This work is a fascinating compendium of information about a neglected aspect of East Asian culture and of the history of timekeeping. Incense timekeeping devices played important roles in early Chinese and Japanese social and technological history in addition to their use for measuring time. They served in rituals in Buddhist temples, as replacements for community water clocks in times of drought, as regulators of the flow of water to farmers for irrigation in agricultural regions, and in palaces and government offices for establishing time schedules. In China they became a favored feature of the studios of poets and scholars, a practice that continued to recent times, while in Japan they were also adapted for use in geisha houses.

This book will not only appeal to students of Chinese and Japanese history, but will also prove useful to museum curators and particularly to the fast growing body of collectors of these exotic devices in the Western world and in East Asia. Excellent illustrations range from early East Asian forms of timekeeping to a large variety of incense time-measures, with their intriguing functional designs, and beautifully executed decoration, demonstrating the highest quality of workmanship. The appendices include a catalogue of examples which have appeared in sales rooms in recent years.

THE TRAIL OF TIME

SILVIO A. BEDINI

THE TRAIL OF TIME

Time measurement with incense
in East Asia
Shih-chien ti tsu-chi

CAMBRIDGE
UNIVERSITY PRESS

CAMBRIDGE UNIVERSITY PRESS
Cambridge, New York, Melbourne, Madrid, Cape Town, Singapore, São Paulo

Cambridge University Press
The Edinburgh Building, Cambridge CB2 2RU, UK

Published in the United States of America by Cambridge University Press, New York

www.cambridge.org
Information on this title: www.cambridge.org/9780521374828

First published 1994
This digitally printed first paperback version 2005

A catalogue record for this publication is available from the British Library

Library of Congress Cataloguing in Publication data
Bedini, Silvio A.
The trail of time: time measurement with incense in East Asia
Shih-chien ti tsu-chi / Silvio A. Bedini.
p.   cm.
Bibliography
Includes index
ISBN 0 521 37482 0
1. Clocks and watches – East Asia – History. 2. Incense seals and
containers – East Asia. 3. Time measurements. I. Title.
II. Title: Shih-chien ti tsu-chi.
TS543.E18B43   1990
681.1111 – dc20
89-35784 CIP

ISBN-13 978-0-521-37482-8 hardback
ISBN-10 0-521-37482-0 hardback

ISBN-13 978-0-521-02163-0 paperback
ISBN-10 0-521-02163-4 paperback

For Joseph J. Kusaila
(1921–1987)

As the curling smoke wafts sinuously aloft from a freshly lighted [incense] stick and the first faint whiff strikes the nostrils, a profound symbolism takes effect – even subconsciously. The glowing coal is the spark of life. The smoke signifies the incorporeality and evanescence of spiritual truths, while the fragrance demonstrates the tangible reality and actual penetration of these spiritual truths.

From unidentified ancient writings, quoted in *Altars of the East* by Lew Ayres (1956)

Contents

Illustrations

Preface

The journey along the trail of time which led to the present publication began more than thirty years ago. At that time a clock collector in Los Angeles with whom I had been corresponding mailed me photographs of two "water clocks" he had unwittingly purchased by mail from an antiquities dealer in Japan. The objects illustrated could not conceivably have been water clocks, nor did they appear to be clocks at all. The collector was curious, however, and wishing to have them identified, offered me one of them in exchange for a satisfactory description. Any curiosity to be so easily acquired without cost was a temptation, and I set to work at once.

Inasmuch as the presumed timepieces had been shipped from Japan, the first sources to be explored were horological publications relating to Far Eastern countries. In due course the solution was found in the catalogue of N. H. N. Mody's collection of Japanese clocks. Therein similar timepieces were illustrated and described as Japanese "incense clocks" (*kōbandokei*), a category little known to Western collectors.

Fulfilling his promise, the collector presented me with one of the two he had acquired, with the consequence that my own curiosity increased, and I was determined to learn more about the measurement of time with incense.

During the next few years the search led to the discovery of another category of "incense clocks," Chinese incense seals. Again the quest for information brought about the development of a considerable correspondence, with horological scholars and collectors in the United States and Europe, curators of Oriental art, and university professors of Far Eastern languages and literature.

The interest generated among Far Eastern specialists in the academic world was most reassuring and rewarding. Each contributed a little more to my knowledge, cooperating by suggesting other early Chinese writings, copies of which I obtained from the Harvard-Yenching Institute and other sources, and rendering translations.

Their patience proved to be endless, many extending considerable effort and time on my behalf. Gari K. Ledyard, then at the University of California at Berkeley, for example, translated major sections of the *Hsin tsuan hsiang p'u* and the *Hsiang ch'eng* of Chou Chia-chou, which formed a vital base for the study on which I was engaged.

My research eventually resulted in a monograph which was published in 1963 by the American Philosophical Society in its series of *Transactions*, with the title *The Scent of Time: A Study of the Use of Fire and Incense for Time Measurement in Oriental Countries.*

During the succeeding years numerous inquiries which came from museums and collectors encouraged continued study, undertaken from time to time during the next two and a half decades. Bit by bit a substantial amount of new data on the use of incense for time measurement came to light, making it possible to produce this entirely new and much more comprehensive work on the subject.

A major windfall was the acquisition for my personal library of the extremely rare work, *Yin hsiang t'u p'u* of Ting Yün (Ting Yüeh-hu). This slim two-volume work consists of designs of one hundred covers and templates of Chinese incense seals derived from archaic forms, compiled or designed by a Chinese scholar in the late nineteenth century. These are accompanied by dedicatory prefaces by his friends and associates. Printed entirely from wood block in a variety of esoteric scripts, several attempts to have the work translated ended in failure due to the difficulties presented by the scripts. Eventually the project was undertaken by Kirby R. Vining, who devoted many months of difficult study to the work. This collection of designs and writings made it possible to add yet another chapter to the story, as incense seals of these designs were found and studied in private collections.

The history of time measurement with incense in the Far East has never heretofore been fully reported even in native writings. Other than occasional brief mentions in modern Japanese works for clock collectors, its history survives in Chinese and Japanese state records and remote documents which were sought out with considerable effort and translated.

At the time this research was first undertaken, incense timepieces were virtually unknown in the Western world, not only to horological scholars and collectors, but also to curators and historians of Far Eastern art and technology.

While engaged in this project, it was quite by accident that I discovered several Chinese incense seals for sale in New England by antiquities dealers. A systematic search of other shops revealed a few more, generally buried in the storerooms, making it possible to acquire a number of Chinese incense seals for study. Generally they were sold to me as hand warmers, opium stoves or incense burners. Never once was their true function identified. In the course of time it was possible to acquire more examples and study others found in private collections and shops on the West Coast.

The publication of *The Scent of Time*, however, brought about a fast growing awareness of the purpose of "incense clocks." The monograph had immediate substantial distribution not only to American and European clock collectors but also to those in Japan. A community of collectors of incense seals developed, with the consequence that prices realized for these devices rose rapidly. In addition to the incense seals that made their way across the Pacific in years past, a substantial number of them have been brought out of mainland China in recent years by traders, after it was opened to Western commerce. The growing number of collectors continues to exceed the number of surviving incense seals available, however.

For the most part, the Chinese incense seals which have appeared on the market in recent years are of nineteenth-century vintage, many of exquisite form and decoration. Earlier examples are virtually impossible to find and are particularly prized. The Japanese *kōbandokei* have been produced in much more limited numbers and consequently are extremely rare. It is believed that a few early incense seals still survive in Buddhist temples and perhaps among the possessions of old families in China and Japan, neglected or used merely as common incense burners; their original function long forgotten and now unknown.

Although in modern times the incense time measurers have been reduced to the status of decorative curiosities and categorized as examples of applied art rarely found in museums, for centuries they nevertheless played a significant role in the ordering of the religious and civil life of Far Eastern countries, and as such deserve a place in the history of the Far East.

Acknowledgments

Over the course of the thirty years and more that I have been engaged in research on time measurement with incense in the Far East, I have accumulated a tremendous debt to the many scholars, librarians, museum curators and collectors who contributed their generous and interested assistance.

For my earlier work, *The Scent of Time*, the contents of which have been incorporated in the present text, vital assistance was provided primarily by Gari K. Ledyard, who translated some of the most important texts. In my subsequent research, I am equally indebted to Anthony J. Cannon and Richard J. McGhee for their translations of Japanese texts, and particularly to Kirby R. Vining, who has translated the dedicatory prefaces of Ting Yün's memorial volume and numerous other writings.

Many others have contributed information and provided assistance in the preparation of this study. Among them are the late Teiichi Asahina of the National Science Museum, Tokyo; Derk Bodde of the University of Pennsylvania; Paul C. Blum of the Charles E. Tuttle Company in Tokyo; Chang Lin-sheng of the National Palace Museum in Taipei; Jonathan Chaves of The George Washington University in Washington, D.C.; James I. Crump of the University of Michigan; Wolfram Eberhard and the late Edward H. Schafer of the University of California at Berkeley; Rev. Louis J. Gallagher, S. J. and Rev. Joseph Sebes, S. J. of Georgetown University; Akio Gotoh, director of the Tezukayama Astronomical Observatory in Nara City, Japan; L. Carrington Goodrich of Columbia University; Dan S. Greenwood of the Harvard College Library; Shojo Honda of the Japanese Section, The Library of Congress; Jack Jacoby of the C. V. Starr East Asian Library, Columbia University; the late Edward S. Jones of Los Angeles, the collector who first interested me in the subject; Oumi Kobayashi of the Kokuritsu Kobunshoken in Tokyo; Thomas Lawton, the late Harold P. Stern and Lily Kecskes, director, former director and librarian respectively of the Freer Gallery of Art, Smithsonian Institution; Sanae Iida Reeves and Ann Yonemura of the Arthur M. Sackler Gallery of the Smithsonian Institution; Kosei Morimoto, Chief Librarian, Todai-ji Temple, Nara City, Japan; Li Di of Inner Mongolia Normal University; Min-chih Chou, curator of the Wason Collection, Cornell University Libraries; Exsei Kawaguchi, chief priest of Tozen-ji Temple, Naku-ku, Yokohama; Dr. Carole Morgan of Equipe de recherche sur les documents de Touen-Houang et materiaux connexes, Paris; the late Derek J. de Solla Price of Yale University; the late Edwin Pugsley of New Haven, Connecticut; Sang Woon Jeon, president of Sungshin Women's University in Seoul, Korea; the late Rev. Georg Schurhammer, S. J. of the Institutum Historicum Societatis Iesu in Rome; Shu Fen of the National Central Library of Taipei; Shoichi Sugiyama of the University of Maryland; M. K. Starr of the Field Museum of Natural History in Chicago; the late William Barclay Stephens, M. D., of Alameda, California; Takeo Tamura of Tsuchiura City, Japan; Anthony J. Turner of Le Mesnil-le-Roi, France; Lydia B. Voorhees of Maitland, Florida; Gun'ichi Wada, secretary of the Shōsōin Imperial Treasury at Nara City, Japan; Glenn Vessa of Honeychurch Antiques, Hong Kong; Masakuni Wada of Yokohama; Kiyoshi Yabuuchí of Kyoto University; Ryuji Yamaguchi of the University of Commerce in Tokyo; Yutakia Ojihara of Kyoto University, and Akira Yuyama, director of the International Institute for Buddhist Studies in Tokyo.

Special acknowledgment is due to Joseph Needham of Gonville and Caius College and founder of the Needham Research Institute at Cambridge University for his ceaseless assistance and encouragement over many years; to John Stevens of Tohoku Fukushi University in Sendai, Japan for having provided and translated the Tantric scripture from the Buddhist Canon of "The [Incense] Seal of Avalokitésvara" and other texts critical to this study; and to Mr. Chester W. Howard of Kamakura-shi, Japan for his persistent efforts on my behalf.

Many have generously permitted examination of their collections of incense seals, and among those who provided illustrations for this work are James M. Kallison of San Antonio, Texas; the Patrimonio Nacional in Madrid, Spain; the Peabody Museum in Salem, Massachusetts; Victor and Linda Renaghan of Hawaii; the Science Museum, South Kensington, London; Carl Szego of Millburn, New Jersey, the Time Museum in Rockford, Illinois, and Lydia B. Voorhees of Maitland, Florida.

Lela Bodenlos and James Roan of the Smithsonian Institution Libraries have rendered invaluable service in furnishing sources and verifying citations. Ricardo Vargas and the late Alfred Harrell of the Smithsonian Photographic Laboratories have photographed many of the items used in illustrations. My wife, Gale, has not only read and helped edit the manuscript, but has also compiled the index. To all of the foregoing I express my grateful appreciation.

Cambridge University Press acknowledges with gratitude a subvention from the Arthur M. Sackler Foundation which has helped towards the production of the photographs in this book.

Permission to use translations, research materials and photographs from the following is gratefully acknowledged: Jonathan Chavez; James I. Crump, Jr.; Akio Gotch; T. Morgan Jones; James M. Kallison; Masuo Kawaguchi; Thomas Lawton; Richard J. McGhee; Carole Morgan; Patrimonio Nacional, Madrid; Nasakuni Nada; The Peabody Museum, Salem, Massachusetts; Linda and Victor Renaghan; Royal Ontario Museum; Ontario, Canada; John Stevens, Tohoku Fukushi University, Sendai, Japan; Carl Szego; Takeo Tamura; The Time Museum, Rockford, Illinois; Kirby R. Vining; Lydia Voorhees.

Conventions

The Wade-Giles Romanization has been used for names of Chinese persons, places, things, institutions, etc., except in instances of direct quotation or traditional spelling of common place names.

Dimensions of artifacts and other dimensions are specified in metric measurements; when Chinese feet (*ch'ih*) and inches (*ts'un*) appear in translations, conversion to the equivalent metric measurement is provided.

In translated passages, question marks used after a word or phrase are editorial, and indicate that the meaning is uncertain. The same is true for question marks following dates, both in the text and within quotations.

For the reader's convenience the dynasties and reigns and their dates are listed in Appendix A, a select glossary in Appendix E, and a Romanization conversion table is found in Appendix F.

I

TIME MEASUREMENT AND
INCENSE
IN EAST ASIA

I

Early East Asian time measurement

The measurement of time has preoccupied mankind from its earliest civilizations in all parts of the world, although not always for the same purpose or to the same degree. As man's needs and knowledge increased, so did his awareness of the meaning of time. In the Far East as in the Western world, man learned from observing the common phenomena of nature, notably the movements of the sun, the moon, and the stars.

Observation of celestial phenomena, such as the periodicity of the sun and moon, brought even greater cognizance. Myths and superstitions relating to time were created, which in turn gave rise to cults and religions evolving from special rites designed to keep deities benign.

The Chinese developed the legend of P'an Ku, for example, the first Adam who devoted eighteen thousand years to creating the universe by chiseling stars and planets from the cliffside of Chaos. It was from the beginning of that period, tradition claims, that the Chinese began the measuring of time.[1]

The duration of day and night was ascribed to the periods of waking and sleeping of the Great Dragon, whose breathing regulated the winds and the seasons. The day's division into hours was derived from the legend of the Ten Chariots of the Sun, drawn by six dragon horses and driven by the mother of the sun, which raced after one another across the sky. Each of the chariots represented one hour of the day, and evil befell anyone who viewed more than one at the same time.

Celestial phenomena were explained by the legends describing them, such as the tale of the Excellent Archer who shot down nine of the ten suns with his magic bow, and the story of the Cowherd and the Weaving Girl which explained the Milky Way. The myths and legends gradually became interwoven with religions that evolved, influenced, and eventually became part of the philosophic considerations of scholars.[2]

These exotic practices led from the construction of megalithic monuments

[1] C. A. S. Williams, *Encyclopedia*, pp. 309–11; S. W. Williams, *Middle Kingdom*, vol. 2, pp. 137–42.
[2] C. A. S. Williams, *Outlines*, pp. 121, 273–74, 347.

relating to the directions of sunrise and moonrise and their settings, to the production of astronomical models which illustrated the sun's annual cycle, all designed to reflect the orderly sequence of celestial events – heliacal risings and settings, the moons, and equinoxes and solstices. After the passing of centuries and not long before the beginning of the Christian era, there occurred almost simultaneously in China and Greece the development of astronomical knowledge which resulted in greater awareness of time and provided the first means for its measurement.

As man first attempted to measure time, the tools at his command were primitive in the extreme – the shadow cast by a sunlit tree or a vertically erect pole led to the invention of the sundial. Later, the realization that the flow of water could serve as measurement brought about the invention of the clepsydra or water clock.

With increasing awareness of the natural world came the realization that time provided a means by which order was furnished to all organic and human experience. Man learned to consider time in terms of two basic units – the interval and the epoch, the first to measure duration and the second to measure location in time. Thus, time became a subject of man's philosophical considerations long before he attempted to harness it for practical use.

Chinese philosophers speculated about time and its content from very early periods, and definitions and discussions of time and its relation to space are to be found in Chinese writings as early as the fifth century B.C. By the first century the Bureau of Astronomy had become an important office of the civil service, concerned with celestial observation and continued development of the calendar. Consciousness of time increased among Chinese intellectuals as they acquired more and more new knowledge about the natural world.[3]

Inevitably from early times man applied his senses for measurement of time, first his sight and then his hearing. Much later, with the advent of the mechanical clock, he applied other senses. The sense of touch was utilized with the eighteenth-century French invention of timepieces designed for the blind so that the current hour could be determined by raised knobs opposite the relevant numerals. In the same period was developed the horological oddity, admittedly without wide application, of a clock on which the hour numerals were identified by the sense of taste.

It was in the countries of the Far East that the fifth sense – the sense of smell – was first applied for the measurement of time by means of burning incense. Incense is believed to have come into use in China as part of Buddhist religious rites which had been transmitted from India, following the translation into Chinese of Indian Tantric scriptures by the Indian Buddhist monk Amoghavajra.

---

[3] Needham, *Time*, pp. 1–52.

In the Western world the use of incense has generally been limited to religious or cult practices, whereas in East Asian countries it served many other practical uses as well. It is consequently not surprising to discover that it was widely utilized also as a means of measuring time in the palace, government office, Buddhist temple and scholar's study.

The awareness of time achieved by Chinese philosophers, scholars, and the court bureaucrats from astronomical studies and activities, however, did not extend to the common people engaged in everyday pursuits. The latter's awareness was related only to their daily activities, and consequently had little meaning for them. It was observed by foreign travelers during the eighteenth and nineteenth centuries that unlike many of the common people in Western countries, those in the Far East remained comparatively unconcerned with the time factor. Punctuality did not appear to be a characteristic of the people, nor in fact was it considered to be a desirable trait. Again and again visitors to China reported that the Chinese appeared to express no need to know the time except approximately, demonstrating a lack of concern that has persisted to modern times. However, the outlook of the man in the street has probably varied considerably from one historical epoch to another.

For the populace time periods were related to common experiences, such as the time required to drink hot tea, to consume a bowl of rice, to travel a specified distance, or to burn a stick of incense.[4]

An expression frequently encountered in Chinese writings is "the time of burning an incense stick" (*i chu hsiang ti shih hou*), a common phrase used to indicate the lapse of a time period.[5]

It is conceivable that because of its wide use, the burning of a stick of incense became in fact a common unit of measure, and that it was in fact a consequence of this practice that eventually led to the marking of an incense stick with graduations of the time divisions required for its consumption. The incense stick came into use at every level of daily life. In the time of Cheng Ho and before, for example, the incense stick was used to time watches kept at sea, and mariners changed their compass direction only when a number of these had burned to the end.[6]

Among those commenting on the apparent lack of concern for time were American residents in China in the mid-nineteenth century. Clocks in public buildings, particularly railroad stations, appeared never to register the time with

---

[4] J. D. Ball, *Things Chinese*, p. 713.

[5] *Chung-shan ta tz'u tien i tzu ch'ang pien*, p. 472. Communication to the writer from Prof. Yang Lien-sheng, November 6, 1959.

[6] Needham, *Science and Civilisation in China*, vol. 4, part 3, pp. 564, 570, 583; these volumes are referred to hereafter as *SCC*.

any degree of accuracy. When one of these public clocks malfunctioned, no attempt was made to repair it; instead a new clock was added to the wall and the inoperative ones allowed to remain. Travelers noted seeing walls of some public offices and railroad stations covered with a number of clocks, most of them no longer functioning, the others inaccurate. An American doctor living in China in the mid-nineteenth century wrote, ". . . clocks are not often met with in China; they are generally confined to the public offices, where it is common to find half a dozen in a row."[7]

Robert K. Douglas, the British Museum's Keeper of Oriental Books and Manu-scripts and Professor of Chinese at King's College, London, writing in the late nineteenth century, also commented on the apparent unconcern with precision time measurement:

For the most part, . . . even at the present day, Chinamen are dependent on the sun for their knowledge of the time of day. Happily for them, punctuality in all matters of daily life is foreign to their social system, and the division of the day into twelve periods, measuring two hours apiece, supplies with sufficient minuteness all that is required for fixing appointments and keeping engagements in that leisurely land. In some cities clep-sydrae are used to mark the progress of time, and occasionally joss-sticks, which are carefully divided by the astronomical board into periods corresponding to the hours are kept burning for the same purpose. The advance from these rough contrivances to clocks and watches is as great as that from the native candles to kerosene lamps.[8]

Early in the twentieth century, another traveler commented that in Hong Kong, Macao and in the vicinity of the Treaty Ports "clocks are found in every shop and watches abound, but in many places there is no standard of correct time, and in places where there is, it is ignored extensively." A Westerner obviously annoyed by the attitude of the Chinese so different from that of his own culture, wrote:

Time – but what idea has a Chinaman of time? Time does not enter into the essence of his ordinary conceptions of a day, or, at all events, the idea is so very vague that the conception of it seems but an inchoate one . . . Life is not such a mad rush as with our feverish pursuit of wealth, a livelihood, or learning. Fix a time for an engagement with a Chinese, and he comes in half an hour late, or even two or three hours after, occasionally a few days later than the day fixed upon, with no idea that he had done anything out of the way.[9]

More recently, Chiang Monlin described the wide dichotomy that existed in the sense of time of the privileged class as compared with that of the peasants:

Clocks were unnecessary – for what is the use of keeping exact time in a village? What

---

[7] Magowan, "Modes," p. 336.   [8] Douglas, *Society*, p. 313.
[9] J. D. Ball, *Things Chinese*, p. 709.

difference would it make to be two or three hours too late or too early? The country folk counted time in days and months, not in minutes or hours.[10]

The same indifference to precise measurement of time prevailed also in Japan. Sir Ernest Satow, who was attached to the British Mission in Japan during the years of its revolution which restored the Mikado in the mid-nineteenth century, wrote:

In those days neither clocks nor punctuality were common; and, as the hour altered in length every fortnight, it was very difficult to be certain about the time of day, except at sunrise, noon, sunset and midnight.[11]

For an understanding of how incense first came to be applied for the measurement of time in the Far East and of how it was developed for both religious and community purposes, it is necessary first to review briefly the evolution of time measurement in East Asia. Then, consideration must be given to the development of incense from its function in primitive religious rites to its pervasive presence in daily life in East Asia, a presence which had no counterpart in the Western world.

The published history of time measurement in the Far East has been fragmentary at best. Research in recent years has produced excellent published studies of scientific time measurement in early China, particularly the remarkable astronomical instruments and clocks of the Chinese imperial court and of Korea. There is, however, room for further study. Meanwhile, the early history of time measurement in Japan remains to be studied and reported. Most neglected has been the story of time measurement among the common people of Japan.[12]

For centuries communal time measurement was achieved in the Far East by three means – sundials, clepsydrae, and incense timekeepers.[13]

The moving shadow cast by a sunlit gnomon (from the Greek word *gnomon* meaning "interpreter" or "the one who knows") has measured the sun's daily journey across the sky and divided the daylight hours from time immemorial in every early civilization. Through the centuries the gnomon has assumed many forms, from the primitive denuded tree trunk or erect pole in an open area, or megalith in Majorca, to the shining metal stylus of a Ming fire-gilt monumental equatorial dial of a Chinese palace (Fig. 1).

In Egypt the sundial appears to have been preceded by the shadow clock, an

[10] Chiang Monlin, *Tides*, pp. 34–35.
[11] Satow, *A Diplomat*, quoted in Robertson, *Evolution*, p. 193.
[12] Needham, The Wilkins Lecture; The Henry Myers Lecture; *SCC*, vol. 3; Needham *et al.*, *Heavenly Clockwork*; Needham *et al.*, *Hall of Heavenly Records*.
[13] For the most comprehensive accounts of Chinese sundials, clepsydrae, sand clocks, and the astronomical clocks of Su Sung, see Needham *et al.*, *Heavenly Clockwork* and Needham, *SCC*, vol. 3.

example of which was recovered from an Egyptian burial of the fifteenth century B.C. Other time-reckoning devices may have existed there even before that period.[14]

A sundial, in the unusual form of a refracting instrument, is described in the Second Book of Kings of the Old Testament. The sundial was made in about 771 B.C. by Uriah, the priest of Jerusalem, for Achaz, the eleventh ruler of Judah. It was mentioned again in the thirteenth year of the reign of Hezekiah, the son of Achaz.[15]

The earliest knowledge of sundials among the Greeks is attributed by tradition to Anaximander of Miletus in the sixth century B.C. This tradition was partially reinforced in the fifth century B.C. in a statement by Herodotus that the concept of the sundial had come to Greece from the Babylonians. The gnomon had in fact been known to the Greeks from an earlier period, used for solar and astronomical observations. The sundial evolved over the centuries in various forms, and continued in use long after geared clockwork timetelling devices came into being.[16]

The earliest form of timekeeper noted in East Asian countries appears to have been the most primitive type of the sundial. The *K'ao kung chi* chapter of the *Chou li* (Records of the Rites of [the] Chou [Dynasty]) attested to "the installation of a straight pole to observe its shadow."[17]

The use of the gnomon is mentioned also in the *Chou pei suan ching*, an ancient work on mathematics attributed to c. 1100 B.C.[18]

Gradually more sophisticated forms were developed. One of the earliest known surviving Chinese examples is an equatorial dial consisting of a stone disk elevated at an angle on a pedestal, with a bronze pin penetrating it at the center and projecting on both sides. The upper side of the pin points to the North Pole and the lower end of the projection to the South Pole.[19]

A sundial (*kuei piao*) is described in the writings of the Shu period of the Three Kingdoms Dynasty. Relics of two examples, probably of the Western Han period, were reported. Of primitive form, they included indications for compass directions. Their primary purpose may have been for the measurement of time only, and the possibility exists that they were introduced from other countries, and were not indigenous to China.[20]

---

[14] Priestley, *Man and Time*, pp. 22–24, 143–48.

[15] The Bible, Book of Kings II, 20:9–11, Book of Isaiah, 38:8; Earle, *Sun-Dials*, pp. 391–95; Sachse, "Horologium," pp. 21–30.

[16] Gibbs, *Greek and Roman Sundials*, pp. 3–11, 66–88.

[17] *Chou li*. Translation by E. Biot, *Le "Tcheou-Li" ou "Rites des Tcheou"*; Needham, *SCC*, vol. 3, pp. 28 b, 290–92.

[18] *Chou pei suan ching* (Arithmetical Classics of the Gnomon and the Circular Paths of Heaven), sixth century B.C. – ante A.D. 80; Needham, *Heavenly Clockwork*, p. 19, fn.

[19] Maspero, "L'Astronomie," pp. 267–356.   [20] Kim Yong-Woon, "Origins," pp. 4–11.

Listed in the bibliography of the *Ch'ien Han shu* is a "Sundial Book" (*Jih kuei shu*).[21] The same work mentioned sundials in relation to a gathering of calendar experts who assembled in about 104 B.C. It described the manner in which the true east and west points were established, after which "gnomons were raised and clepsydrae activated (*li kuei i hsia lou k'o*)." The calendar experts marked out the twenty-eight *hsiu*, and fixed the first and last days of each month, the equinoxes and solstices, the phases of the moon, as well as the movements and positions of the heavenly bodies.[22]

Two surviving sundials (*ts'e ching jih kuei*) attributed to the Han period have inscriptions which appear to be identical. In each instance the inscription appears only upon the upper surface and is in the form of a diagram consisting of two circles with the annular space divided into equal segments of hundredths along the circumference. At the center is a socket for a fixed gnomon, and at the intersection of the lines with the outer circle are sixty-nine sockets for the insertion of a movable gnomon, the shadow of which would coincide with that of the fixed gnomon.[23]

In later periods monumental sundials were erected in palace courtyards and other public places for the use of the court and the public. A number of these have survived (Fig. 2).[24]

Sundials were mentioned also in literary works. As an example, the T'ang poet, Li Ho (A.D. 790–816), at one time Supervisor of Ceremonies in the Court of Imperial Sacrifices in the T'ang capital of Ch'ang-an, became dissatisfied with his position. In his poem, "After Days of Rain in the Ch'ung-i District," he wrote longingly of his distant mountain home, describing how

> The Southern Palace is darkened by ancient blinds,
> Its sundials blank beneath a watery sun.[25]

That the time measured by the sundial was not always the same due to the irregularity of the apparent motion of the sun through the sky was known from ancient times to the Babylonians and the Greeks and to the Chinese. The sun moves through the sky sometimes faster and sometimes slower than its average motion. The irregularity is due to two circumstances, the eccentricity of the earth's orbit, and the difference between the planes of the equator and the ecliptic.

---

[21] Yin Hsien (first century B.C.) assisted in editing and classifying the books brought together by the emperor from all parts of the Chinese empire. Needham, *SCC*, vol. 3, p. 302; Giles, *Dictionary*, pp. 945–46.

[22] *Ch'ien Han shu*, (History of the Former Han Dynasty), ch. 21A, p. 16 b. Needham, *SCC*, vol. 3, p. 302; Needham, *Heavenly Clockwork*, pp. 200, 204.

[23] Yetts, *Cull Chinese Bronzes*, pp. 150–65; Neeham, *SCC*, vol. 3, pp. 302–9.

[24] Needham, "Astronomy," pp. 67–82; Needham, *SCC*, vol. 3, pp. 302–7.

[25] Frodsham, *Goddesses*, p. 110.

As a consequence, the greatest positive difference between clock-time and sundial-time is 14½ minutes occurring in February and the greatest negative difference of 16½ minutes occurs in November, which is known as "the equation of time." Li Ch'un-feng (fl. A.D. 620–680) developed algebraic methods for determining it in preparing the Lin Te calendar of 665. The consequent inequality of the four seasons – the time passed between the solstitial and equinoctial points – was realized by the ancients. In the eleventh century the discrepancy, of approximately one Chinese quarter (*k'o*) undoubtedly became apparent in the use of the clepsydra or water clocks. It is possible also that the discrepancy could have been detected in the use of the "perpetual lights" or oil lamps used as timekeepers in temples. As reported by Fan Shun-ch'en in the *Shan chu hsin hua* of c. 1360, in these the oil supply was measured by *k'o* revealing a discrepancy between the time measured by the oil lamp and the sundial.

Because sundials functioned only in fair weather and in sunlight, it was inevitable that other forms of timekeeping would be provided for measurement of the hours in which sunlight was absent. Thus was born the water clock (clepsydra). It was not a Chinese invention, having existed in both Egypt and Babylonia for centuries before the early Shang period. The advent of the clepsydra in China is attributed by some sources to the legendary Yellow Emperor, Huang Ti, who began his reign in 2698–2695 B.C. Tradition claims that he had twelve bells cast to correspond to the twelve moons to indicate the seasons, months, days and hours.[26]

The *Chou li*, the oldest surviving record of institutions of the period and concerned primarily with customs and ceremonies, assigns the invention of this primitive form of timekeeping to the year 2356 B.C. in the reign of Yao. It reports that efforts were then being made to mark the revolutions of the planet Jupiter. It was ordered that the hours of the night be divided into intervals, and that the people be informed of them by public display of wooden tablets, each inscribed with the character for the current hour. It was further noted that the intervals were measured by means of water clocks, each of which had a large hollow basin into which water dripped. The interior of the basin was marked with the divisions of the hours and their parts, and was illuminated by lamps so that the divisions remained visible throughout the night.

In the Early Chou Dynasty a court functionary named Ch'ieh Hu Shih was designated to maintain the clepsydrae of the Duke of Chou, brother of the first ruler of the family, who was said to have been the first to provide a means of heating for clepsydrae to prevent the water from freezing during the winter months.[27]

[26] Needham, *SCC*, vol. 3, pp. 123, 202; Giles, *Dictionary*, pp. 338–39.
[27] Gaubil, *Histoire*, cited in Planchon, "L'Heure," p. 194.

Generally the clepsydrae were made in the form of a series of water tanks ranging in number from two to four of varying shapes and sizes. An early description occurs in the *Lou shui chuan hun t'ien i chih* (Apparatus for Rotating an Armillary Sphere by means of Clepsydra Water) Chang Heng.[28]

Early allusions to clepsdrae in Chinese writings occur also in the *Shih chi* in connection with the career of a military figure in the period of the Warring States. Mei Yao-ch'en, a poet of the Northern Sung period, wrote of the simple form of the inflow clepsydra with two vessels, in common use in rural districts:

> The astronomers know of the rising and setting of stars,
> No error they make in predicting the heat and the cold;
> But the farmer too keeps time with his rustic pots
> And grudges the loss of a single inch on the dial.
> The drops of his sweat match those of the dripping clepsydra,
> The swatches of his cutting advance like the shadow itself.
> Who could disdain or regret this life-giving labor?
> Who can snatch back an hour from the realm of the past?[29]

In time clepsydrae were improved by using four tanks with spouts shaped like mouths of dragons and other animals from which the water was emitted. In the Later Han period, about A.D. 164, clepsydrae were constructed also in other forms, including some made as spheres featuring signs of the zodiac to mark the time divisions. In the Northern Wei Dynasty as noted by the Taoist Li Lan in his *Lou k'o fa* preserved as part of the encyclopedic *Ch'u hsüeh chi*, water, the common fluid in clepsydrae, was replaced with mercury.[30]

Yet another form of the water clock incorporated a wooden figure attached to a rod which floated in a water tank, with its outstretched hand pointing to the current hour on a vertical scale. There is considerable similarity between this device and the water clock described in the Western world by the first-century Roman architect and engineer, Marcus Vitruvius Pollio.[31]

The earliest known depiction of a Chinese clepsydra is found in *Liu ching t'u* (Illustrations of Items Mentioned in the Six Classics), a twelfth-century work by

---

[28] Needham, *SCC*, vol. 3, p. 320; Needham, *Heavenly Clockwork*.

[29] Wang Chen, *Nung Shu*, ch. 19, 20a *et seq.*; Needham, *SCC* vol. 3, pp. 315–20. Mei Yao-ch'en's dates are also given as 1002–1060. A native of Wan-ling in Anhwei, he inherited official rank. After presenting an account of his poetic abilities at the Imperial Academy, he rose to the position of second-class secretary, and was appointed to the commission to prepare a new history of the T'ang period. Giles, *Dictionary*, p. 579.

[30] Planchon, "L'Heure," p. 194; Needham, *SCC*, vol. 3, pp. 326–27.

[31] Vitruvius, *Architecture*, 1673, 2nd edition, 1684. Although there are many editions of this work from the fifteenth century to the present, this is considered to be the earliest of merit. Clepsydrae are described in Book 4, ch. 4; Diels, *Antike Technik*, p. 213.

Yang Chia (fl. 1155) which was subsequently enlarged by Mao Pang-han (fl. c. 1170) in his encyclopedic *Shih lin kuang chi* (Fig. 3).[32]

Several of the earlier Chinese clepsydrae have survived. One is the brass clepsydra in Canton, which was installed by Tu Tzu-sheng in A.D. 1316, the third year of the reign title Yen Yu of the Yüan Dynasty. It was called *t'ung lu ti lou* and was featured in a pavilion under a double arch that traversed a road leading to the city's south great gateway and to the palace of the treasurer of the province. A small brick stairway was provided for the keeper. The device consists of four brass tanks, one of which contains a rule marked with the hours mounted upon a float (Fig. 4).

Surmounting the installation was an altar dedicated to P'an Ku, the Adam of the Chinese, and at the right of it is another altar dedicated to the goddess Kuan Yin. At the beginning of each hour the keeper was required to strike the number of the hour upon a drum during the day and upon a gong during the night, and to display outside the door of the room a wooden tablet bearing the character of the current hour. He utilized his free time to mark incense sticks with graduations for the hours. These he sold to the public and others to augment his modest salary.[33]

The remarkable astronomical clocks of Su Sung have been described in detail in several recent studies. The earliest timekeeper devices separate from astronomical apparatus were a series of thirteen apparatus devised by Kuo Shou-ching for reforming the calendar in 1276, the thirteenth year of the reign title Chih Yüan of the Yüan Dynasty. One of these was the so-called "lamp clepsydra" or illuminated clock in the Ta Ming Palace, which was used exclusively for measuring time.[34]

Clepsydrae were frequently mentioned in literary works. In the early T'ang period, for example, Tu Shen-yen, the Archivist in the Imperial Academy and poet, wrote:

> The wind clear – the waterclock's drip can be heard at dawn.
> We sit hand in hand till the last of our pleasure is gone,
> And it seems once again we had never left the crowd.[35]

Later a reference to the clepsydra is found also in "The Song of the Crows Roosting at Night" by the celebrated T'ang poet, Li Po (A.D. 701–762):

---

[32] *Liu ching t'u*, quoted in Needham *et al.*, *Heavenly Clockwork*, p. 91.

[33] Planchon, "L'Heure," pp. 193–95; Liu Hsien-chou, pp. 330–35; Magowan, "Modes," p. 430. See also Needham *et al.*, *Heavenly Clockwork*, pp. 85–94.

[34] *Yüan shih*, ch. 48, pp. 7a and 7b; Needham *et al.*, *Heavenly Clockwork*, pp. 135–36.

[35] Tu Shen-yen, "On An Autumn Night: Banqueting at the Cottage of Magistrate Cheng of Lin-chin," in Owen, *Early T'ang*, p. 331. Living during the late seventh and early eighth centuries, Tu Shen-yen was a native of Hsiang-yang in Hupeh and the grandfather of the famous poet Tu Fu. Giles, *Dictionary*, pp. 783–84.

> Half only of the sun projects from the jaw of the green hills,
> The silver arrow on the water-clock has marked many hours;[36]

In the same period Wang Ch'ang-ling, in his "An Autumn Song for the Ch'ang-hsin Palace," wrote about that most famous of "deserted consorts," Pan Chieh-yü. The chief favorite of the Han Emperor Ch'eng Ti, she had the title Chieh-yü conferred on her as the imperial concubine having the greatest literary ability. When she was replaced in the Emperor's affections by Chao Fei-yen, another lady of the seraglio, Pan Chieh-yü sent the emperor a fan on which she had bitterly described how, like a fan in autumn, she had been

> ... laid, neglected, on the shelf,
> All thought of bygone days, like them, bygone.

She retired to another palace and thereafter attended her friend the Empress Dowager. The term "autumn fan" became part of the language to describe a deserted wife. Wang Ch'ang-ling wrote of how

> Censer for her robes, pillow of jade lack all loveliness.
> She lies, listens to the clear dripping of the water-clock stretch
> on.[37]

Water as a means of time measurement was eventually replaced by the use of sand. Sand clocks, in which wheelwork was operated by the movement of sand, were first developed at the end of the fourteenth century by Chan Hsi-yüan (fl. c. 1370). In about 1360, as the Yüan Dynasty was drawing to a close, he devised the "five-wheeled sand clepsydra" which appears to have been the first attempt to produce an independent mechanical wheelwork for time measurement. It was described in the *T'ien wen chih* (Astronomical Records), a chapter of the *Ming shih*, the official history of the Ming period.

Sand clocks were revived again in the final years of the Ming Dynasty to replace the clepsydra when the latter had become inoperable in winter. To prevent the sand, which provided the motive power, from running through too quickly, the

[36] Li Po, despite a lifetime of dissipation, was a celebrated Chinese poet recognized by the imperial court and appointed to the Han-lin Academy. Giles, *Dictionary*, pp. 455–56. Waley, *Li Po*, p. 48. In his compilation, *The Great Age*, p. 121, Owen has translated this passage as follows:

> From waterclock more and more drips away.
> From the basin of gold with its silver arrow.

[37] Wang Ch'ang-ling, "An Autumn Song for Ch'ang-hsin Palace," in Owen, *The Great Age*, p. 102. Wang Ch'ang-ling was a native of Chiang-ning who, after employment at the capital, fell into disfavor and was sent to Hunan. Upon his return to his native city, he was slain by the Censor Lü Ch'iu-hsiao. Giles, *Dictionary*, pp. 808–9.

orifice was slightly enlarged, and at first four and later six wheels were added to the main driving wheel with scoops. Eventually the wheelwork was designed in such a manner that its rotation coincided with the movement of the heavens.[38]

Sundials and clepsydrae were introduced into Japan from China in the seventh century. The first clepsydra in Japan was recorded in A.D. 671. It was installed by order of Imperial Prince Tenchi (626–671), who, not only produced the first water clock but also devised a system of time measurement for the country.[39]

The use of water clocks persisted in Japan, and in the Heian period a water clock called the *rōkoku* was maintained in operation by one or two designated officials and their twenty assistants.[40]

Three horary systems existed in China from the most ancient periods. In two of the systems, the Chinese day began at midnight instead of at sunset as in the Hebrew and Greek tradition, utilizing a division by "double-hours" that had existed since at least the Babylonian culture, and from which it may have come to China. Reckoning a full day to be from midnight to midnight, it was divided into twelve "hours" (*shih*), each of which was twice the length of a Western hour. This division into twelve double-hours was already well known and stabilized in China by the Former Han period, but it probably had existed even earlier, possibly the Chou period.[41]

Prior to about 1270 B.C. in the early Shang period, and before the adoption of the sub-division by equal double-hours, the Chinese divided the day into major segments defined by terms such as pre-dawn (*mei*), first light (*hsi*), morning (*chao*), midday (*jih*), pre-dusk (*mu*) and post-dusk (*hun*). Later for a time the division of the day consisted of six unequal and varying periods which were given special terms, in addition to another for midnight.

In another of the systems prevalent in China, each of the double hours was divided into two parts, providing a total of twenty-four divisions. The first half of each of these sub-divisions was called the "beginning" (*ch'u*) and the second part the "mid-point" (*cheng*).[42]

In one system of time measurement, the day was divided into 100 quarters or *k'o*, each equivalent to 14 minutes and 24 seconds of Western time, so that consequently each *shih* consisted of $8\frac{1}{3}$ *k'o*. It seems likely that although the use of double-hours was borrowed from another civilization, the division into 100 *k'o*

---

[38] Chang T'ing-yü *et al.*, *Ming shih*, ch. 25, pp. 18b–19; Liu Hsien-chou, *Chung-Kuo*; Needham, *Heavenly Clockwork*, pp. 154–61.

[39] Asahina "On the Jikōban," p. 24.

[40] Oda, *Dai Bukkyo jiten*; Fumio Masutani, *Amidakyō*, and *Shinsen Bukkyō daijiten*, cited in Asahina, "On the Jikōban," p. 24.

[41] Needham *et al.*, *Heavenly Clockwork*, pp. 199–200; Sang-woon Jeon, *Science and Technology*, pp. 86–93; Needham, *SCC*, vol. 4, part 3, p. 461.

[42] Needham *et al.*, *Heavenly Clockwork*, pp. 199–201; Needham *et al.*, *Hall of Heavenly Records*, pp. 8–9.

originated with the Chinese. Efforts were made from time to time to change the number of *k'o* to either 96 or 120, but were unsucessful. Eventually in some calendars a further subdivision was made of each *k'o* into fractions called *fen*, 100 in some while in others it was divided into 60. With the adoption of Western timekeeping in China in the seventeenth century, the *k'o* was established as the Western quarter-hour, or 15 minutes, and the minute as a *fen*.

By the third or fourth century B.C. the Chinese had begun to link the *shih* to the twelve cyclical chararacters, each *shih* associated with a symbolic animal equated to the signs of the zodiac. Following is the astronomical succession:

11:00 P.M.–1:00 A.M.	Tzu (Rat)
1:00–3:00 A.M.	Ch'ou (Ox)
3:00–5:00 A.M.	Yin (Tiger)
5:00–7:00 A.M.	Mao (Hare)
7:00–9:00 A.M.	Ch'en (Dragon)
9:00–11:00 A.M.	Ssu (Snake)
11:00 A.M.–1:00 P.M.	Wu (Horse)
1:00–3:00 P.M.	Wei (Sheep)
3:00–5:00 P.M.	Shen (Monkey)
5:00–7:00 P.M.	Yu (Cock)
7:00–9:00 P.M.	Hsü (Dog)
9:00–11:00 P.M.	Hai (Boar).[43]

Another system of timetelling measured the night from sunset to sunrise, and divided the period of darkness into five "night-watches" (*keng*). The "night-watches" might vary with the season according to the length of the period of darkness and with the latidude of the region. Each *keng* was further subdivided into five equal parts, called *ch'ou*. There were 25 *k'o* from "Beginning of night watches" period and the "Waiting for dawn" period, occurring between *chu' keng*, which occurred 10 *k'o* after dusk, and lasted until *tai tan*, 10 *k'o* before dawn. These were divided into 5 equal parts, with midnight placed at the exact center of the third watch. Accordingly, the middle *keng* straddled midnight so that in European reckoning, the *ch'u* was the period from 11:00 P.M. to 12:00 midnight and the *cheng* was the period from 12:00 midnight to 1:00 A.M. In the order of their appearance the terminal points of the system of "night-watches" were named

> *jih ju* – sunset
> *hun* – dusk – 2½ *k'o* after sunset
> *ch'u keng* – 10 *k'o* after dusk
> *tai tan* – period of "waiting for dawn" – 10 *k'o* before dawn
> *hsiao* – dawn – 2½ *k'o* before sunrise
> *jih ch'u* – sunrise.[44]

---

[43] E. Chavannes, "Le Cycle turc," p. 31. Needham *et al.*, *Heavenly Clockwork*, pp. 200–1.

[44] W. S. Williams, *Middle Kingdom*, vol. 2, pp. 67–80; Needham *et al.*, *Heavenly Clockwork*, pp. 190–95.

The hour was announced to the public in the daylight period by the sound of the drum and during night by that of the gong, a signal which served two purposes. It informed the people of the hour and at the same time provided assurance that the guards were indeed actively circulating through the streets of the city, where they questioned anyone found wandering about after curfew. In their left hand some of the guards carried a hollow bamboo cylinder which when struck with their right hand made a loud noise. Others carried instead a piece of wood in the form of a fish, which served the same purpose. Guards also were equipped with a rattle attached to their arm, which they operated as they moved from one post to another. Fellow guards upon hearing the rattle replied by striking their bamboo cylinder or wooden "fish" to indicate that they were actively making their rounds. The officers of the guard were often mounted on donkeys and preceded by a soldier carrying a lantern (Figure 7).[45]

In the account of his travels, the Venetian adventurer Marco Polo (1254?–1324?) noted having seen stone towers in every Chinese city. To these the inhabitants removed their possessions for security in the event of a fire in the city, which apparently occurred frequently. By imperial edict a guard of ten watchmen was stationed under cover, five on duty during the daylight hours and five others during the night:

Each of the guard rooms is provided with a sonorous wooden instrument [drum] as well as one of metal [gong], together with a clepsydra [*horiuolo*] by means of which latter the hours are ascertained. As soon as the first hour of the night has expired, one of the watchmen gives a single stroke upon the wooden instrument, and also upon the metal gong [*bacino*] which announces to the people of the neighboring streets that it is the first hour. At the expiration of the second, two strokes are given, and so on progressively, increasing the number of strokes as the hours advance . . .[46]

To pay the salaries of watchmen maintained in towns of the Chinese interior, subscriptions were collected from householders in the wards nightly patroled by them. At each period the guards assembled at a designated central point and began their vigil by the sound of a tattoo beaten on a drum and voiced by a clarinet. The drumming began with a slow beat which increased gradually until it became a brisk rattle. In Peking the time for the neighboring wards was determined by a clepsydra in the Drum Tower in the northern part of the Tartar City, and the change of beat swept quickly to the adjacent districts.[47]

[45] Magalhaens, *Relation*, pp. 150–55; Planchon, "L'Heure," pp. 247–50.
[46] Yule and Cordier, *Marco Polo*, vol. 1, pp. 375, 378 no. 6, vol. 2, pp. 187–89; Burkhardt, *Chinese Creeds*, vol. 2, pp. 13–14; Planchon, "L'Heure," pp. 247–48.

Sulaiman al-Tajir (fl. ninth century), also known as "Sulaiman the Merchant," a Muslim traveler in the Far East, described the drum towers of Chinese cities he visited. He reported that five trumpets at each of a city's four gates were blown at specified hours of the day and night. To announce the time to the populace, ten drums maintained by the city were beaten at the same time as the trumpets were sounded.

Sulaiman noted that the Chinese were equipped with sundials and "clocks with weights," by which he meant steelyard clepsydrae. The account of Sulaiman's journeys, entitled *Akhbār al-Sin wa'l-Hind* (Information about China and India), was recorded by an anonymous author and published in 851. This is the earliest Arabic account of China and many of the coastlands of the Indian Ocean.

Each morning the imperial palace gate and the "Locked Gate" were opened one-quarter of an hour after the middle of the double-hour *mao*, or approximately 6:15 A.M. A drum was beaten at each *k'o*. After the eight drum beats, the official in charge publicly displayed an ivory tablet inscribed in gilt with the character for the hour. There were seven of these. The termination of the remaining time was signaled by a "cock-crow" (which may have been the sound of a trumpet) and thereafter the drum was beaten fifteen times to indicate the middle of the double-hour.[48]

References to the community time signals occur frequently in Chinese poetry. Writing early in the T'ang Dynasty, Tu Shen-yen noted:

> But when the bells and drums of Lo-yang reach me,
> Carriage and horses turn back slowly, unwilling.[49]

The sounds of the night watch were mentioned in a poem recalling a "Night on a tower," by Tu Fu (A.D. 712–770) a poet of the high T'ang,

> Drums and horns of night's fifth watch, notes both strong and
>   sad,
> In the Three Gorges the river of stars, reflections stirring,
>   shaking.[50]

---

[47] Ferrand, *Voyage de Sulayman*, cited in Sarton, *Introduction*, vol. 1, pp. 571–72 and in *Isis*, vol. 6, p. 146.

[48] Wang Ying-lin, *Yü hai*, ch. 11, abridged in Toktaga and Ou-yang Hsüan, *Sung shih*, pp. 3b ff. Quoted in Needham *et al.*, *Heavenly Clockwork*, p. 90.

[49] Tu Shen-yen, "On A Summer Day, Passing By the Mountain Library of Mr. Cheng," in Owen, *Early T'ang*, p. 331.

[50] Tu Fu, "Night in the Tower," in Owen, *The Great Age*, p. 213. Tu Fu was born in Tu-ling in Shensi. Although showing great promise in his youth, he failed his examinations and became a professional poet. He was appointed to a position by Emperor Hsüan Tsung, but the revolu-

It was not until the advent of Jesuit missionaries to China in the early seventeenth century, particularly under the influence of Rev. Johann Adam Schall von Bell, S. J. (1591–1666), that the Chinese adopted a new division of the day, of twenty-four hours each consisting of sixty minutes and each minute of sixty seconds, thus making each *k'o* of fifteen minutes' duration.[51]

Undoubtedly the least accurate of the common methods of time measurement used by the Chinese was one reported by the French Lazarist missionary and traveler, Père Evariste Régis Huc (1813–1860), observed during his travels in China in the mid-nineteenth century.

One day as he was passing through a rural region with several companions, he stopped to ask a farm boy whether it was yet noon. The sky was overcast and the sun was hidden by clouds. The boy asked them to wait, left his work and returned moments later holding a cat in his arms. No, he reported, it was not yet noon; one could tell from the cat's eyes. Unwilling to display their ignorance, Huc and his companions accepted the information, without further questioning the boy. They thanked him and continued on their travel.

Later, when Père Huc and his companions had an opportunity to inquire of others, it was explained to them that observing a cat's eyes to tell the time was a common practice among the rural Chinese. They were aware that the pupil of the cat's eye kept diminishing more and more as daylight increased. By noon the pupil was reduced to a very thin line, and then the pupil gradually dilated and became larger and larger as the afternoon advanced.[52]

Chinese sources stated "the pupil of the cat's eye marks time; at midnight, noon, sunrise and sunset, it is like a thread; at 4 o'clock and 10 o'clock, morning and evening; it is round like a full moon; while at 2 o'clock and 8 o'clock, morning and evening, it is elliptical like the kernel of a *tsau* [*tsao*] or date."[53]

Studies of cat vision indicate that a cat's eyes are sensitive to the ultra-violet rays of the spectrum, and the pupils are capable of considerable expansion and thus admit all the light that is available. Cats, as well as owls, can see better than humans in partial darkness, such as at dusk, and consequently would react in the same manner even on cloudy days.

tion that drove the emperor from his throne brought Tu Fu into exile. Under the successor, Emperor Su Tsung, he became Censor. Falling into disgrace at court, he was appointed governor of Shensi. He resigned and retired to Szechuan. Later he became secretary of the Board of Works, then returned to wandering. During an inundation he sought refuge in an abandoned temple, living for days on nothing but roots. He was eventually found and died almost immediately thereafter, having just partaken of roasted beef and white wine to excess after such a long fast. Giles, *Dictionary*, pp. 780–82.

[51] Planchon, "L'Heure." pp. 193–95; Needham, *SCC*, vol. 4, part 3, p. 461.

[52] Huc, *Journey*, vol. 2, pp. 303–4; Stimpson, *Nuggets*, p. 79.

[53] K. Ball, *Decorative Motives*, p. 151; Couling, *Encyclopedia Sinica*, p. 84; C. A. S. Williams, *Encyclopedia*, pp. 57–58.

The tradition of considering the cat as a "timepiece" had its origin in ancient times in China, and the Egyptians, who had a superstitious reverence for cats, also claimed that its eyes changed with the course of the sun across the sky. They compared the cat's eyes with the moon, which waxed and waned from a thin crescent to a full moon and back to a crescent once more.[54]

According to an early history of Japan, the *Nihon shoki*, Prince Tenchi who later became emperor, in 660, in May of the sixth year of the reign of Emperor Saimei, devised a type of clepsydra called the *rōkoku*, the first known in Japan. He is credited also with having created Japan's first time system, in which April 25 of the lunar calendar, which was equivalent to June 10 of the solar calendar, was designated "Time Day."

In accordance with the Guchū calendar, the natural day was divided into 50 *koku*, each of which contained 6 *bu*. The day was divided into 12 double-hours (*shinkoku*), each of which contained 4 *koku* and 1 *bu*. Although this system of time division was generally used by the public, the official astronomers and calendar makers utilized a more accurate division of the day into 100 *koku*. By this method 1 *koku* consisted of 84 *bu*, and 1 *shinkoku* or double-hour was made up of 8 *koku* and 28 *bu*.

The Japanese name for the hour was *toki*, each of which was divided into 10 *bun*. The *bun* in turn was divided into 10 *rin*.[55]

A number of references to time periods occur in the *Nihon shoki*, such as *Jūyokka tora no toki* (The Tiger's Hour of the Fourteenth Day) and *Jūrokunichi no ne no toki* (The Mouse's Hour of the Sixteenth Day) for example. Although few details are known of the time system Tenchi devised, it appears to have been the first system of time measurement used in Japan. However, the time periods were twelve in number and designated by the Chinese signs of the Twelve Terrestrial Branches.[56]

In Japan, during the Tokugawa shogunate, time measurement was based upon the Chinese system of double-hours varying with the seasons, with the exception that the Japanese natural day began at dusk instead of at sunset. The Japanese utilized two systems, the equinoctial system used by the astronomers, and the temporal hour system in civil life. In the equinoctial system, each hour was equivalent to two Western hours as in China, and remained the same length throughout the year.

The Japanese continued to use the Chinese symbols of the Twelve Terrestrial Branches, which have often been mistakenly described as signs of the zodiac.

[54] *The Chinese Repository*, vol. 7, March 1839, Article 4, p. 598, quoted in C. A. S. Williams, *Encyclopedia*, p. 57.

[55] Hirayama Kiyotsugu, *Rikihō oyobi jihō*.

[56] Asahina, "On the Jikōban," pp. 19–34.

Whereas in China each sign represented the middle point of the *shih*, the Japanese reckoned from the beginning of the hour. As an example, while in China the "Tiger" hour was the period from 3:00 to 5:00 A.M., in Japan it designated the period from 4:00 to 6:00 A.M.[57]

The Japanese counted the hours backward – the highest number, 9, was allotted to midnight and mid-day, and the progression of the hours was represented by the numerals 9, 8, 7, 6, 5, 4 inasmuch as six numerals were required for each of the two periods of the full day. The numbers 1, 2 and 3 were reserved for use by the temples. This curious system of numeration was derived from the ancient system of beating the drum to announce the current hour. Several theories exist concerning the derivation of the system. All are based on the assumption that the magical number 9 be assigned to the one invariable moment of the day when the sun was in the meridian. The number 9 was selected for the purpose from the Chinese classic, "The Book of Changes."

One theory proposes that the numerals 1 through 6 were multiplied by the mystical number 9 and that the second resulting digit in each instance was adopted:

$$
\begin{array}{ccccccc}
 & 1 & 2 & 3 & 4 & 5 & 6 \\
\times & 9 & 9 & 9 & 9 & 9 & 9 \\
\hline
 & 9 & 18 & 27 & 36 & 45 & 54
\end{array}
$$

Another version suggests that inasmuch as the hours were divided on a decimal basis, the numbers 1 through 6 were derived by subtracting each of them from 10:

$$
\begin{array}{ccccccc}
 & 10 & 10 & 10 & 10 & 10 & 10 \\
- & 1 & 2 & 3 & 4 & 5 & 6 \\
\hline
 & 9 & 8 & 7 & 6 & 5 & 4
\end{array}
$$

Yet another possible explanation was that inasmuch as the primary occupations of Japan were agriculture and fishing, and work began at dawn, the period remaining before the time for rest at noonday became less and less as the day progressed and was represented by the diminished number of the hour.

In Tokyo and presumably elsewhere in Japan, the passing hours were first announced by one stroke of the bell, followed a minute later by a second stroke, and then quickly by a third. Then came a long pause, and then the number of strokes corresponding to the hour were tolled at ten second intervals, except for the last stroke which followed quickly to indicate that the striking was completed.[58]

[57] Bayer, *De Horis Sinica*, pp. 1–32. Cited also by Robertson, *Evolution*, p. 205n.
[58] Hoffman and von Siebold, cited by Robertson, *Evolution*, pp. 200–1; Fraissinet, *Japon contemporain*, pp. 161–69.

Table 1

	Number	Sign	Equinoctial hours	Temporal hours
9	Kokonotsu	Rat (*ne*)	XI–I	XII–II midnight
8	Yatsu	Bull (*ushi*)	I–III	II–IV
7	Nanatsu	Tiger (*tora*)	III–V	IV–VI
6	Muttsu	Hare (*u*)	V–VII	VI–VIII dawn
5	Itsutsu	Dragon (*tatsu*)	VII–IX	VIII–X
4	Yotsu	Serpent (*mi*)	IX–XI	X–XII
9	Kokonotsu	Horse (*uma*)	XI–I	XII–II midday
8	Yatsu	Goat (*hitsuji*)	I–III	II–IV
7	Nanatsu	Ape (*saru*)	III–V	IV–VI
6	Muttsu	Cock (*tori*)	V–VII	VI–VIII dusk
5	Itsutsu	Dog (*inu*)	VII–IX	VIII–X
4	Yotsu	Boar (*i*)	IX–XI	X–XII

Source: Yamaguchi, "Japanese Clocks," pp. 73–74.

Table 1 compares the two Japanese time systems based on the equinoctial and on the temporal hours.

When mechanical clocks first came into use in Japan, they were made to conform to the practice of the temples, which informed the community of the time by the striking of their great bells.

Another Japanese time system was known as *rokuji*, which meant literally "six hours" (i.e. *roku* – 6 and *ji* – hours). In this system each day was divided into the following six periods:

1. Shinchō – morning
2. Nitchū – afternoon
3. Nichibotsu – evening
4. Shoya – early night
5. Chūya – midnight, or
   Hanya – 'half night'
6. Kōya – after night.

According to some sources, this system was derived from the Buddhist religious practice of praying to Amida, at six intervals throughout the day, a practice which was known as *rokuji no raisan* and literally means "to pay homage at the six hours." Amida Nyorai (Amitabha Tathagata) was the "Buddha of Infinite Light and Lord of the Western Paradise."[59]

---

[59] Yamaguchi, "Japanese Clocks," p. 71; Robertson, *Evolution*, pp. 191–92.

With the end of the civil rule by the Tokugawa shogunate following the revolution of 1867 and the restoration of the Mikado, Japan was thrown open to Western influence. The ancient and complicated methods of timetelling were abandoned, and on January 1, 1873 the Western calendar and system of time measurement were officially adopted.[60]

An early account of the Chinese and Japanese horary systems was provided by the Jesuit missionary, João Rodrigues Tçuzu, S. J. (1562–1633). He observed that

Both the Chinese and Japanese count the hours by the names of animals. The artificial day begins at six o'clock in the morning when the sun rises in the middle of their hour of the Hare. Then follows the hour of the Dragon, which begins at seven o'clock and ends at nine; next comes the hour of the Serpent, which begins at nine and ends at eleven. Then follows the hour of the Horse, which begins at eleven, reaches its midpoint at noon and finishes at one. The hour of the Goat lasts from one to three, the hour of the Monkey from three to five. The last hour of the day, that of the Cock, begins at five, with the day finishing at six o'clock with the setting of the sun in the middle of this hour.

Thus the artificial day is divided into six of their hours or twelve of ours. After these six hours the night begins from six o'clock and continues in the same fashion until midnight and thence to six in the morning when the night comes to an end. Thus the night also lasts six of their hours, twelve of ours. They divide each of their hours into eight quarters which they call *koku*, or, in Chinese, *ke* [*k'o*], although in older times they used to make ten divisions, that is, two short ones, and eight long ones. The first four quarters, lasting until the middle of the hour, are called the first prime, the second prime, the third prime, and the fourth prime. Then comes the half-hour, followed by four quarters called the first, second, third and fourth divisions after the half-hour. After that comes the name of the following hour.[61]

Few records remain relating to time measurement in Korea during its early periods. The Chinese calendar was adopted probably in the sixth century A.D., in the ancient kingdom of Silla in southeast Korea, and the sundial was in use in Silla from early times. A fragment of a monumental granite sundial of that period has been preserved. The first clepsydra in Silla was constructed at a monastery in Whang Ryong in A.D. 718, and special officials were appointed to supervise it. Tradition relates that the great bell of Silla was struck twenty-eight times at curfew to honor the twenty-eight constellations, and thirty-three times in the morning to acknowledge the thirty-three Buddhist heavens. During the Koryŏ period, time was measured by the sundial during daylight hours and by graduated candles and the clepsydra at night. During the Yi period, sundials were installed in public areas and clepsydrae were shipped to the provinces to be used there. A system of marking time by the striking of bells was promulgated in A.D. 1414.[62]

---

[60] Yabuuchi in "Chūgoku no tokei" gives the source as T. Oda.

[61] Rodrigues Tçuzu, *Historia*.

[62] Rufus and Won-Chul Lee, "Marking Time," pp. 252–57; Sang-woon Jeon, *Science and Technology*, pp. 42–64.

The regulation and recording of time was entrusted to certain petty officials who utilized what was called the "dew clock," probably a clepsydra. The period of night hours until twelve o'clock midnight was divided into five parts (*giung*), and each of these in turn was divided into five smaller units (*jium*). The *jium* were announced by the sound of a gong and the beating of a drum. At midnight the palace gates were opened.

An early surviving clepsydra, now in the National Museum in Seoul, was constructed in 1536, the thirty-first year of the reign of Choong Jong, the eleventh ruler of the Yi Dynasty. It was originally installed in the "Time House" (Boo Roo Kak) inside the west gate of the Kyŏng Bok Palace. It consisted of three containers from which water flowed through a tube to elevate a floating arrow in the form of a tortoise which indicated the time.[63]

Maintained in the king's palace was a timekeeper in the form of a lantern of oiled paper which enclosed a hempen rope soaked in nitre. This "fire rope" (*kwa-sung*), which was protected from the wind by the lantern, was kept burning continuously and at a constant rate. Each hour was divided into four parts indicated by cords tied at these junctures to the rope. The time was announced by means of a lantern placed before the palace window, its transparent sides marked with the different *giung*. This timekeeping device was the particular charge of an official and served as a check on the "dew clock."

Another type of Korean lantern timekeeper was the perpetual or eternal lamp kept alight in Buddhist temples (*changmyŏngdŭng*), also known as a "jade lamp" (*okyungjan*). Soybean or sesame oil served as the fuel and the amount of oil consumed indicated the passage of time.[64]

A device said to have been used in Korean Buddhist temples for time measurement in the Yi Dynasty was the *ok-dungjan*, a ceramic lamp in which a special type of vegetable oil served as the fuel.[65]

As in China, incense sticks inserted in a bed of wood ash in a bronze container also served the purpose of measuring the time in Korea.

An interesting account of timetelling in Burma observed by His late Royal Highness Prince Damrong during a visit in 1944 was featured in his volume of *Nithan Borankhadi* (Tales of Other Times). He noted that the procedure was quite similar to that current in Thailand. Like all royal palaces in Burma of earlier epochs, the Royal Palace included a very tall tower. The bottom level contained a room for storing clocks. At the top of the tower was a hallway in which were stored the drums and gongs used to strike the hours. He noted that they had remained in disuse for several decades. Unable to obtain information about the

---

[63] Sang-woon Jeon, *Science and Technology*, pp. 57–60.
[64] *Ibid.*, pp. 64–65.
[65] Communication to the writer from Dr. Sang-woon Jeon, March 2 1960.

schedule for striking the drum and gong, His Highness conjectured that the gong was used in daylight hours and the drum during the night. He noted that the same system had been used in ancient times in Thailand. It was reflected in the Thai language, in which the Thai word for "hour" during the daytime was represented by the word "*mong*," the sound of the gong, such as "*si mong*" and "*haa mong*." During the night the sound of the drum, the word "*thum*," was used for "hour," such as "*si thum*" and "*haa thum*."[66]

The common forms of time measurement prevalent in the countries of East Asia, by means of the sun's shadow, the flowing of water and sand and fire consumption of cord, were to be supplemented by yet another, relatively accurate and more exotic – the burning of incense in a measured trail or path.

[66] Damrong Rajanuhab, *Nithan Borannakhadi*. Courtesy of Lydia B. Voorhees.

2

Incense and the populace

Incense has played such an important role in so many aspects of daily life in East Asian countries from the most ancient times, and formed part of so many functions, that it is not surprising that it was eventually utilized for time measurement.

Incense is defined as the material used to produce a fragrant odor, as well as both the perfume itself and the fumigation derived from its consumption. Materials used included certain resins, barks, woods, seeds and fruits. The origin of the use of incense is laden with an infinity of tradition and detail. From the earliest times incense was associated with death, and tradition claims that it originated from the practice in ancient societies of burning fragrant woods in funeral pyres in order to cloak offensive smells emitted in cremation. In time the exudations of such fragrant woods as frankincense and myrrh also were burned in temples to camouflage the odor of the burning flesh of animals and human sacrifices. Gradually incense assumed religious significance and was used to achieve spiritual enlightenment.

It is believed that incense was first used in the countries of the Near East. In Egypt, for example, incense was known as early as the Eighth Dynasty (1580–1350 B.C.). It was next introduced into India in a very early period, and is mentioned in both of the ancient Indian epic poems, the "Ramayana" and the "Mahabharata."[1]

From India the use of incense spread simultaneously into Nepal, Tibet, Ceylon, Burma, China, Korea and Japan. There is evidence that prior to the introduction of incense from India, aromatic drugs were already known in China as early as the late Chou or early Ch'in period, between about 250 and 200 B.C.

According to the *T'u shu chi ch'eng*, as derived from the thirteenth-century work *Tung t'ien ch'ing lu*, the Chinese ancients did not burn incense as such at

---

[1] Atchley, *Use of Incense*, pp. 6–28; Nakamura, "Burning Tradition," p. 70. The *Ramayana*, ascribed to the poet Valmiki, consists of 24,000 stanzas in seven books which recount the life of the young hero, Rama. He is the reincarnation of the deity Vishnu and heir to the throne of Ayodhya. After winning Sita as his bride he is forced into exile but is eventually restored to his throne. The *Mahabharata* details the long struggle between the five Pandavas, sons of Pandu, and the Kauravas, the family of Dhritarashira, who refuse to surrender the throne to rightful nephews. Best known of the numerous episodes is the story of Nala and Damayanta.

first, but attempted to communicate with spiritual beings by means of burning "southernwood" and mugwort. It was burned in a powder form, neither as paste nor as incense sticks. Pine resin was added to their elixirs and pills of longevity because it was considered to be an ingredient conducive to immortality inasmuch as it was full of "soul essence." From this practice grew the concept that the burning of incense wafted one's prayers to the gods; another version was that incense was the "Messenger of Earnest Desire" which brought down Buddha's mercy to the supplicant.[2]

The Chinese also used "incense water," prepared by immersing aromatic wood and plants in water, in the same manner that it was used in India. The solution was rubbed on the skin to provide relief from intense heat.[3]

In the early period, little distinction was made between perfume, aromatics, fragrances, drugs, spices and incense. These were burned to produce a pleasant fragrance to be used on the person while bathing, on clothing, in places of worship, in the home, and in government offices. Incense became pervasive, for it was believed to sustain the spirit and to attract the auspicious as well of gods and lovers. Fragrance was featured in state business as well, particularly in the presence of the emperor and in civil offices, for it was deemed necessary to provide and maintain a pleasant frame of mind.[4]

The burning of incense in Far Eastern countries served religious, secular, and recreational purposes. In religious rites, it was an act of consecration similar to anointment; in secular use it served to freshen the air, to cool the body, to scent clothing and accoutrements; and in recreation it was utilized in games as an elegant test of the accuracy of the participants' sense of smell.[5]

The oldest known official pharmacopoeia, *Hsin hsiu pen ts'ao* of approximately A.D. 659, lists six prominent constituents of incense: aloes wood, frankincense, cloves, patchouli, elemi and liquidambar, which then were considered to be the most important ingredients of aromatics.[6]

A discussion of aromatics also forms part of the *Ch'ing i lu* (Records of Unworldly and Strange Things) compiled in about A.D. 950 by T'ao Ku.[7]

---

[2] Casal, "Incense," pp. 48, 54; Hearn, *Ghostly Japan*, pp. 27–28.

[3] *T'u shu chi ch'eng*, vol. 1126 (*k'ao kung tien*, book 250, *hui kao*, p. 4), vol. 1124, (*lu pu tsa lu* Book 236, p. 2a); *Tung t'ien ch'ing lu* (thirteenth century), quoted in Laufer, pp. 178–79, fn. 4; Casal, "Incense," p. 47.

[4] Waseda, "Incense Ceremonies." pp. 14–15; Needham, *SCC*, vol. 5, part 2, pp. 133–34.

[5] Needham, *SCC*, vol. 5, part 2, p. 148.

[6] *Hsin hsiu pen ts'ao*, ch. 12, pp. 12b, 13a; Needham, *SCC*, vol. 3, p. 143.

[7] T'ao Ku was in the imperial service of the Chin Dynasty from an early age. When Emperor Yeh-lü Te-kuang wished him to accompany the court in the north, T'ao hid in a Buddhist temple until the emperor's death shortly thereafter. Subsequently he was appointed Supervising Censor under the new dynasty. Under the Later Chou and Northern Sung Dynasties he

The subject occurs from time to time in literary works, a notable example of which was the essay *T'ien hsiang chuan* by Ting Wei (A.D. 969–1040). Entries on aromatics and incense are to be found also in early encyclopedias, such as the *T'ai p'ing yü lan* produced in about A.D. 983.[8]

Many of the earlier writings on incense have been lost, but several works on the subject produced in the T'ang and Sung periods have survived. Possibly the earliest is the *Hsiang p'u* (List of Aromatics) written by Shen Li in about 1074, of which only fragments exist. Another is the twelfth-century work with the same title written by Hung Ch'u in about 1115. It contains forty-three entries providing history, description, place of origin, and uses. Somewhat later, in about 1151, a catalogue of incense entitled *Hsiang lu* (Record of Aromatics) was produced by Yeh T'ing-kuei.[9]

A compilation of previously existing works about incense was assembled in the twelfth or thirteenth century by Ch'eng Ching, also entitled *Hsiang p'u* (List of Aromatics). It included not only the *Hsiang p'u* of Hung Ch'u but a substantial number of excerpts from the work of Shen Li as well. Somewhat later a provincial professor named Hsiung P'eng-lai was the author of another *Hsiang p'u*, which appeared in about A.D. 1322.[10]

Of the writings on incense that appeared in the Ming period, most notable was the *Hsiang chien* of T'u Lung in the sixteenth century, and the more extensive *Hsiang ch'eng* (Records of Incense) produced by Chou Chia-chou in the early seventeenth century. Another important work was the *Hsiang kuo* by Mao Chin, in which was listed more than fifty varieties of plant aromatics used for incense as well as a number more for other uses.[11]

The advent of new aromatics in the Han period, imported by Emperor Wu (140–87 B.C.) as the consequence of extensive commercial relations developed

rose to the position of president of the Boards of Punishment and Revenue. Giles, *Dictionary*, pp. 720–21.

[8] Ting Wei was a native of Ch'ang-chou in Kiangsu who entered official life at an early age. He distinguished himself in a campaign against the aborigines of Szechuan, and became president of the Board of Civil Office. Later he became minister of state but because of his oppressive rule he was degraded and banished. Giles, *Dictionary*, pp. 737–38.

[9] Hung Ch'u possibly may be identified with Hung Chüeh-fan, the distinguished poet and calligrapher of the eleventh and twelfth centuries. Grandson of Hung Hao and native of Hsing-ch'ang, he became a member of a trio, consisting of his uncle P'eng Yüan-ts'ai and a townsman named Tsou Yüan-tso, which became known as the "Three Wonderful Men of Hsin-ch'ang." He later took orders as a Buddhist priest assuming the name Hui Hung. Giles, *Dictionary*, p. 343.

[10] Ch'en Ching, *Hsiang p'u*; *Hsiang p'u* by Hsiung P'eng-lai; Needham, *SCC*, vol. 5, part 2, pp. 134–46.

[11] T'u Lung, *Hsiang chien*; Chou Chia-chou, *Hsiang ch'eng chou*; and Mao Chin, *Hsiang kuo*; Needham *SCC*, vol. 5, part 2, pp. 134–37, 147.

outside the empire, resulted in the invention of a vessel for burning them, the incense burner. The talents of the bronze worker and the potter were enlisted, and of the gem carver as well, to produce the earliest form of what was known as the "hill censer" (*Po-shan lu* or *Po-shan hsiang lu*).

Made of pottery or bronze, its first form was of a goblet narrowing below into a short foot supported on a tray or basin. Its separate lid was conical, moulded into a shape resembling a hill surrounded by two rows of waves and generally perforated at four or more points. During the same period the "hill censer" was produced also in jade. Variations included the heads of zoomorphic beasts and birds.[12]

The later forms of the incense burner were copied from the various types of vessels used for ancestral worship of the Shang and Chou Dynasties. A favorite form was the ancient *ting*, found in Taoist, Confucian and Buddhist temples. It was a bronze vessel shaped like a cauldron supported on three or sometimes four feet, which was originally used for containing meat offerings in ancient rituals.

An account of how the incense burner known today evolved from the "hill censers" of the Han period occurs in the *Tung t'ien ch'ing lu*:

Therefore they had no censers [incense stoves]. The vessels which nowadays are called censers are all made after models derived from sacrificial vessels in the ancestral temples of the ancients. The censer in the shape of a "goblet" (*chio* [*chüeh*]) took its pattern from the goblet of the ancients. The "lion" censer is made after the ancient tazza with a single stem. The censer in the form of a ball is copied from a cauldron of the ancients, and there are many others of this kind. There are also new casts, but in shape resembling forms that existed in times of old, with the only exception of the *Po-shan lu*, which was used in the palace of the heirs-apparent of the Han time. It is from this type that the manufacture of censers first commenced. There are counterfeits among them which can be discriminated by the color [patina] of the object.[13]

The invention of the *Po-shan lu* is attributed to a mechanician named Ting Huan, who first produced it in about 180 B.C. His "nine-storied hill censers," (*chiu ts'eng Po-shan lu*), were said to incorporate figures of strange birds and beasts which were activated by the hot fumes from the burning incense. He also invented the *pei chung hsiang lu*, a censer with gimbals used for perfuming bedclothes, although some form of the latter may have existed half a century earlier.[14] Although several ancient examples in jade were known, generally they were made of bronze or pottery. The earliest censers were used exclusively in the

---

[12] Laufer, *Pottery*, pp. 187–92; Needham, *SCC*, vol. 3, pp. 580–81.

[13] *Tung t'ien ch'ing lu*, quoted in Laufer, *Pottery*, pp. 178–79.

[14] Needham, *SCC*, vol. 4, part 2, pp. 233–35; vol. 5, part 2, pp. 134–46; *Hsi ching tsa chi*, ch. 1, p. 7b.

palace of the imperial heirs-apparent and served as wedding gifts when the rulers married[15] (Fig. 5).

It was recorded in the *Hsi ch'ing ku chien* that the people of the Chou period burned artemisia mixed with animal fat, but that when the *Po-shan hsiang lu* were first produced in the Han period, they were used for burning *lan hui*, a fragrant species of orchid, and presumably artemisia as well. The use of artemisia and orchids was abandoned after the importation of aromatics new to the Chinese in the reign of Emperor Wu (140–87 B.C.). These included camphor and cloves brought from Annam Baros (North Viet Nam and Sumatra), and Parthian incense and attar of roses from countries in the West.[16]

By the T'ang period, the use of incense had become an established part of secular as well as religious life, especially among the gentry. In addition to its use by both sexes of the upper classes for toiletry, for perfuming their homes, as well as their offices and temples, it was popularly featured in love potions and lovemaking. In China, incense served also for the purification of dwellings, a ceremony known as *kuan huo* or "changing the fire," which took place from at least the seventh century A.D.[17]

From very early times, incense or aromatics were blended from many essences. A combination that was frequently mentioned in early poetry was "hundred-blend aromatics" (*po ho hsiang*). Writing in the T'ang period, the statesman and poet Ch'üan Te-yü (759–818)[18] noted its use in describing a beautiful girl in her boudoir:

> At the green window, pearl screened, embroidered with Mandarin ducks,
> An attendant slave-girl first burns a "hundred-blend aromatic."[19]

Although in China a large number of aromatics had been produced from indigenous plants and animals from an early period, those imported from other lands during the T'ang period were more highly prized. The ports of Canton and Yangchow became major centers of commerce in incense.

Among the aromatics imported by the Chinese in earliest times were aloes wood (*ch'en hsiang*), anise (*hui hsiang*), basil (*lo le hsiang*), benzoin (*an hsi hsiang*), ambergris (*lung hsien hsiang*), camphor (*lung nao hsiang* and other varieties of camphor), cassia (*kuei hsiang*), civet (*ling mao hsiang*), clove (*ting*

---

[15] Laufer, *Pottery*, pp. 174–90; Needham, *SCC*, vol. 3, pp. 580–81; vol. 5, part 2, p. 133.

[16] *Ko chih ching yüan*, book 57, p. 1a; quoted in Laufer, *Pottery*, p. 180.

[17] Needham, *SCC*, vol. 5, part 2, p. 148; Tiffany, *Canton*, pp. 180–81.

[18] Ch'üan Te-yü rose to the highest state offices in the public service and was recognized as an eminent scholar. Giles, *Dictionary*, pp. 201–2.

[19] Ch'üan Te-yü, "Ku yüeh fu," as cited in Schafer, *Golden Peaches*, p. 159.

hsiang), frankincense (*ju hsiang*), jasmine (*mo li hua hsiang*) and storax (*su ho hsiang*).[20]

The legendary frankincense, also known as Olibanum, an aromatic well known to Westerners, is a gum resin derived from the tree *Boswellia thurifera*; the tree is indigenous to southern Arabia and Somalia, from whence the aromatic was brought to India. From India it was exported into the countries into which Buddhism was introduced.[21]

Benzoin, also known as "Arsacid aromatic" or gum guggul, was imported from Indonesia and Indochina. It was already known in China from the fourth century A.D. and in the next two centuries it was brought from Turkestan and associated with Gandhara, a source of Buddhist doctrine. Benzoin from Sumatra, called the "frankincense of Java," replaced the earlier bdellium in the mid T'ang period. Benzoin continues to be widely used in India in the present time. Incense sticks of benzoin sold throughout the country are called *ud-buti* or "benzoin lights" as well as *aggar-ki buti* or "wood aloe lights."[22]

Undoubtedly, the most popular ingredient in Far Eastern incense was sandalwood (*Santalum album*). This is a fragrant wood native to India, where its use has been traced to at least the fifth century B.C. Tradition relates that the first image of Buddha imported into China from northern India by the earliest Buddhist pilgrims, soon after the beginning of the Christian era, was carved of white sandalwood. The archaic name for sandalwood in China was *chan t'an*, and it has been used for religious purposes in various forms. As chips it is heaped before the Buddhist altars and ignited. In a pulverized form it is moulded into incense sticks to be burned before altars in sacred shrines. It is a wood favored also for carving the rosaries worn by Buddhist monks, and in general it is widely used wherever Buddhism exists. In modern times it is known as "sandal perfume" or "sandal incense" (*t'an hsiang*). The wood is still imported into China from India and used for carving Buddhist images.[23]

A favorite aromatic, particularly during the T'ang period, was aloes wood, known in Sanskrit as *agaru*. It is a product of several trees of the genus *Aquilaria* native to southeast Asia. The wood used for incense is of that part of the tree made dark and heavy by a saturation of resin, in contrast to the surrounding softer lighter wood. Because these saturated portions were heavier than water, the Chinese referred to the product as "sinking aromatic." It was an important component of incense in medieval China.[24]

---

[20] The components of many early Far Eastern perfumes and varieties of incense are extensively described in Schafer, *Golden Peaches*, pp. 155–94.
[21] Watt, pp. 173–74; Dastur, *Useful Plants*, pp. 51–52; Van Beek, "Frankincense," pp. 99–126.
[22] Schafer, *Golden Peaches*, pp. 169–70.
[23] Yabuuchi, "Chūgoku no tokei," pp. 19–25; Schafer, *Golden Peaches*, pp. 134–38, 157–60.
[24] Schafer, *Golden Peaches*, pp. 158–65.

Popular in the same period was lakawood (*kayu laka*), known also as purple liana aromatic, which consists of the heartwood of a rosewood liana brought from Indonesia.

Camphor is an aromatic of crystalline form derived from the wood of a large tree indigenous to China, Japan and Tonking (North Viet Nam). A type preferred in early times but not found in these particular regions was laevo-borneol (*kapur Baros*), which the Chinese called "dragon brain aromatic" (*lung nao hsiang*). It was obtained from tall trees and brought from Sumatra and Borneo. One of the scents derived from camphor and used in the T'ang period was known as "auspicious dragon brain."

The dark purple storax was brought to China from Rome and Parthia prior to the T'ang period. This substance was Malayan balsam and called "God's tallow." A strong drug, some Chinese believed it was lion's dung.[25]

Another source of incense favored by the Chinese was wood aromatic, also known as costus root or putchuk, which emanated the odor of violets. Chiefly used in medicine, it also served a minor role in the preparation of incense.

The Chinese also used the grey, light substance known as ambergris, a secretion in the intestines of the sperm whale which is still used by perfumers to render flower scents permanent. In medieval times Arab traders obtained it from natives of the Nicobar Islands and possibly from Somalia as well, and sold it to other countries. When it was first imported into China in the tenth or eleventh century, it became known as "dragon's spittle."[26]

Onycha, an aromatic derived from the operculum of a gastropod mollusk found on the shores of China south of the Yangtze, was called "plate aromatic," and was mixed with other ingredients to form the incense "plate decoction."[27]

Oleoresins derived from the trees of the genus *Canarium* were called *elemis* or *breas*. In the T'ang period the *elemi* of the copaliferous *kanari*, found in parts of Lingnan in the form of a white granular substance like sugar, was mixed with pulverized carbon to create "tram-sugar aromatic."[28]

The production of the incense stick (*hsiang pang*) in eastern Asia dates from at least the medieval period. It was occasionally mentioned in the literature, such as in a poem by the T'ang writer, P'i Jih-hsiu (c 833?–883), who wrote:

> I block out the midday brightness with a
> screen depicting dark woods,
> Burn a stick of heavy incense, nursing my hangover.[29]

To produce incense, a prescribed variety of woods and dried gums were pulverized into mixtures with pestles in stone mortars or by means of longitudinal

---

[25] *Ibid.*, pp. 165–69.   [26] *Ibid.*, pp. 174–75.   [27] *Ibid.*, p. 175.
[28] Needham, SCC, vol. 5, part 2, pp. 136–44; Schafer, *Golden Peaches*, pp. 163–75.
[29] P'i Jih-hsiu, "Impromptu on a Hangover," in Watson, *Columbia Book*, p. 290.

travel edge runner mills (*yen nien*). Certain elements were required to provide the qualities necessary for setting as well as to provide the tinderlike characteristic of the product. Generally these were elm root, sawdust of cypress, juniper, myrtle, and cedar, dried leaves of Perilla, refuse of nutmeg, pine resin and gum arabic of various kinds.

The powder was rendered into a paste by mixture with water or distilled wine for thinning it; rhubarb and saltpetre may also have been added. The dried leaf powder of *Perilla ocimoides* (*frutescens*) favored slow even burning of timekeeping incense, according to the formula proposed by the eleventh-century engineer, Shen Li, who also recommended the withered and dried flowers of the pine tree carefully powdered. The source of pine resin was presumably the red pine, *Pinus massoniana*.

The paste was formed into sticks by extruding it through small holes in a drawplate by means of a simple pump or syringe. The process was similar to that used for making noodles. The resulting extrusion was cylindrical and could be made straight or shaped into coils of several sizes before drying.[30]

The method of manufacture of incense sticks in ancient China appears to have remained much the same to recent times, as observed and described by Western visitors. Three quarters of the length of the stick consists of wood dust derived from grinding wood in stone mortars with primitive pestles. The remaining quarter is compressed from a combination of the bark of elm root powdered and thinned with water, to which is added scented powders of incense, clove, camphor, and odoriferous woods such as cypress. The scented powders are thinned in Chinese wine. The elm root bark and the powders are all ground together and made into a binding paste placed in a type of pump which crushes the mass and from which it emerges through round holes of a drawplate in the form of sections molded like thin wire. These are dried and then cut into ritual lengths, usually 15 to 18 cm. The dried sticks are then sold in batches of 19, 37, 61 or 91.

These apparently haphazard quantities of incense sticks are the result of tying the sticks into cylindrical bundles. The bundles are started with a single stick at the center, 6 sticks ringed around it, 12 more around those. The next larger addition accumulates another 18 sticks in the outer layer. The two final layers add 24 and 30 respectively, as follows:

$$1+6+12=19$$
$$1+6+12+18=37$$
$$1+6+12+18+24=61$$
$$1+6+12+18+24+30=91.$$

For centuries the custodians of the temples were traditionally the principal

---

[30] Planchon, "L'Heure," pp. 249–50; Needham, *SCC*, vol. 5, part 2, pp. 136–46.

purveyors of incense sticks. Their primary functions were to light the sticks in the morning and at night, in addition to maintaining the temple in good order. They lived on the premises, and derived additional income from the sale of incense sticks, the length and size of which were determined by the Astronomical Board.[31]

Western observers frequently noted in their writings the prevalence of incense sticks or "joss sticks" at every level of Chinese life. The most common of these were "about as large as macaroni stems," and made of a mixture of sawdust and a type of gum processed through molds. Some were perfumed and others were colored, and their consumption was so considerable that enormous sums were expended annually for their purchase by the population. As one nineteenth-century American observer in China wrote:

in every house, boat, street, and garden the traveller, after a little observation beholds signs of religious import, principally in the innumerable joss sticks that are forever smoking ... Throughout the length and breadth of the vast empire; through cities and villages; in enormous temples, and solitary roadside shrines; in districts where the eye can reach over leagues of green culture, and on barren crags by the salt sea; in the labyrinthine palace of the monarch, and in the hut of the beggar; in the tenements of the living, and by the tombs of the dead, appear the silent but everlasting signs of adoration.[32]

Incense continues to be used profusely throughout the countries of the Far East, even among the poor who spend their last few coins for it. It is burned so extensively day and night, often several sticks at a time, in the homes as well as in public places, that foreign visitors writing about their travels commented that they found the resulting pervasive stench, particularly in the temples, to be most irritating.[33]

The use of incense was as prevalent in Japan as in China. The scent of flowers formed part of Shintō worship, their fragrance intended to purify the sacred premises from molesting evil spirits, so that the beauty of the flowers could be enjoyed. The practice of burning incense was probably initiated in Japan in the sixth century, during the reign of Emperor Kinmei (540–571), introduced with the religion of Buddhism. It became a necessity of life for priests and laity alike.

The classic Japanese novel, *Genji monogatari* (The Tale of Genji) written between 1001–1021, and other literary works of the Heian period make constant mention of incense for perfuming clothing, and of games to identify incense scents, the use of perfume bags (*nioi-bukuro*), incense burners (*hitori*), scent cages (*kōkago*) and decorative fragrant rosettes (*kusuri-dama*). Incense became particularly popular among the members of the imperial court, who enthusiastically

---

[31] De Poncins, p. 94 n; Needham, *SCC*, vol. 5 part 2, pp. 144–46.
[32] Douglas, *Society*, p. 313.
[33] Tiffany, *Canton*, pp. 180–81; Magowan, "Modes," p. 340.

adopted the customs of the Chinese court aristocracy at Ch'ang-an, the T'ang capital of China.

It is a matter of interest to note that the family that served as purveyors of incense exclusively to the imperial family in the sixteenth century survives to the present as the firm of Nippon Kōdō, with branches in many major cities throughout the world.[34]

Several classifications of incense in Japan were made by the tenth century, during the Enki and Tenryaku periods. Later, in the late thirteenth century, Kimitaka-Sangi visited China and returned with information about other previously unknown aromatics, of which he compiled descriptions. Later still Takauji devised a classification of incense, and the shogun Ashikaga Yoshimasa (1435–1490) developed a nomenclature with the 130 varieties he had collected.[35]

Preserved in the Shōsōin, the Imperial Treasury at Nara, is an interesting artifact relating to the history of incense in Japan. It is a piece of wood called "Yellow Fever Perfume" (*ōjukkō*). It is reputed to be the wood known as kyara or aloes, and traditionally described as "some Chinese wood that was buried." When last reported, the piece, which originally was much larger, had been reduced to somewhat more than 152 cm in length and almost 122 cm in width at one end, tapering to only 13 cm at the other.

This historic piece of wood is generally known as the *Ranjatai*. It was a gift from the Chinese court to Emperor Shōmu, who ruled from 724 to 748, and at first it was kept in the treasury of the Tōdaiji, the imperial temple, and later moved to the Shōsōin.

Small bits were removed from it from time to time, but only on the most auspicious occasions. A piece measuring one square inch was cut from it in 1465 and presented by Emperor Gotsuchimikado (1465–1500) to the shogun, Ashikaga Yoshimasa (1435–1490), in recognition of his eminent service. From time to time subsequent shoguns made requests for pieces of the *Ranjatai* but without success.

It was more than a century later, in 1574, after the general and statesman Oda Nobunaga (1531–1582) had succeeded in pacifying the country and made a beginning towards its unification, that another piece was cut from the slab. In a great ceremony the door of the Imperial Treasure House was opened in the presence of the emperor, courtiers and warriors and the famous wood was brought out. A small bit was removed and presented to Oda by Emperor Oogimachi. Characteristically, Oda kept only one third for himself, and parceled out small bits to several of his generals. In this manner incense was brought even into politics.[36]

---

[34] Ayres, *Altars*, p. 51; Van Beek, "Frankincense," pp. 99–126; Waseda, "Ceremonies," pp. 15–17; Tiffany, *Canton*, pp. 180–81.   [35] Hearn, *Ghostly Japan*, p. 23.   [36] *Ibid.*, pp. 22–23.

In 1602 the shogun Tokugawa Ieyasu (1542–1616) had achieved sufficient power to procure a piece of the *Ranjatai*. Thereafter for a time each Tokugawa Shogun appears to have succeeded in acquiring a small bit of the treasured wood. The last piece was removed in 1877 by Emperor Meiji for his own use. The remainder is carefully preserved in the Shōsōin.[37]

Although the wood's provenance as a Chinese gift to Emperor Shōmu appears to be well established, it has also been suggested that it may be identified with one mentioned in the *Nihongi*. In its chronicle for A.D. 595, it was noted that in the fourth month, "lign-aloes wood drifted ashore on the island of Ahaji. It was stated to be a fathom round. The people of the island, being unacquainted with aloes wood, used it with other firewood to burn in their cooking range, and the resulting smoky vapor spread a perfume far and wide. Wondering at this, they presented it to the Empress Suiko."[38]

Aloes wood incense achieved popularity after the late sixteenth century following its importation from the south by Portuguese and Spanish traders. Its rarity and costliness made it particularly desirable as a staple in gifts to important personages.[39]

Secularization of incense burning began in the Nambokuchō period, after the Japanese imperial court adopted incense parties as a courtly pastime. These gatherings became as popular as the tournaments in poetic composition (*uta-awase*), and a continuing demand for new fragrances evolved. The ports of Shimonoseki and Sakai developed a flourishing trade in aromatics as a consequence, and a new profession of specialists in incense was created, who graded aromatics according to their perfuming qualities and established prices.

Incense was classified according to region of origin – Manaban (Malacca), Rakoku (Thailand), Kyara and Manaka (India), Sumotara (Sumatra) and Sasora (Sassori). By the middle of the eighteenth century, however, importation of incense woods to Nagasaki came to an end.[40]

Several important writings on the subject were produced in Japan, one of which was the Kun shu rui sho (Incense Collector's Classifying Manual). It is a comprehensive work, containing the teachings of the Ten Schools of the Art of Mixing Incense. It provides advice on the seasons preferred for making incense; instructions concerning the various types of fire to be used in burning incense, such as "literary fire" and "military fire"; the use of incense bags; religious uses; and legends about incense.

One of these legends was about a radical figure called Sue Owari-no-kami. He

[37] Casal, "Incense," pp. 69–71.
[38] *Nihongi*, for the year A.D. 595; cited by Takigawa Seijirō.
[39] Schafer, *Golden Peaches*, pp. 163–65.
[40] Waseda, "Ceremonies," pp. 14–23.

built himself a palace entirely of incense woods and on the night of his revolt he set fire to it, with the consequence that the smoke from its burning perfumed the regions for an area of twelve miles.[41]

Ceremonials of the old "incense assemblies" became popular, and in time a vast related literature was developed, including poems, tales, dramas and love songs. Specialized dictionaries and manuals were published in China and Japan in which the varieties of incense were classified and which included recipes for preparing them. Special varieties of incense were reserved for the use of princely families and recipes were handed down from one generation to another. The variety of mixtures as well as materials was infinite.[42]

Very much in vogue among the Japanese aristocracy of the Heian period was the blending of incense, a hobby which was refined to an art. One of the elegant pastimes of cultured cognoscenti were incense guessing parties, such as those reported in the *Genji monogatari*. It is believed that this pastime may have originated in China among Taoists and literati in T'ang or pre-T'ang times.[43]

"Incense sniffing" developed in Japan as an aristocratic form of social entertainment at least as early as the eighth century. Mention of it first appeared in Japanese literature after the ninth century, in the Heian period. The game was intended primarily for the creation of improved and more permanent perfumes, and prizes were awarded to the most successful competitors. By the beginning of the Muromachi Shogunate (1392–1490), the Ashikaga Shoguns had brought art appreciation and love of splendor, of which "incense sniffing" formed a part, to a high level. "Couplet tournaments" were featured at court, where the tea ceremony had already become extremely popular.

The practices of "incense blending" (*kōawase*), the "incense assembly" (*kōe*), and the "incense or *bunkō*" (*kikikō* sniffing), were all entertainments devised by Japanese aesthetes during the Muromachi period and were particularly popular during the Tokugawa reign which followed.

The practice of "incense blending" evolved gradually from competitions held to develop new aromas. It was a literary pursuit that tested not only the player's abilities to recognize numerous scents of incense, but also revealed his knowledge of ancient literature. A participant's success depended not only on his familiarity with classic poetry, but also on his ability to apply this knowledge. The scents were identified with euphemistic allusions and lyrical names, and the contestants were required to judge their qualities and guess their names from their odor alone. "Literary fancies" or puns were popular, and the devotees of the game claimed that it liberated them from worldly distractions and purified their spirit.[44]

[41] Hearn, *Ghostly Japan*, p. 24.
[42] Bushell, *Chinese Art*, vol. 1, p. 85.
[43] Needham, *SCC*, vol. 5, part 2, pp. 143–44.
[44] Waseda, "Ceremonies," pp. 17–18; Casal, "Incense," p. 61.

Among those who particularly relished the game was Ashikaga Yoshimasa, and his palace at Higashiyama became the principal center for such activities during his reign. It was said that the Shogun collected 130 varieties of incense, a number of which included unblended fragrant woods. Yoshimasa was initiated in the recreation of incense comparison, according to contemporary claims, by Nishisanjō Sanetaka, an acknowledged authority in "the way of incense" (*kōdō*). Nishisanjō was the founder of the school of incense-burning that became known as Oie-ryū.[45]

The originator of the true "incense comparison" (*kōawase*), however, is considered to have been Shino Sōshin, a bonze who lived in the late fifteenth and early sixteenth century and who was undoubtedly greatly encouraged in this endeavor by Yoshimasa. With the assistance of Nishisanjō, Shino developed the pastime into a form of philosophy, with endless strict rules and occult terminologies. In accordance with his rules, for example, the meetings had to be conducted with the strictest etiquette, making the sniffing of incense not only an art but also a science.[46]

Compound incense had begun to be manufactured for the first time in Japan, and disciples of the "Shino" or "Shino-ryū" school established by Shino Sōshin recognized as many as sixty-six distinct blends, each distinguished by names based upon literary allusions. It was alleged that the use of compound incense was restricted to the court aristocracy, and furthermore that the military invariably preferred simple varieties. The members of the imperial household and court nobility favored the Oie-ryū "school," while the feudal lords and the general populace were said to prefer the Shino-ryū version.[47]

In time these "schools" of incense entertainment branched out to form other schools of incense burning, known as Tatebe-ryū, Yonekawa-ryū, Hachiya-ryū, Satomi-ryū, and Sono-ryū.

The incense sniffing party – *kōkai, kōe* or *kikikō* – ranked second only to the tea ceremony as a pastime among individuals of prestige and taste, and was always conducted by formal rules. In a simple form of the game, called *jitchū-kō* (ten sticks of incense), the host who served as chairman of the event (*kōmoto* or *himoto*), sat at the head of the gathering with a tray before him containing one or two porcelain censers.

The utensils he would use were at his right, and the incense container and a shelf for other supplies were at his left. The utensils required for the game were extremely expensive. Wrought of the finest gold lacquer or made of precious metals, many were rare works of art. The records were kept by the *kiroku*, seated at the host's right, and the contestants sat in a circle before them.

---

[45] Brinkley, *Japan Its History*, vol. 3, pp. 1–9.
[46] Waseda, "Ceremonies," p. 17; Hearn, *Ghostly Japan*, pp. 30–41.
[47] Brinkley, *Japan Its History*, vol. 3, pp. 1–9.

The host selected three varieties of incense, of which three portions of each had been separately wrapped in elaborate paper folders. An additional folder or packet contained a fourth variety, which was designated as the *kyaku* (guest) and which was generally the most costly.

To each was assigned a name based on a literary subject, a reference to the season, or perhaps a poetical phrase. In the spring season such names might be "cuckoo and rain," or "aged plum tree," while "iris" or "Weaver Princess" were used in the summer, "maple" or "brocade [of leaves]" might suggest autumn, and "hoarfrost" might refer to winter. Each of the three kinds was assigned a number from one to three. To assure that the varieties could not be identified by color, the samples were covered with a thin layer of charcoal dust which had been specially prepared and maintained in a pottery container.[48]

Before the game began, the participants were divided into competing parties. At the beginning there was a *tameshi* or tryout, during which the host lighted each of the incense varieties in irregular order and passed them around "to be sniffed" and memorized. The participants then presented the *kyaku* or "guest" packet of incense, which, however, was never burned. After the preliminary trial, portions were taken from each group indiscriminately and the players were required to identify them by means of their aroma only, and to give each a literary name.

The identifications of the scents were recorded by each player with its number on a tablet called a *kōfuda*. The tablet was then placed in a case made for the purpose which was maintained by the *kōmoto*. The victor was chosen on the basis of his erudition and cleverness in choosing the names. Although each side may have made the correct identification, for example, one may have chosen the phrase "moonlight on a couch" and the other "water from the hill." The former was derived from the poem

> When autumn's wind breathes
> Chill and lone my chamber through
> And night grows aged,
> Dark shadows of the moonlight
> Cast athwart my couch,
> Sink deep into my being;

while the second was taken from

> Stream with scented breast
> From flower-robed hills that flowest,
> Here thy burden lay,
> Thy freight of perfumed dew-drops
> Sipped from sweet chrysanthemum.

[48] Waseda, "Ceremonies," pp. 17–20.

In a comparison of the two names, the first would be selected as superior, because it conveyed the forceful concept of the penetrating influence of a fine aroma even though it had no material allusion to incense-burning. The second would be judged to be the more commonplace, inasmuch as the scent of flowers was an ordinary simile frequently used in praising incense. The pastime involved much more than the mere smelling of incense, for as the *uta-awase* (couplet composing) tested literary originality, the other determined literary knowledge.[49]

Often the notations made on the *kōfuda* were replaced by the *Genjimon* (linear signs) representing the chapters of the *Genji monogatari*. In the incense-sniffing game known as *Genjikō*, five varieties of incense were used, each divided into five packets, making a total of twenty-five. After mixing the packets together, they were submitted to the participants in groups of five to identify. A special form of notation was devised, consisting of fifty-two symbols, which were allotted to the respective chapters of the *Genji monogatari*. Inasmuch the novel contained fifty-four chapters, two were left without symbols.

When it had been established that the contents of a set of five packets were judged to be different, a symbol consisting of five vertical lines corresponding to the second chapter was noted. When a participant considered that a repetition had occurred in the packets of incense, he connected the five lines with a horizontal line at the top. Five vertical lines connected in this manner indicated that the five packets were of the same incense. If the first two and the last two were judged to be the same variety, and that the third was different, the third line was stepped down and the first and second and fourth and fifth lines were connected. Symbols used for this purpose might have included representations of fireflies and iris, a nightingale and prunus blossoms, for example.

In another of the incense games, called *Ujiamakō*, the identifications were given in the form of the stanzas of a short poem, such as one by Hoshi Kisen from the anthology *Hyakuninshu*.

Waga iwo wa	My home is near the Capitol,
Miyako no tatsumi	My humble cottage bare
Shika zo sumu	Lies southeast on Mt. Uji; so
Yo wo Ujiyama to	The people all declare
Hito wa yūnari.	My life's a "Hill of Care."

Five varieties of incense were burned in the preliminary stage of the game, but only one in the final stage or "real" game, on the assumption that the Muses preferred quality to quantity. The means by which contestants identified the single incense being burned with one of the five burned in the preliminary exercise would be by quoting any one of the five lines of the poem; if it was judged to be

---

[49] *Ibid.*, pp. 20–22; Needham, *SCC*, vol. 5, part 2, pp. 143–44.

the same as the second incense burned, a participant would quote, "*Miyako no tatsumi*," for example.

Often the contestants were divided into two camps called Gen-Pei, named after the two combatting clans of the Kamakura period. Each side was assigned a different color and given tablets in that color. A variety of incense was burned in a censer and circulated among the contestants. After each participant had sniffed the censer as it came to him, he recorded his guess on a tablet, which would be placed in a tablet case. When everyone had taken his turn, the master (*kōmoto*) counted the tablets, and the side having the greater number of correct guesses would be declared the winner.

An elaboration of the game featured a novel method of scoring, with scoring tracks and two toy horses, one for each side. The appropriate horse was moved forward as its team won points, and the first to reach the winning post was declared the victor. The horses and scoring tracks were often veritable works of art, and there was as much competition in providing the most elaborate figures as there was in winning the game.[50]

For advanced participants, varieties of incense were burned in combinations, and then words having poetical significance were used in which certain syllables recurred to indicate the two "numbers" recognized in a double scent. The markers were collected after each round, and when the sitting ended, they were sorted and referred to the host's master list. The one who had submitted the most correct scores won the prize honors.[51]

Depending on the erudition of the participants, further complications might be introduced by providing each incense with a descriptive appellation suggestive of its fragrance. The game demanded politeness and restraint at all times, and provided training of one's mind in apperception and subtlety at the same time that it enhanced one's memory of literature.

Another version of the incense pastimes required "listening" to incense with "ears of the spirit", said to have been invented by Sasaki Dōyo. The word for "to sniff" in Japanese is the same as that for "to listen," hence this development from "incense-sniffing" to "incense-listening." It originated as a ritual and leisure game among the Japanese nobility and the warrior class. In this game one did not refer to "sniffing" the incense, only "listening" to it, expressive of the thoughtful concentration required in making identification. The ceremony began by first sniffing the incense and then "listening" to the fragrances, using the aromas to create mental images that were then translated into poetic forms. It was said to

---

[50] Waseda, "Ceremonies," pp. 18–22.
[51] *Ibid.*, pp. 18–23; Brinkley, *Japan Its History*, vol. 3, pp. 3–7; Casal, "Incense," pp. 65–66; Nakamura, "Burning tradition," p. 70.

calm the mind, and to constitute an effort to find meaning in a materialistic world while engaged in communication with friends.

The ceremony generally required one and one-half hours. To sharpen their senses, contestants generally fasted or avoided a heavy meal before the game. Conversation had to be reduced to a minimum to avoid disturbing the air. No one was allowed to leave the room from the time the game began until it terminated. During the game no flowers were permitted to be in the room, nor could there be open doors or windows, nor any change in the air of the room.

The fixed formula for "listening" was indicated to be "an even number of inhalations to the right and to the left" (*yucho sahan*). Participants were to inhale an even number of times, either twice or four times, through the right nostril, and to make an odd number of inhalations or sniffs, either one or three times, through the left. When the nose became insensitive, as frequently happened due to excessive exposure to the scents, its sensitivity was restored by inhaling vinegar. As part of the ritual, there was often a *kumikō* or contest based on a literary theme using the scents of two or more kinds of incense.

Although the game of *kōdō* might appear to be no more than a casual pastime, Japanese warriors, savants and priests devoted much time to its pursuit. A considerable literature exists on the subject delineating the exact principles to be observed and describing the numerous modifications that the game underwent in the course of time.[52]

A remarkable modernization of the practice of *kōdō* took place in 1987 in Carnegie Hall in New York City, when Seiji Ozawa conducted the Boston Symphony orchestra in a rendition of a new work entitled *Kō o kiku* (Listen to the Incense).

As previously noted, the equipment required for the incense pastimes was almost as elaborate as that for the tea ceremony. The paper folders in which the incense was packed were made of gilt decorated with paintings by notable artists, the containers often lacquered in gold and enclosed in rich brocaded bags. The incense censers were masterworks of the arts of the potter, lacquerer or metal-worker. They were most often made of gold, or iron inlaid with gold or silver, or of the finest porcelain; among the most highly prized were those made of celadon. They were carefully maintained in boxes almost as elaborate as the censers. Equally elaborate were the miniature chopping blocks, mallets, knives, tongs and spatulas of gold and silver; and marking boards of silver, gold, vermilion lacquer, or mother-of-pearl. The pastime of incense-smelling lost its vogue after the Restoration of 1868, although it is still popular among certain groups.[53]

[52] Waseda, "Ceremonies," pp. 22–23.
[53] Brinkley, *Japan Its History*, vol. 3, pp. 1–9; Casal, "Incense," pp. 61–69; Nakamura, "Tradition," p. 72.

One type of incense said to have been known in China from the Han Dynasty was *fan hun hsiang* (spirit-recalling incense). It was imported into Japan where it was named *hangon-kō*. A curious innovation which became popular in Tokyo at the end of the nineteenth century was a new type of cigarette named *hangon-so* (Herb of Hangon), so named to suggest that the smoke functioned in the same manner as spirit-summoning incense. Its popularity was due not to its success in summoning spirits, however, but to the appearance of a photographic image of a dancing girl on the cigarette's mouthpiece caused by the smoke.[54]

[54] Hearn, *Ghostly Japan*, pp. 41–44.

3

Incense in religious rites

The use made of incense for religious rites varied in Far Eastern countries. In China, for example, it was employed in several religions, forming part of the rituals in ancestral worship, with Taoist and Confucian rites and in the ceremonies of the Buddhist religion. In Japan, however, the only religious use made of incense was in relation to Buddhism.

Odoriferous gums and resins have been used in several Chinese religious rites from a very early period. This has been and continues to be particularly true of Taoist rituals. Taoists used incense for exorcism, incantations, and magical rites because it was considered to render the gods indulgent and because it scattered evil. The officiant's commands to the spirits were borne aloft on the tenuous tendrils of smoke, which were compared to the wrigglings of an ascending dragon and could compel the dragon into bringing fructifying rains, for example.[1]

In Taoist temples incense is burned in large standing stoves or burners made of bronze or cast iron (*ch'ing lu*) maintained in the courtyard in front of the temple hall, as well as in others placed upon the altar. Clusters of incense sticks are inserted upon an ash base in the large ornamental cauldrons. Powdered incense is sifted upon an ash base on flat metal surfaces. It is occasionally burned also in long handled basins (*shou lu*).[2]

The incense burner on the altar is the most important feature in the Taoist ritual. The "furnace master" (*lu chu*), chosen annually by lot, and his assistants (*lu hsia*), begin each liturgy with the ceremony of the lighting of the incense burner (*fa lu*), and end the ritual with a return to it (*fu lu*). Part of the liturgy includes a representation of the forces of evil attempting to steal the hand-held incense burner and its recovery by the "furnace master." The ritual has not changed since the fifth century.[3]

The Taoist altar generally holds not only the missal, flowers, fruit, tea, water, rice, wine, bread, candles, the wooden drum (*mu ku*), and the gong (*ch'ing lo*), but also an incense burner (*ch'ing lu*).[4]

---

[1] Casal, "Incense," pp. 53–56.   [2] Needham, SCC, vol. 5, part 2, pp. 129–34.   [3] *Ibid.*
[4] Chin Yün-chung, *Shang ch'ing ling pao ta fa*; Chu Ch'i-feng, *Tz'u t'ung*; Needham, SCC, vol. 3, p. 129; SCC, vol. 5, part 2, pp. 128–31.

The possibility exists that the incense burned in the rituals of the ancient Taoists may have generated hallucinogenic smoke, in view of the fact that the incense of Taoist liturgy began shamanistically as much as a means of fumigation and purification as it was a religious offering to the gods.[5]

Stimulation of the sense of smell by means of psychotropic drugs undoubtedly could substantially enhance the flagellation and violent prostration which were featured in the ritual of the early Taoist church, as described by Yang Lien-sheng, and similar quasi-orgiastic rites. In early China hemp (*Cannabis indica*) was used, and its characteristics were well known.[6] Incense was liberally used also by the Confucians from at least the Sung period.[7]

With the transmission of Buddhism from India, which occurred in the first century A.D. or possibly earlier, new fragrances became available for use in Chinese temples. This led to the development of new customs which supplemented the old traditions, and new beliefs about incense evolved. As a consequence, incense became an increasingly important element in the imagism, liturgical observances as well as the literature of Buddhism.[8]

According to one belief, the smoke of burning incense conveys one's faith in Buddha up to his heaven, and according to another, it brings down Buddha's mercy. The curling smoke of incense, according to the *Soshi-ryaku* (Epitomized History of Priests) was "the Messenger of Earnest Desire," and as a burnt offering, symbolized the pious desires of the faithful, forming a link between the supplicant and divinity. The Sanskrit word *gandha* (aromatic) was frequently used to mean "pertaining to Buddha." A temple was called a *gandhakuti* (house of incense), and the gods of fragrance and music (*gandharvas*), lived on Gandhamadana (incense mountain).[9]

Incense had been used in Japan for religious purposes from as early as the sixth century. Although at least twenty-four varieties of fragrant wood were then known and used for secular purposes, its use in religion was introduced with Buddhism, in A.D. 551 with the gift of temple materials from Korean King *Song-myŏng* of Paekche to the Japanese Emperor Kinmei. Included in the gift were a collection of sutras, a bronze image of Buddha, and a complete set of furniture for a Buddhist temple. The Buddhist doctrine had already been preached many years earlier in Japan by enterprising missionaries who wandered through the country, but without making significant impact.

---

[5] Needham, *SCC*, vol. 5, part 2, pp. 149–52.
[6] Suggested by Professor Joseph Needham, *SCC*, vol. 5, part 2, pp. 150–54.
[7] *Ibid.*, pp. 131–34.
[8] S. W. Williams, *Middle Kingdom*, pp. 188–89.
[9] Schafer, *Golden Peaches*, pp. 156–57; Cobbold, *Religion*, pp. 30–66; Hearn, *Ghostly Japan*, pp. 27–29.

The first Japanese emperor to become interested in the new faith was Kinmei. He was charmed by the mystery and loftiness of the religion, and particularly impressed by the Korean ambassador's assurance that Buddhism had become the faith of civilized Asia. When, in his enthusiasm, the emperor proposed that the new creed be adopted, however, only the premier, Soga no Iname, espoused it. The imperial court as a whole remained cautiously unresponsive on the subject at first, however, possibly because Japan already had a great number of existing deities, and the other ministers feared that the adoption of a new god would insult the old.

As a compromise, the emperor postponed the decision to adopt or reject the doctrine and named Soga keeper of the sutras and the statue. Soga's family persisted in its support of Buddhism, with the consequence that it was subsequently severely persecuted. Later Prince Shōtoku Taishi (574–622), prince regent to his aunt, the Empress Regnant Suiko (593–628), became a fanatically zealous follower, and the religion became firmly established in Japan. He popularized Chinese art and undertook to develop a religious center. He built many fine temples at Nara and its vicinity, among them the Hōryō-ji, which survives.

It was not until Empress Shōtoku (765–769) became a patron of the new religion, however, that Buddhism was adopted by the court and the nobility. Until the fifteenth or sixteenth centuries many features of Japanese Buddhist temples, including idols, altar furniture and libraries, were of Korean origin.[10]

Buddhism in Korea achieved its greatest period in the seventh and eighth centuries due to the migration of many priests from T'ang China, who established numerous temples. It became the state religion in the fifteenth century and survives in its Chinese form to the present day.[11]

In Japan the inexpensive common incense burned by the poor and pilgrims before Buddhist icons was known as *ansokukō*, as contrasted with more expensive varieties burned in temples of the rich. Notable among the latter was *ranjatai*, imported in quantity for the temple Tōdai-ji by Emperor Shōmu, who reigned from A.D. 724 to 749 and died in 756.[12]

The use of incense was greatly influenced by other superstitious and religious beliefs which originated prior to the establishment of a formal religion. For example, incense is still burned in the presence of a corpse so that the fragrance will shield the body and the liberated soul from evil demons. It is also used to summon spirits or the souls of the absent, and for this purpose the incense is called *fan hun hsiang* in China and *hangon-kō* in Japan. According to Japanese legend,

[10] Brinkley, *Japan Its History*, vol. 1, pp. 89–195; Chamberlain, *Handbook*, pp. 4–96.
[11] Griffis, *Corea*, pp. 330–36.
[12] Hearn, *Ghostly Japan*, pp. 26–27.

any form of burning incense will summon the *jikikōki* or "incense-eating goblins," a class of hungry demons (*pretas*) recognized in Buddhism. These are the souls of men who in their lifetime had been guilty of selling incense of inferior quality for profit, and who are now compelled to seek their only food in incense smoke.

Cheating the purchaser of incense was considered to be a crime equivalent to cheating the gods, and Buddhism condemned dishonest dealers and manufacturers to become *jikikōki*. They are described as demons belonging to the *gaki* class of *pretas* and are distinguished by large bellies, huge mouths, and tiny throats. They suffer unappeasable hunger and live either in hell in the service of Yama, in the air, or among men but are visible only at night.[13]

A huge bronze or iron pot, usually decorated with dragons, lotus flowers and other Buddhist symbols, stands before each Buddhist temple. It is half filled with ashes, into which are inserted burning incense sticks (*hsiang pang* in Chinese, *senkō* in Japanese). The sticks must be ignited from a pure flame which is available nearby. Smaller and plainer censers are placed on the altars before the idols. According to one source, in Japan the incense stick used was little more than 0.16 cm in diameter and 20 or 23 cm in length and burned for approximately forty minutes.[14]

Incense plays an important role in death ceremonies in both China and Japan. Incense sticks in jars are featured during the burial ceremony. In China, after the corpse has been laid in a bed of lime or cotton or covered with quicklime in the wooden coffin, which is sealed with mortar and then varnished, the coffin sometimes remains in the house for months and sometimes years, until a "lucky" place for burial is found, or until the family can afford suitable burial. During this period incense is burned before the coffin morning and evening.[15]

Incense also forms an important part of the offering made during the Buddhist funeral in Japan. Standing before a temporary wooden tablet bearing the name of the deceased, a Buddhist priest delivers the customary incantation:

In my heart's core I respectfully request that the scent of this stick of incense offering from the heart may pervade the regions where the Law prevails and that the Messenger of Hades may conduct the soul thither.[16]

Incense also plays a role in the memorial services held for Buddhist dead, which take place several times a year. The ancestral tablets are exposed and an incense bowl is placed in front of them, as well as food offerings, flanked by a candle on one side and a jar of shikimi leaves on the other. The shikimi tree (*Illicium*

---

[13] Eitel, *Hand-Book*, p. 125; Hearn, *Ghostly Japan*, p. 45; Casal, "Incense," pp. 51–54.
[14] Casal, "Incense," p. 56.
[15] S. W. Williams, *Middle Kingdom*, vol. 2, pp. 243–55.
[16] Casal, "Incense," pp. 53–54.

religiosum) is considered by the Buddhists to be sacred; not only are the branches of the tree displayed, but the bark is used as an ingredient for making incense.[17]

Many similes in Buddhist literature were provided by incense, often in prayers, such as in an example from the *Hōjisan* (Praises at Pious Observances):

Let my body remain pure as a censer! – let my thought be ever as a fire of wisdom, purely consuming the incense of *sila* [observance of the rules of purity in act and thought] and of *dhyana* [or *zenjō*, one of the higher forms of meditation] – that so may I do homage to all the Buddhas in the Ten Directions of the Past, the Present, and the Future![18]

The prevalent use of incense in religious ceremonies was noted repeatedly by Western travelers to the Orient. Marco Polo returned from China in the thirteenth century with one of the earliest records in the West of the use of incense. In his account of a visit to the palace of the Great Khan at Shandu, he described the position of the Baksi, a religious order from Tebeth and Kesmir (probably Tibet and Kashmir) that frequented the palace. When the festival days of their idols drew near, they asked the Great Khan for appeasement of their idols in the form of

. . . a certain number of sheep, with black heads, together with so many pounds of incense and of lignum aloes, in order that we may be enabled to perform the customary rites with due solemnity.

Marco Polo then went on to relate that the kingdom of Kanan or Tana

. . . produces a sort of incense, in large quantities, which is not white, but on the contrary of a dark colour. Many ships frequent the place in order to load this drug.[19]

The "drug" was probably benzoin, which was not produced in India, but was imported from Sumatra to supply the markets of Arabia, Persia, Syria, and Asia Minor. In his account, Marco Polo noted that the Tartars were idolators, and that they burned incense in front of certain tablets inscribed with names, and before an image of Natigai.[20]

Other early Western travelers in China also made particular note of the use of incense. William of Rubruck, traveling in East Asia during the years 1253 to 1255 wrote of it, as did another famous traveler in Asia, the Italian Franciscan missionary Odoric of Pordenone (1286–1331). Near the end of his life he related that when the Great Khan traveled through any country the subjects kindled fires before their doors, and cast spices thereupon to perfume the air.[21]

---

[17] *Ibid.*; R. C. Armstrong, *Buddhism*, pp. 32–44; S. W. Williams, *Middle Kingdom*, vol. 2, pp. 243–55.   [18] Hearn, *Ghostly Japan*, pp. 27–28.
[19] Yule and Cordier, *Marco Polo*, vol. 1, book 1, ch. 61, p. 327; vol. 2, book 3, pp. 445–49.
[20] *Ibid.*, vol. 2, book 3, pp. 445–49.
[21] Rockhill, *Rubruik*, pp. 143–49; S. W. Williams, *Middle Kingdom*, vol. 2, pp. 422–24.

While in Japan, Saint Francis Xavier (1506–1552) made little note of the use of incense, except when he encountered it on his voyage aboard a Chinese ship bound from China to Japan over the South China Sea. In a letter to the Jesuits at Goa, he described a Chinese sea-god's shrine aboard the junk and how he had observed that

Then there was more casting of lots, preceded by many sacrifices, offerings and adorations, that the idol might tell whether we should have favourable winds or no. The lot foretold good weather and advised that we should delay no longer, so we weighed anchor and hoisted sail in a happy mood, the heathen confiding in that idol of theirs on the poop, which they worshipped with lighted candles and incensed by burning before it sticks of sweet-smelling wood, while we put our trust in God the Creator of Heaven and Earth . . . [22]

Father Matteo Ricci, S. J. (1552–1610), founder of the first Catholic mission in China in the early seventeenth century, described the memorial shrines constructed to honor magistrates when they left the cities over which they had presided, noting that

A yearly allowance is granted for these temples for incense and to pay servants for attending to perpetual lights. The large incense bowl, made of bronze, used in these shrines is similar to that employed for the same rite in the veneration of the idols.[23]

Ricci also described the incense offerings brought to the seasonal worship at the Temple of Confucius, as well as in other Confucian temples honoring the titular spirits of the cities.[24]

Friar Sebastião Manrique (c. 1600–1669), the Portuguese Augustine missionary who voyaged to Portuguese Asia in the seventeenth century, described a festival in a temple on the island of Saugar. Inside the temple the music and "hot wafts of scent from flowers or incense worked on minds open to bewitchment."[25]

Accounts by later travelers provided additional details. The American lecturer, John Lawson Stoddard (1850–1931), observed that in the Temple of Five Hundred Arhats at Canton the worshippers burned a stick of incense in a small jar of ashes in front of each of the idols.[26]

The Swedish geographer and explorer, Sven Hedin (1865–1952), noted that in the temple of Ta-fo-ssu at Jehol, "a lama stepped forward bowed and lighted sticks of incense in a bronze bowl half full of ashes."[27]

The Victorian author, Lafcadio Hearn (1850–1904), commented on the same use of a bowl of ashes in Japan in an account describing a shrine adjacent to a

---

[22] Schurhammer, *Epistolae*, vol. 2, pp. 145–51. Quoted in Broderick, *Saint Francis*, p. 336.
[23] D'Elia, *Ricciane*, vol. 1, pp. 31–33, 164.
[24] *Ibid.*, pp. 40, 118–20.   [25] Collis, *Experiences*, p. 76.
[26] Stoddard, *Lectures*, vol. 3, pp. 298–99.   [27] Hedin, *Jehol*, p. 81.

Buddhist temple. There he looked "through the blue smoke that curls up from half a dozen tiny rods planted in a small brazier full of ashes."[28]

Hearn also described the statues of the *roku-jizō* in a Buddhist cemetery. Each held a symbolic object, such as a lotus flower, a pilgrim's staff, and a Buddhist incense box, among other items. In his account of a pilgrimage to Enoshima, he reported that, "at the right hand of Shaka enthroned on a lotus some white mysterious figure stands, holding an incense-box."[29]

George Nathaniel, Marquis Curzon of Kedleston (1859–1925), observed that when he visited the Buddhist monastery of Ku-shan, there arose from the altars below the images of the three Buddhas "a thin smoke curling upwards from the slow combustion of blocks of sandal-wood, or from sheaves of smouldering joss-sticks standing in a vase."[30]

After having made a study of modern religious uses of incense in Far Eastern countries, the American traveler Lew Ayres recently wrote:

Cautiously used, incense can contribute greatly to a highly satisfying liturgical experience. Excessively used, it simply becomes a revolting, irritating stench. Throughout the Orient incense is used with utter abandon. One's last few coins are spent for it. In many places it has become the custom to use not fewer than three sticks at a time. It is burned profusely day and night. The temples reek of it. One wonders if there is any significance in the fact that since ages past, the temple owners have been the leading incense concessionaires.[31]

Various writers have described the altar which forms the center of the family's religious life in every Japanese household of old fashioned upbringing. From the time that Shintōism became the national religion of Japan, every home has had a Shintō altar. If the family was Buddhist, every such household had a second, Buddhist, altar as well.

The Shintō altar was plain, made in the form of a wooden shelf called "God's shelf" (*kamidana*) shaped somewhat in the form of a Shintō shrine. It was placed over a doorway. However, incense played no part in the Shintō ritual and was not included among the offerings placed on the shelf.

The Buddhist altar (*butsudan*), was invariably built against the outside wall of the inner back room of a domestic dwelling. It was made of hardwood or of rich lacquer and decorated in gilt. It was shaped in the form of a miniature Buddhist shrine in which an image of Buddha was housed, as well as the "soul com-memoratives" (*ihai*). These were mortuary tablets dedicated to the memories of deceased members of the family. A lamplet, an incense cup, and a water vessel formed part of the altar furniture placed before the shrine in addition to candles,

---

[28] Hearn, *Miscellany*, pp. 117–18. [29] *Ibid.*, p. 234.
[30] Curzon, *Leaves*, pp. 264–74. [31] Ayres, *Altars*, p. 51.

bundles of incense sticks, and often a pair of bamboo cups containing sprays of the sacred shikimi tree.[32]

In Japan the incense burner (*kōro*) was usually placed upon a polished hardwood or lacquered stand in a recess in the main wall of the room (*tokonoma*) at the side of a hanging scroll (*kakemono*) with a vase of flowers. The *kōro*, made of bronze or lacquer, was usually global in shape with two ear handles, three-footed, and lined with silver or copper, with a cover terminating in a knob. The sides were plain or worked in relief with an auspicious poetical theme. *Kōro* made of pottery were generally cylindrical or hexagonal in form. *Kōro* made of carved hardstones were often to be found in the homes of the wealthy. Incense was burned before the altar in the daylight hours, while at night a small floating wick lamp was kept alight.[33]

The American writer and diplomat, Bayard Taylor (1825–1878), noted that in front of the image of the Buddha on the *butsudan*, "in the porcelain bowl of ashes, stand glowing bundles of fragrant wood, irreverently styled 'joss-sticks' by the infidel, which waft little clouds of incense before great Buddha." The reference to a bowl made of porcelain is noteworthy, because they were generally made of metal.[34]

To foreigners in East Asia the incense stick is commonly known as the "joss-stick." This name came into general parlance after the advent of the early Portuguese traders to China. "Joss" is derived from the mispronounced Latin word *deus*, or the Portuguese *deos*, meaning "god" and applied to the figure of Buddha which the Chinese worshipped. The Chinese rendered the word as "joss," with the consequence that the Buddha became generally known as a "joss." As a result, sticks of incense offered to the native deities became known among foreigners in the Far East as "joss-sticks," and temples became known as "joss-houses."[35]

Buddhism reached Tibet in the seventh century, spread by traveling preachers from India, and it eventually developed into a peculiarly Tibetan form. The number of its priests or lamas multiplied rapidly, and in time managed to usurp authority in matters of state to such a degree that the kingship of Tibet was eventually assumed by the priesthood in the form of the Dalai Lama.

The role which incense has played in Tibet is comparable to that in China and Japan. It is offered daily in the temples and homes. An incense burner is part of the home altar furniture, placed on a lower shelf or in front of it. When incense is not readily available or too expensive, juniper spines (*shūka*) serve as a substitute. An

[32] Hearn, *Miscellany*, pp. 243–45, 256–57; Hearn, *Glimpses*, vol. 2, pp. 399–405; Kates, *Furniture*, pp. 41, 75.
[33] Hearn, *Glimpses*, vol. 1, pp. 399–405; Holland, *Old and New*, pp. 44–59, 67–68.
[34] Taylor, *Japan In Our Day*, pp. 261–62.
[35] Casal, "Incense," p. 52.

incense kiln is frequently present beside the entrance to the homes; juniper spines are burned in this kiln each morning and evening. Tibetans traveling about the country place lighted incense sticks in rocky clefts where evil spirits were believed to lurk.

In the Tibetan temples, the lamas burn incense at dawn, in elaborate censers of gold and silver, as offerings to the several classes of divinities. Incense is among the Eight Essential Offerings which form part of every rite. On the anniversary of Buddha's death, incense is burned on every hilltop, and in every temple, home shrine, and lamasery.[36]

The same religious use of incense occurs also in Mongolia and Manchuria. The Russian painter, Nicholas Konstantin Roerich (1874–1947), who traveled through central Asia on painting expeditions, noted that at sunset in Ladak, smoke arose slowly above every house on the plain as incense was ignited during the hour of prayer. He reported also that the shops of Urga sold large stocks of objects used in temple ceremonies and that among the most popular were incense sticks of Tibetan and Chinese manufacture. The incense sticks made at the Sera Monastery at Lhasa were particularly prized in Mongolia.[37]

Despite the prevalent use of incense in ceremonial functions in Manchuria, Mongolia, and Tibet, there is no evidence that it was at one point used also for time measurement. However, an unusual incense burner which may have served as a time keeper is made of copper with a cover of "white metal" (*paktong*), and measures 38 cm in length, 14 cm in height and 8 cm in breadth. Bosses of paktong in repoussé are affixed to both sides of the burner by means of copper rivets. On one side two cloud dragons face each other with the pearl of wisdom between them, and the other side is decorated with the Buddhist symbols of the canopy, twin fish, sacred base, lotus, conch shell, sacred knot, umbrella and Wheel of the Law. Elaborate grotesque animal masks are attached to each end and the cover is decorated with a floral design perforated to enable the smoke and heat to disperse. Featured atop the cover are the lotiform Wheel of the Law flanked by two resting antelopes. Although the burner was acquired in Outer Mongolia, it may be of Tibetan or Chinese craftsmanship, possibly of the nineteenth century.

According to the catalogue information, presumably derived from the source from which the burner was obtained:

When used as an incense burner this long, deep traylike receptacle was filled with sifted wood ash on which were placed glowing bits of charcoal. Pellets of incense were then laid on the glowing coal. The ash served as an insulation so that the burner never became overheated.[38]

---

[36] Rockhill, *Diary*, pp. 123, 130, 132, 198, 201–2.　　[37] Roerich, *Trails*, pp. 25–29.
[38] *Northeastern Asiatic Art*, Item 142. From the collection of the Newark Museum, Newark, New Jersey.

"Joss perfumery," the use of incense sticks, has always formed an integral part of the Buddhist ritual in China but not in Japan. Incense is also featured in the ceremony of "receiving the fire" (*pul-tatta*), the rite undergone by young bonzes upon taking the vows to enter the Buddhist priesthood. After the hair of a professing young man has been shaved completely, several cones of incense (*moxa*) are laid upon his head and then ignited. They produce painful sores, the scars of which serve not only as a record of initiation but also as a mark of dedication and holiness. If the vows are later broken, the torture is repeated, thus maintaining ecclesiastical discipline.[39]

The incense used in Buddhist temples in Korea was in the form of sticks known as "longevity incense" (*man-su-hyang*) made from powdered mushrooms that grew on oak trees. After the mushrooms had been boiled in lye, they were dried in the shade and then ground into powder in mortars with pestles. The powder was then placed in incense trays, kept in the "tray room" (*nojon*) which was situated off the "main hall" (*taeungjon*) of Buddhist temples.[40]

The pervasion of incense in Far Eastern life was most vividly expressed by Lafcadio Hearn:

It was almost ubiquitous, – this perfume of incense. It makes one element of the faint but complex never-to-be-forgotten odor of the Far East. It haunts the dwelling-house not less than the temple, – the home of the peasant not less than the yashiki of the prince. Shintō shrines, indeed, are free from it; – incense being an abomination to the elder gods. But wherever Buddhism lives there is incense. In every house containing a Buddhist shrine or Buddhist tablets, incense is burned at certain times; and in even the rudest country solitudes you will find incense smouldering before wayside images, – little stone figures of Fudō, Jizō, or Kannon. Many experiences of travel, – strange impressions of sound as well as of sight – remain associated in my own memory with that fragrance; – vast silent shadowed avenues leading to weird old shrines; – mossed flights of worn steps ascending to temples that moulder above the clouds; – joyous tumult of festival nights; – sheeted funeral-trains gliding by in glimmer of lanterns; – murmur of household prayer in fishermen's huts on far wild coasts; – and visions of desolate little graves marked out by threads of blue smoke ascending, – graves of pet animals or birds remembered by simple hearts in the hour of prayer to Amida, the Lord of Immeasurable Light.[41]

[39] Griffis, *Corea*, p. 335; Douglas, *China*, pp. 359–60.
[40] Sang-woon Jeon, *Science and Technology*, pp. 64–65.
[41] Hearn, *Ghostly Japan*, pp. 19–21.

4

Incense for time measurement

Incense was well suited for time measurement, because it burned constantly, at an even rate, and without flame. Since it was flameless, it was relatively safe and could be used without the fear that it would cause a fire. It was first adapted for the purpose of time measurement in Far East countries in several simple forms, primarily incense sticks and powdered incense, with which it was possible to establish the burning rate of specific units of incense.

The date of introduction of incense in the form of a stick or rod is not known; but it may have been preceded by primitive forms of the incense seal. Incense sticks were already in common use by the T'ang Dynasty, however.

The manner in which incense sticks were used for measuring time, not only for official needs but also for the individual, was quite simple. The burning time for a stick was noted and then marked off along its length to indicate time segments.[1] One or more sticks marked with graduations for the time periods were inserted vertically into a bed of wood ash contained in a suitable receptacle, such as a *ting*, a deep three-footed container originally used for ritual purposes. The passage of time was easily noted by the progress of burning of the stick. An early illustration is included in an eighteenth-century work on Chinese buildings, furnishings and customs by Sir William Chambers (Fig. 6).[2] A later representation by a nineteenth-century French writer depicts the arrangement in greater detail (Fig. 7).

Among the poorest classes in China the inexpensive knotted match-cord was used as a time alarm. Consisting of a length of specially prepared cord made of punk, it was similar to the incense stick in that punk burns slowly and evenly.[3] The speed with which a specified length of the match-cord burned was readily established and consequently it was possible to determine the spacing between knots. Knotted at measured intervals, each knot indicated the passage of a time period by the progress of the burning.[4]

[1] Magowan, "Modes," p. 431; Needham, SCC, vol. 3, pp. 329–30.
[2] Chambers, plate 14; Needham, SCC, vol. 5, part 2, p. 147.
[3] Hough, "Timekeeping," p. 208.
[4] Punk is variously described as rotted wood or a fungus growth on wood which in a dry state is used for tinder. It forms a composition that smoulders when ignited and is used to ignite other combustible materials. *The Oxford English Dictionary*, vol. 8, p. 1604.

Chinese messengers, who by the nature of their work had limited periods for sleeping, were said to awaken themselves at designated times by putting a lighted bit of joss-stick or strip of knotted match-cord between their toes, and igniting it before falling asleep. When the cord had burned down to his skin, it would awaken him, and undoubtedly he would inadvertently awaken the rest of the household (Fig. 8).[5]

Candles inscribed with hour graduations (*k'o chu*) were used in China for time measurement as early as the Six Dynasties, in about the fourth or fifth century, and they were mentioned several times in the *Nan shih* (History of the Southern Dynasties).[6] Such candles may have been first devised by Buddhist monks to mark their vigils. That they were already well known to the Chinese is confirmed in a couplet by the sixth century poet, Yü Chien-wu (fl. A.D. 520):

> By burning incense [we] know the o'clock of the night,
> With graduated candle [we] confirm the tally of the watches.

Chinese candles of the period were made of beeswax by the dipping method, with a thin, hollow bamboo wick. From the T'ang Dynasty and later, other materials were also used, including insect wax and vegetable tallow. The candle was adapted for time telling by having the body marked into equal segments, each labeled with the appropriate character for time division, and each of which required a specific time unit to burn.

The incense candles favored by Emperor I Tsung (860–874), were not graduated for time periods inasmuch as they were only 6 cm in length, but nevertheless they were reported to have burned through the night.[8]

Graduated candles and incense sticks were in common use for time measurement during the Sung period, and both are mentioned in the literature of the period. The *Dai Kanwa jiten* describes the graduated candle and notes also that five incense sticks were burned, one for each of the watches of the night. This suggests that the period of burning for each stick was from 3 to 7 *k'o*, depending on the season and the length of the night hours.[9]

Incense sticks were used not only for measuring the hours of the day and night, but for other purposes as well. During his travels along the Yellow River in western China, the American diplomat William Woodville Rockhill (1854–1914),

---

[5] Kunitomo, *Tokei no hanashi*; Yabuuchi, "Clocks of China," cited by Seijiro Takigawa, "Kawachi," pp. 8–29.

[6] Morohashi, *Dai Kanwa jiten*, vol. 2, p. 246b; Needham, SCC, vol. 5, part 2, p. 147.

[7] Yü Chien-wu, "Feng ho ch'un yeh ying ling." Quoted in Schafer, *Golden Peaches*, p. 160.

[8] Po Shou-i, "Sung shih I-szu-lan chiao-t'u-ti"; Schafer, *Golden Peaches*, p. 160.

[9] Morohashi, *Dai Kanwa jiten*, vol. 2, p. 264b, vol. 5, p. 963a.

observed that water was raised by huge wheels of 17 to 18 m in diameter. Generally these wheels were communal property and belonged to the villages. In rare instances, however, they were privately owned by individuals who would, for a small consideration, sell water to the peasants. The price was calculated by the quantity of water which flowed from the wheel during the burning of a given length of an incense stick.[10]

Even until recent times farmers in Kansu have used incense sticks marked in *pouces* (French; inches, equivalent to the size of the thumb or big toe) to measure the time necessary for the flow of water to irrigate their lands.[11]

Chinese coal miners, working underground continuously for consecutive periods of approximately three hours each, carried with them what they called a "timepiece" in order to know the time and to measure the period of their labor. This was an incense stick which glowed for three hours without flame. This practice continued until recent times, and perhaps even to the present.[12]

An unusual use of incense sticks was made by Chinese physicians, who adapted them as a simple time check. The physician made several breaks in an incense stick, carefully rupturing it without totally separating it so that it formed several angles. After igniting one end, he instructed his patient to take the first dose of his medicine when the burning had reached the first angle, the second dose when the burning reached the second angle, and so on.[13]

Another practice of former times, which may persist to the present, was for older cultured men, or scholars, to sit together and compose poetry upon a mutually selected theme. The poem had to be completed within a designated period of time. The time elapsed was measured by an incense stick to which a small bell was affixed by means of a string or thread at a designated point along its length. When the incense had burned to the string, it parted, making the bell fall and indicating that the time for the composition was up.[14]

During the medieval period the incense stick served Chinese navigators on shipboard in the same manner that the time-glass or "mariner's dyoll" was used in the Western world. The time-glass did not come into use in China until after its introduction by the Dutch or Portuguese at the end of the sixteenth century. From at least the twelfth century, incense sticks, and later a more refined form of incense time measurement, were undoubtedly maintained in the ship's shrine where the compass was kept, and served for timing the watches. On shipboard at night and

---

[10] Rockhill, *Land*, p. 42.

[11] Wins, "L'Horloge," p. 23.

[12] Hommel, *China At Work*, p. 4.

[13] Mason, in Seely, "Development of Timekeeping," p. 49; Hough, "Timekeeping," p. 208.

[14] Communication from Dr. K. M. Starr, Curator of Asiatic Archaeology and Technology, Field Museum of Natural History, Chicago, Illinois, October 29, 1959.

in times of storm, the watches were kept by means of incense sticks which were graduated with the hours.[15]

The incense stick figured also in Chinese imperial functions. When the emperor was about to take a bride, he was instructed concerning his new relations by four young ladies, and the Astronomical Board consulted the stars to determine the most auspicious time for the supreme ceremony. The prospective bride, and the four young ladies who accompanied her, were housed in a palace specially prepared for them. One day before the wedding was to take place, a gold scepter, a seal, and a gold tablet were sent to the bride; on the tablet was inscribed the edict elevating her to imperial status. At eleven o'clock on the evening selected, a procession was formed with the bride seated in the "Phoenix Chair," and accompanied by a number of officials. The entourage made its way slowly to the imperial palace. Troops guarded the procession along its length and the houses all along the way were closed. At the side of the bride's chair marched an official of the Astronomical Board carrying "a lighted joss-stick, so marked as to indicate portions of time, by means of which he regulated the pace of the procession, in order that the imperial palace might be reached at the fortunate moment of two in the morning."[16]

The graduated incense stick was also prescribed as the means of timing the passage of the royal procession during the coronation of the emperor. As the entourage moved from the imperial palace to the temple, a graduated incense stick of a specified fixed length was carried by a functionary.[17]

Incense sticks and graduated candles continued in common use for time measurement during the Sung Dynasty and both are mentioned in the literature of the period. The length and size of the incense sticks were established by the Astronomical Board.[18]

The graduated candle and its use for official purposes was described in the *Dai Kanwa jiten*. The "time incense" (*gong heung* or *keng hsiang*) used by the Chinese for timetelling at night consisted of five sticks of pressed wood dust, to be burned during the night hours, which were divided into five watches. The sticks provided for the purpose were of two sizes – long for the winter nights, and short for the summer nights.[19]

---

[15] Needham, *SCC*, vol. 3, p. 330; vol. 4, part 2, pp. 127, 462, 526, 570; part 3, pp. 569–70; vol. 5, part 2, pp. 146–47.

[16] Douglas, *China*, pp. 71–75.   [17] *Ibid.*, pp. 73–75; Hough, "Timekeeping," p. 208.

[18] Mason in Seely, "Development of Timekeeping," p. 49; Hough, "Timekeeping," p. 208; Douglas, *Society*, p. 313.

[19] Morohashi, *Dai Kanwa jiten*, vol. 5, p. 963a. A bundle of these "time incense sticks" from Canton was donated to the Smithsonian Institution in the nineteenth century by Stewart Culin. They are one-fourth inch in diameter and 16 inches in length. Hough, "Time-keeping," pp. 207–8.

In the *Hsiao hsüeh kan chu*, the thirteenth-century encyclopedia in ten *chüan* by Wang Ying-lin (1223–1296), the compiler quoted Hsüeh Chi-hsüan (1125/1134–1173) as having written,

Nowadays time-keeping devices are of four different kinds. There are the bronze vessels [clepsydrae], the [burning] incense stick, the sundial, and the revolving and snapping springs.

The "revolving and snapping springs" referred to the steelyard clepsydra, an advanced form of the balancing inflow water clock.[20]

Chang Hsüan, writing in the Ming Dynasty, commented in his *I yao* on the use of incense sticks:

In olden times Chang Chung-ting . . . would set up an incense stick in his bedroom and sit up all night. If the watch drums in the commandery tower were off by one *k'o*, those in charge were reprimanded if they were late by as much as a single notch . . .[21]

Another form of incense for time measurement was the spiral incense coil. In addition to other purposes, it served for time telling when longer periods of burning were necessary than could be measured by means of incense sticks. These coils were used during certain periods for keeping time for the "night watches" of the community; the length of the spirals varying with the length of the months during which they were to be employed. The incense coils were made so that the burning of each spiral lasted for a single night, and were marked off evenly into intervals for each of the watches to designate the reliefs of the five night watches.[22]

Incense coils are used in the present day as offerings in Buddhist temples, particularly in Southeast Asia. They are suspended from the ceiling rafters in great numbers and confusion (Fig. 9). In the early twentieth century, they were commonly used also as an insect repellent in the home, as Monsignor Alphonse Favier illustrated in a work on Peking.[23]

A modern work on Chinese timekeeping provides a summation of the subject of Chinese combustion clocks, as an inexpensive method for telling time that continued to be in use in the early years of the Chinese Republic:

As a measure of time, the Chinese rather used combustion clocks [*horloges á feu*] made by means of the same mixture causing combustion of which great use was made in China in the form of odoriferous sticks which are continuously burned before the tablets of ancestral worship or their idols. This mixture, with a base of clay, is composed of the

---

[20] Wang Yang-lin, *Hsiao hsüeh kan chu*, ch. 1, p. 42b. Quoted in Needham *et al.*, *Heavenly Clockwork*, p. 163; Needham, *SCC*, vol. 4, pp. 462, 526.

[21] Needham, *SCC*, vol. 4, part 3, p. 127. Chang Hsüan, *I yao*, p. 95.

[22] Jünger, *Das Sanduhrbuch*, pp. 49–52.

[23] Favier, *Peking*, p. 364.

sawdust of the wood of various species of aromatic trees from Tibet, to which are added musk and gold dust.

These sticks, which consume themselves slowly without ever igniting, are occasionally marked with graduations to serve as time measurers; their burning can last for several days and they indicate the time with sufficient accuracy. These combustion clocks assume several forms and can also serve as alarms [i.e., the dragon boat alarm].[24]

In addition to the various other types of time measurers utilizing incense, the Chinese also devised several forms of alarms. Although they might not be particularly audible to the Western ear, accustomed to normal cacophonies of sound, the subtle sounds of these alarms appear to have been adequate for the Chinese. An early form consisted of a large spiral incense coil suspended from a special bracket placed upon a table. The base was in the form of a bronze or brass basin. A small weight or bell was attached to the point on the curve desired, measured for the passage of time. The spiral was ignited at the top and as the burning eventually reached the weight, it fell into the basin, and the sound it made in doing so alerted the sleeper as to the time (Fig. 10).[25]

The widespread use of the suspended incense coil alarm was noted by the Spanish Dominican monk and apostolic prefect of Chekiang Province, Martin Fernández Navarrete (1610–1689?). In the account of his travels in China in the mid-seventeenth century, he wrote:

They make small perfumed pastilles in conical form, for lighting them at each hour of the night. They are marked to indicate each hour as it burns. [The missionary] Magalhaens observed that these pastilles are composed of sandalwood or of other odoriferous woods reduced to powder, with which is made a form of paste and which is shaped in molds . . . Their length has somewhat the size of two or three palms [the size of a hand] and no more. They last for one, two, or three days, depending on their size. Some are also made for the temples, that burn twenty or thirty days . . . This method of measuring time is so accurate as never to result in any considerable error. Those who wish to be awakened at a certain hour suspend a small weight [on the coil] at [the desired hour] mark. When the burning reaches that point, the weight falls into a brass basin and the sound awakens the sleeper.[26]

The incense alarm was developed in yet another and more intriguing form in China, known as the "dragon boat" alarm. This consists of a sculptured container carved of wood in the form of one of the ancient dragon boats, with figurehead and stern representing the head and tail of a fiery dragon. The center of the sculpture was hollowed and the outer surfaces finished in black lacquer with gold trimming to represent the scales of the beast. The head was gilded and the tail was

[24] Chapuis, *Relations*, pp. 16–17.
[25] Planchon, "L'Heure," pp. 249–50.
[26] Navarrete, *Account*, quoted in Guitton, *Quand sonne*, pp. 96–97.

either gilded or appropriately painted a flaming red inasmuch as the tail represented flames. The device was supported on four carved feet, again representative of the dragon. The interior was fitted with a pewter liner pierced at intervals with nine openings along its length into which are inserted V-shaped wires which serve as a rack for supporting an incense stick.[27]

The dragon boat rested upon two pedestals approximately 15 cm in height. A metal pan or basin having a high resonance was placed between them on the table surface. The "alarm" consisted of a pair of small bronze bells tied at the terminals of a silk thread. The thread was draped over the incense stick and the sides of the dragon boat at the point at which the sleeper wished to be awakened. When the burning of the incense stick had reached that point, the silk string burned and parted, dropping the bells into the basin. The sound that resulted presumably awakened a very light sleeper. Some of the dragon boats were equipped with more than one pair of bells, which serve as repeaters (Fig. 11).

The "dragon boat" was designed as an alarm primarily for use during the night. The incense stick which it accommodated was graduated with the hours, each segment requiring one double-hour to burn.

Surviving "dragon boat" alarms were generally made in the eighteenth or early nineteenth century, of wood or of metal, the greater number being of the former. The vessel was constructed from three blocks of wood mortised and glued together, and consisted of the head and neck, the body, and the tail. Each was carved and modeled to the final form, and the assembled unit was lacquered and gilded. For the most part they measure approximately 69 cm in length, 6 cm in width and 15 cm in height. In some examples the gilded feet concealed tiny wood or brass wheels which enable the dragon boat to be moved easily along the surface of a table.

A comparison of surviving examples indicates that a standard form existed, from which several variations were made, generally exaggerating the features of the head and tail, and often made with a shallower body.

A small number of dragon boat alarms made of bronze or brass have survived, having round openings on both sides of the bow through which pins or pegs were inserted and by means of which an incense stick was supported inside the body (Fig. 12).

Another unusual form, of which occasional examples have been noted, was fashioned as a steam paddleboat. The superstructure lifts off revealing a pewter-lined body with traverse wire struts for supporting an incense stick. The nature of this form suggests that they were produced, probably in China, in the late nineteenth century.[28]

---

[27] Guitton, *Quand sonne*, pp. 86–87.

[28] The examples made in the form of paddle boats are generally about 50 cm in length. Several are noted in auction sales catalogues; see Appendix D.

The standard dragon boat alarm takes its form from the vessels used in the *Wuyüeh chieh*, or Fifth Month Festival of ancient origin, known to foreigners in China as the "Dragon Boat Festival." It is the second of the four annual festivals during which the Chinese settle their accounts; the first is New Year's Eve, and the others are the Moon Festival in September or October, and the Winter Solstice Festival which occurs in November or December.

The Dragon Boat Festival is celebrated during the first five days of the fifth lunar month. This period, occurring at about the time of the Summer Solstice in June or July, includes the three settlement days for accounts.

The festival also marks the turn of the year in agriculture. It is the time when primitive peoples attempted to propitiate the gods. These rites have survived in China, particularly in the region south of the Yangtze which has many rivers and lakes, as part of the cult of the water gods upon whom the livelihood of the water people depended. Among the early water divinities were ghosts of those who had drowned and who were forced to wander. Offerings of rice were cast upon the waters and lighted lanterns were set adrift to guide them to the feast in order to placate these ghosts, and to divert their attention from those who made their living from the water, such as the fishermen.

The dragon was the controller of the waters and the dispenser of rain, which was due in China each year in the last week of the month. Consequently it was necessary to propitiate the dragon to ensure an adequate supply of rain for the crops which were ripening at this time.

On the first day, after decorating doorposts and windows with leaves of sweet flag and artemisia, the river people dressed in their finest garb and made their way to the lake or river to observe the gala event, the race of the dragon boats.

According to some semi-historical sources, the Festival originated in the year 295 B.C. (some say 278 B.C.), on the occasion of the death of Ch'ü Yüan, also known as Ch'ü P'ing (332–295 or 278 B.C.), a virtuous minister of the Ch'u state, who drowned himself in the Mi-lo River in grief over being slandered at court. His death took place on the fifth day of the fifth moon. Greatly beloved by the people for his virtue and poetry, search parties went out in boats in an effort to find his body, but without success.[29]

The search has been continued on succeeding anniversaries throughout the centuries. It is said that after his death, Ch'ü Yüan appeared to a fisherman and gave specific instructions for the method of parceling rice which was to be set adrift in bamboo tubes as part of a festival celebration. Accordingly, the people next prepare a rice-cake (*tsung tzu*) wrapped in bamboo leaves or leaves of a plant

---

[29] Giles, *Dictionary*, pp. 200–1.

having the appearance of iris, and bind it with raffia before cooking. In the Peking area the rice cake is triangular in shape while in the south it is square.[30]

As the festival begins, the boatmen in the dragon boats, which are decorated with triangular banners, all depart from the same site amidst a beating of gongs and loud encouragement from spectators, racing to be the first to reach the traditional site of the drowning and to make a sacrifice to the drowned man's spirit. The boats participating in the event each year are propelled by paddles, and are approximately 38 m in length, 76 cm deep, and 1.7 m wide, just broad enough to seat two rowers side by side. They are extremely costly to build. The bow of the dragon boat is made in the form of a carved dragon's head or is decorated with such a carving, while the stern represents the dragon's tail. In addition to the rowers, the boats carry passengers who wave flags and beat gongs to encourage the rowers (Fig. 13).

Impromptu preliminary races are arranged, building up the excitement of the participants and observers before the major events between competing villages. So many accidents have occurred during these races over the years that they have been forbidden in some parts of China, and restrictions have been placed upon them in others.[31]

In his account of his travels in China, Martin Fernández Navarrete described the annual event as he had observed it in the seventeenth century:

One general solemnity is kept throughout the whole empire on the fifth day of the fifth moon. This day they go out upon the rivers in boats finely decked and adorned, to solemnize the festival of a certain great magistrate, who was very zealous for the publick good. They report of him, that an emperor refusing to take his advice, he cast himself into a lake and was drowned. Against this festival they provide a sort of cakes, and other meat, which they throw into the water in honour of that magistrate. Others say they do it, that he may have something to eat. I have before made mention how one year above five hundred vessels went out from *Nan King*, upon the river they call *the son of the sea*, but a sudden gust of wind rising, they all sunk to the bottom, not one escaping.[32]

The actual date of origin of the dragon boat alarm is not known, but that it already existed in the Sung period is suggested in the writings of the Northern Sung poetess, Li Ch'ing-chao (1084-1151), China's most noted woman writer and claimed to be one of its greatest literary geniuses.[33]

---

[30] Douglas, *China*, pp. 360–63.

[31] Morgan, *Chinese Symbols*, pp. 162–64; Burkhardt, *Creeds*, pp. 26–29; S. W. Williams, *Middle Kingdom*, vol. 1, pp. 816–17; Douglas, *China*, pp. 360–63; J. D. Ball, *Things Chinese*, pp. 27, 222–23.   [32] Navarrete, *Account*, Book D, p. 46.

[33] A native of Chi-nan Li, Ch'ing-chao was married in 1101 to the scholar and antiquarian Chao Ming-ch'eng. They escaped during the dynastic collapse of 1126 but he died shortly there-

In a poem entitled "To the Air of *Feng-huang t'ai shang i ch'ui hsiao*" she wrote:

> Incense
> > cold
> > in bronze lion boat
> > Bedcover rumpled
> > in crimson seas
> Getting up listless
> > I comb my hair
> > alone
> Neglected
> > dressing table
> > veiled with dust
> Daybreak climbs
> > the bed-curtain bar . . .[34]

Although the dragon boat alarms generally followed a standard design, there were substantial variations, as a consequence of the artistic inclination of the individual carver, or possibly in imitation of actual forms of vessels participating in the festival (Fig. 14). A study of surviving examples suggests that generally the dragon's head is raised belligerently with its mouth open, and its red tongue exposed. In other examples the head is lowered and thrust forward. On occasion the dragon's head and neck are represented as being extremely hairy, which may have given cause for describing it as a lion's head, and sometimes the dragon's eyes are made to protrude excessively (Fig. 15).

Particularly elaborate examples are decorated along the sides with paintings of floral and other motifs over the black lacquer, sometimes depicting landscapes or family groups executed in gilt. Notable among these elaborate dragon boats is a pair which belonged to the kings of Spain. They are preserved in the Palacio Aranjuez in Madrid, where the sovereigns prepared to undertake sea voyages (Fig. 16).

Occasionally the elaboration led to the addition of human figures and various decorative motifs. An unusual example includes two sculptured figures instead of a dragon figurehead (Figs. 17a and 17b).

In the nineteenth century, dragon boat alarms were occasionally utilized for other purposes, possibly because their original function had long been forgotten. A nineteenth-century work by an American in Canton described one of these later uses:

after. She fled alone along the coast until she reached Chin-hua, where she settled and later died in 1150. Her eminence as a writer resulted in many attempts to smear her reputation in her lifetime. Most of her writings, except for some of her *tz'u* poetry, are lost.

[34] Kwock and McHugh, *Old Friends*, p. 52.

In the middle of the 19th Century the hongs of Canton were exceedingly busy places. The clerks arrived at the counting rooms very early and seldom left before midnight, and a number of what we would term now "breaks" were permitted to relieve the tedium of the long hours. The clerks were allowed to smoke, and a light lunch of bread, plantains, and Calcutta beer was served at noon. But dinner, which took place around four or five o'clock, was the main event of the day.

Owners and clerks customarily partook of this meal together, apparently in an early attempt at good employer-employee relations. There were usually fifteen or twenty of the staff present, though the number might be doubled by the addition of captains and supercargoes from ships in the harbor. The best of foods was served at these meals: lamb, chicken, game birds, vegetables, even potatoes grown near Macao; and a wide variety of fruits and nuts were served as dessert. While the wine was being poured a long narrow tray shaped like a dragon boat, about 50 cm long, was pushed up and down the table top. At one end was a beautifully carved and gilded dragon head, with a red tongue, the tail was edged in red-painted flame. The four low legs, which were gilded, concealed small brass wheels. There was a pewter liner inside the black lacquer body with its gilded scales, across it were nine V-bent wire seats which served as racks for lighted joss-sticks. As the dragon went up and down the polished board the joss sticks were used to light the fine Manila cheroots which ended the dinner.[35]

Surviving examples of the wheeled dragon boat resembling the foregoing description to any degree are relatively rare. Only several are known, including one so curious that it deserves attention. That it was originally a dragon boat alarm is evidenced along the inner sides by holes for the supporting wires. Wooden wheels are concealed under the dragon's feet, and the mid-section is fitted with a box which may have been intended to contain cigars to be passed around the table.

However, the outer surfaces show no jointure between the sides of the vessel and the box, and the vessel may have been relacquered. The exterior of the box is decorated on all four sides with figures that appear to be children at play. The fore and aft hollowed interior sections of the vessel are lined with removable pewter receptacles, which may be fragments of the original inner liner (Fig. 18).

The "dragon boat" alarm came to notice again in the late nineteenth century and in even more recent times in French writings on timepieces, in which it was referred to as the "repeater junk" (*la jonque à répétition*). One writer described it as a Chinese implement and the earliest example of a morning alarm. He reported that one of these devices which he had seen was approximately 80 cm in length with twenty straight rods along the open center upon which the lighted incense stick was laid. Thin threads were laid at spaced equal intervals from edge to edge in contact with the incense stick. To the ends of each of these twenty threads were attached tiny brass bells which hung below the vessel. As the burning of the

---

[35] Tiffany, *Canton*, p. 229.

incense stick progressed, the threads parted one by one and the little bells dropped into a bronze platter over which the vessel was suspended. This version appears to have served the function of a repeater alarm.[36]

Other forms of the dragon boat alarm, possibly earlier, did not include the dragon head and tail, but were carved in the form of river vessels, wide at the bow and narrowing to the stern (Fig. 19).

The relatively rare examples of the "dragon boat" alarm formed to represent river-steamers or paddle boats are generally not functional as alarms. They appear to be late elaborations of the traditional form, possibly produced as conversation-pieces. Nonetheless, such examples have consistently commanded impressively high prices in auction sales during the past decade.

Several examples of a device originating in Thailand similar to the Chinese dragon boat have come to light. These may have been used for the same purpose, or they may be examples modified from a common candle-holder of this form to imitate the Chinese dragon boat's timetelling function. The container is made of brass and is in the form of the *rua hong*, the huge gilt and red royal swan-barge used by the King of Thailand of the Bangkok period in state ceremonies, to the present time. The figurehead is carved in the likeness of the *hansa* or sacred bird, bearing a bell in its beak instead of the traditional flower, suspended from a silken thread. Six pairs of brass balls (not bells) are suspended over the sides of the vessel, the interior of which is fitted to accommodate an incense stick. The Thai vessels are supported on two pedestals rising from a rectangular metal base. Examples exist of a similar device incorporating nine candleholders along the vessel's body instead of supports for an incense stick, suggesting that one or the other purpose was a modification.[37] It is conceivable that in Thailand, with its large Chinese population, the Chinese alarm was duplicated in a native form (Fig. 20).[38]

The methods of employing incense for time measurement in Japan closely paralleled those which evolved in China, from whence they were borrowed together with many aspects of Chinese culture between the sixth and ninth centuries. In Japan incense was used for time measurement from at least the Nara period but it is possible that it was introduced from China even before then.

One form of time measurement used in Japan was the match-cord. Another was

---

[36] Guitton, *Quand sonne*, pp. 86–87.

[37] The *hansa*, according to proverbial lore, has the characteristics of both the swan and the goose, and is a symbol of erudition. Its name implies "the self-existent being," and is derived from the mythical lore of Brahmanism perpetuated in Buddhism. K. M. Ball, *Oriental Art*, pp. 238–40.

[38] *Introduction to Thailand*, p. 51; Bhirasri, *Thai Wood Carvings*, pp. 12–13.

the graduated candle marked in time segments. It was reported, however, that inasmuch as the time of consumption varied with the materials of which the candle was composed, it was not successful as a standardized timekeeper. According to one source, in Japan the graduations along the candle's length were grooved and the channels filled with a white, bone-like powder. Such candles were reported to have been in use until A.D. 901.[39]

The Japanese adopted the use of incense sticks for timetelling as a matter of course. They were introduced into Japan from China by Zen Buddhist priests during the Kamakura period and were said to have replaced the earlier devices utilizing powdered incense in incense timepieces. They were also actively used during the Edo period in the gay quarters, and were found in geisha houses. Known in Japan as "Chinese matches," the most common type were made of powdered sandalwood reduced to paste and hardened. At first the time of consumption required by a stick was observed and used as the time interval. Later the process was refined by marking the sticks at measured intervals with time designations, as in China.

The early history of time measurement in Korea is fragmentary at best. Incense was used from early times, particularly in Buddhist rituals. The use of the incense stick for timetelling in Korean Buddhist temples appears to have been introduced in the later Silla period, during the seventh and eighth centuries, and continued to be used for this purpose in Taoist and Buddhist temples during the Koguryŏ and Yi Dynasties.[40] Mention of the use of incense sticks for time measurement in Korea is found in the collected writings of Nam Pyŏng-ch'ŏl concerning astronomical instruments.[41]

In Korea time measurement with incense was employed only in the larger Buddhist temples, in which the community clepsydrae were also maintained, in a part of the temple called the *ro-jeon*. The temple incense burners were usually made of bronze or brass, and filled to just below the rim with powdered wood ash. Into the ash were inserted several of the *man-su-hyang* sticks graduated along their length with the divisions of the time units. As the incense burned, its rate was observed by temple attendants. When a prescribed period of time had elapsed, as indicated by consumption of the incense, the attendants rang the great temple bell to summon the faithful to prayer. Two types of incense were used for the purpose, one called "30 minute incense," and the other called "one hour incense." The "incense clock" in Korea was known as the *hyang-jeon*.[42] In smaller temples, a

[39] Takabayashi, *Tokei hattatsushi*; Morohashi, *Dai Kanwa jiten*, vol. 2, p. 264b.
[40] Communication from Dr. Sang-woon Jeon, January 20, 1963.
[41] Nam Pyŏng-ch'ŏl, *Uigi chipsŏl*, vol. 2, pp. 31 ff.
[42] Communications from Dr. Sang-woon Jeon, November 29, 1962.

particular vegetable oil was used for the same purpose. It was burned in a receptacle called *ok-dung-an*, made of quartz or similar stone.[43]

Thus far only the simple incense stick and the incense coil were adapted for the measurement of time, all primitive devices at best. The measurement with greater precision of longer periods of time became feasible only after the development of the "incense seal" (*hsiang yin*), which is believed to have originated from a Tantric Buddhist ritual.

[43] Communications from Dr. Sang-woon Jeon, December 31, 1962 and January 20, 1963.

II

THE INCENSE SEAL
IN CHINA

5

The incense seal of Avalokitésvara

It was inevitable that the extensive use of incense for elementary forms of time measurement would eventually lead to the development of a more sophisticated form of the "incense clock." The result was the "incense seal" (*hsiang yin*) which appears to have first come into being in China in the eighth century. It is probable that it was derived from part of the Tantric Buddhist ritual which originated in India. What is believed to be the predecessor of this innovative means of time measurement was featured in a Tantric scripture entitled "The [Incense] Seal of Avalokitésvara Bodhisattva" (Fig. 21).

As the religion of Buddhism flourished, a number of sects developed. The higher forms included the Mahayana and the Hinayana schools, as well as the Ch'an (Zen) and T'ien-t'ai among others. One of the forms most popular with the populace was the Tantric, in which magic played an important role. By the seventh century Tantric Buddhism in India had begun to be systematized and formed into a philosophical basis.[1]

The sect was based on Buddhist sacred texts called *tantra* (*ta chiao* or *shen pien*), accompanied by practices which were sometimes open and at other times esoteric, with the worship of personal gods one of its prominent aspects.

The tantric system of thought had much in common with the magical aspects of China's ancient Taoism. It became known as "the way of the thunderbolt" (*vajrayana* in Sanskrit; *chin kang ch'eng* in Chinese). The *Tantra* invariably reflected a strongly magical element, containing "words of power" (*mantra* and *dharani*), and including hand gestures (*mudra*), talismans (*yantra*), amulets (*kabaca*) and other charms.[2] Tantrism was officially introduced into China in the eighth century by three monks who came from India, Subhakarasimha, Vajrabodhi, and Amoghavajra.[3]

---

[1] Chou Yi-liang, "Tantrism," pp. 241–47; Benoytosh Bhattacharyya, *Introduction*, pp. 28–31, 43–61, 104–8; Needham, *SCC*, vol. 2, pp. 425–31.

[2] Bhattacharyya, *Introduction*, pp. 43–54.

[3] In addition to late sacred texts, the word *Tantra* also means a textile web with its warp and weft. Needham, *SCC*, vol. 2, pp. 425–31; Chou Yi-liang, "Tantrism," pp. 241–47; Bhattacharyya, *Introduction*, pp. 28–31, 43–61, 104–8.

Subhakarasimha, known as Shan-wu-wei in Chinese (A.D. 636–735), a priest of Nalanda, arrived at Ch'ang-an, the Chinese capital, in A.D. 716 and during the next eight years translated five important Buddhist works into Chinese. Among them was the basic text of the Mantrayana in China, the *Mahavairocana-sutra*, known in China as the *Ta jih ching*, the "Sutra of the Great Sun" (Fig. 22).[4]

Vajrabodhi, who became known as Chin-kang-chih in Chinese, was the son of King Isanavarman of central India. He became a monk at Nalanda in A.D. 680. After traveling extensively in India, he went on to Ceylon, and from there to China in 719. He brought with him an extract of the *Vajrasekhara-sutra*, known to the Chinese as the *Chin kang ching* (The Diamond Sutra), which was the counterpart of the *Mahavairocana-sutra*. He translated the extract into Chinese in 723.[5]

Amoghavajra (A.D. 705–774), also known as Amogha, whose name in Chinese was Pu-k'ung chin-kang, was called the "Master of Tantric Buddhism," and became the head of the Yogacara school in China. A Singhalese of northern Brahmanic descent, he lost his father at an early age and accompanied his maternal uncle, a merchant from Samarkand, to China. There he acquired a considerable knowledge of its language, and in 718 he became a disciple of the Buddhist leader Vajrabodhi.

After the latter's death in 741, and at his mentor's wish, Amogha returned to India and then traveled to Ceylon in search of Tantric writings. Upon his return to China in 746, he brought with him more than 500 *sutra* and *shastra* previously unknown there. He was a follower of the mystic teachings attributed to Samantabhadra, one of the four Bodhisattvas of the Yogacara school, and published more than 100 works, chiefly translations of the new Mantrayanic texts.

Having received permission in 749 to return to his home, he was making his way to India when he was stopped in south China by imperial orders. He was later recalled to the capital, where he spent the next few years translating and editing 120 books of Tantric texts. When he died in 774, he was given great honors.[6]

Amogha is considered to be the patriarch of the "Secret Sect" (Mi-tsung) or of the "Doctrine of the True Word" (Chen-yen), called "a sinified Tantrism," which achieved considerable success in China during the T'ang period and for a time obscured all other sects.[7]

---

[4] Van Gulik, *Siddham*, pp. 48–49; Chou Yi-liang, "Tantrism," pp. 247–50.

[5] Van Gulik, *Siddham*, p. 49.

[6] Soothill and Hodus, *Dictionary*, p. 108; Van Gulik, *Siddham*, pp. 49–50.

[7] Amoghavajra was patronized by three successive Chinese emperors and given the honorable title of "Tripitaka Bhadanta." When he died he was appointed minister of state and given a posthumous title. Eitel, *Hand-Book*, pp. 9–10; Chou Yi-liang, "Tantrism," pp. 284–306.

Among Amoghavajra's greatest achievements was the creation of a new system by means of which each Sanskrit sound was replaced by one or more Chinese characters. It ended a long-existing confusion in transliteration of Buddhist texts from Sanskrit to Chinese. He and other Tantric masters of his time wrote in a variation of the Brahmi script which was then widely used in India. The new script, which was named "Siddham," achieved great favor in China and Japan for Sanskrit *mantra* and *dharani*. Over the centuries the script, which was developed two thousand years ago in India, was preserved in virtual secrecy, and its use was limited to esoteric Buddhist rites by the few adept in producing it (Fig. 23).[8]

Amoghavajra advocated the esoteric use of Siddham *mantra* and seed-syllables. His translation of the text *Usnisa-vijaya-dharani* was particularly popular. By imperial decree it was printed on prayer sheets and engraved on pillars in Siddham script throughout the Chinese empire, and all Chinese Buddhist monks and nuns were required to repeat it twenty-one times each day.[9]

After Amoghavajra's death in 774, the influence of the doctrine of the "True Word" gradually declined in China, and it was transmitted to Japan by the Japanese monk Kūkai (A.D. 774–835), known also as Kōbō Daishi.[10]

Kūkai had studied in China at about the same time as his fellow countryman, the monk Saichō (Dengyō Daishi, A.D. 767–822). Together they were responsible for introducing Chen-yen into Japan in the ninth century. There it became known as the Shingon sect (the Japanese pronounciation of Chen-yen), and it has remained one of the most important Buddhist sects in Japan. Kūkai was responsible also for having transmitted the use of the Siddham script to Japan, where it was known as *bonji*. Although the use of Siddham had virtually disappeared in India and China, in Japan it remained an obscure branch of Japanese esoteric science until several decades ago, but is presently undergoing a revival.[11]

The incense seal can be traced to a feature of the Tantric ritual described in one of the Tantric scriptures, rendered by Amoghavajra into Siddham script from the

---

[8] Bhattacharyya, *Introduction*, pp. 55–61; Van Gulik, *Siddham*, pp. 49–56; Stevens, "People's Buddhism," pp. 2–4.

[9] Stevens, *Sacred Calligraphy*, pp. 2–12.

[10] Kōbō Daishi is probably the most famous of Japanese Buddhist saints. He was distinguished as a preacher, painter, sculptor, calligrapher and traveler. Born at Byōbugaura near the modern shrine of Kōnpira in Shikoku, he entered the priesthood in A.D. 793. He was sent to China as a student in 804 and became a favored disciple of the abbot Hui-kuo, who charged him to take back to Japan the tenets of the Yogacara, or Shingon sect at it was known in Japan. He returned home in 806 and was installed as abbot of Tō-ji in Kyoto. Several years later he founded the monastery of Kōyasan in Kishu and remained there until his death. The name "Kōbō Daishi," meaning "the Great Teacher Spreading Abroad the Law," was conferred on him after his death by Emperor Daigo in 921. Van Gulik, *Siddham*, pp. 114–17; Chamberlain and Mason, *Handbook*, p. 77.

[11] Stevens, *Sacred Calligraphy*, pp. 6–11; Stevens, "People's Buddhism," pp. 2–4.

Sanskrit, entitled "The [Incense] Seal of Avalokitésvara Bodhisattva." This part of the Tantric ritual originated in India probably between the fourth and sixth centuries A.D., and was transmitted to China in the eighth century, possibly among the texts Amoghavajra brought back with him from his journey and later translated.

Avalokitésvara, to whom the text is dedicated, was an Indian male divinity adopted by the Mahayana School in India and was highly revered, particularly from the third to the seventh centuries. He was well known as a Bodhisattva who appeared in many places on earth, in many incarnations as a savior of the faith and as the tutelary deity of the Tibetan nation.[12]

In China the Buddhists adopted the female deity Kuan Yin, the Goddess of Mercy, as an incarnation of Avalokitésvara. Kuan Yin may have been a local goddess before the advent of Buddhism. In Japan, Avalokitésvara took many forms. He is known as the female deity Kanzeon Daibosatsu, who contemplates the world and listens to the prayers of the unhappy. In another version, Avalokitésvara is known as the Amida Buddha or as Kannon Bosatsu, a male deity, sometimes many-headed, and other times horse-headed, with forty hands holding various Buddhist emblems; or he is known as the thousand-armed Senju Kannon Bosatsu.[13]

The Tantric text describes the purpose of and manner in which a trail of incense is formed in the shape of a Siddham seal-character and burned. Following is a translation of the complete text, taken from the Buddhist Canon, *The Tripitaka in Chinese*:[14]

The True Method of Perfuming Sentient Beings of All Realms By the Saving Grace of the [Incense] Seal of Avalokitésvara Bodhisattva, The All-Compassionate and All-Wise

The Supreme Holy One Vairocana Buddha thus revealed: great insight and perfect

---

[12] A Bodhisattva is a Buddha-elect and "one whose essence is perfect knowledge." He is described as a type of savior who, in order to help others to gain salvation, has voluntarily postponed his own salvation and entrance into Nirvana. Avalokitésvara is one of the three most important Bodhisattvas. Eitel, *Hand-Book* p. 24; Soothill, p. 108; Liebert, p. 29; De Lubac, pp. 104–54.

[13] Chamberlain and Mason, *Handbook*, p. 52; Eitel, *Hand-Book*, pp. 23–25. Early appearances of the name of the Bodhisattva Avalokitésvara in Chinese Buddhist writings occurred in the works of Hsüan Tsang between 629–645 and of I Ching between 671 and 695. Bhattacharyya, *Introduction*, p. 122.

[14] *The Tripitaka* literally means "three collections." Following the Brahmanic distinction of *Mantra*, *Brahmana*, and *Sutra*, the three divisions of the Buddhist canon are the *Sutra* (Books of Doctrine), *Vinaya* (Statutes on Ecclesiastical Discipline), and *Abhidharma* (Works on Philosophy). A fourth category added by the Chinese is the *Samyukta pitaka* (Miscellaneous Canonical Works). Soothill, *Dictionary*, p. 467; Eitel, *Hand-Book*, pp. 180–81.

freedom can be attained by the utilization and knowledge of secret seals [*mudra*].[15] Those yogis[16] who wish to practice [*Tantra*], be reborn in the Paradise of the West, and facilitate the salvation of sentient beings should follow a guru [*acarya*] possessed of superior wisdom and talent and receive initiation into the Lotus Flower Diamond Tantra.

The mystic rites must be conducted properly and for that an incense burner must be placed in the center of the altar. The incense burner must have a special form, that is, one based on the [esoteric principles] of perfect freedom and total utilization. The incense grid [template] is to be formed according to the secret teachings [of Siddham script] and will thus be all-enveloping.

[On the template or grid pattern] the four letters *ha*, *ra*, *i*, and *ah* are combined into one character *hríḥ*. The meaning of each letter is as follows: *ha*, the origin of all being is ungraspably empty; *ra*, undefiled purity; *i*, ungraspable freedom; *ah*, innately unborn and undying. As the incense smoke wafts up through this pattern it flows back and forth thereby encompassing all aspects of existence. Schematically, this [flow of incense smoke] is:

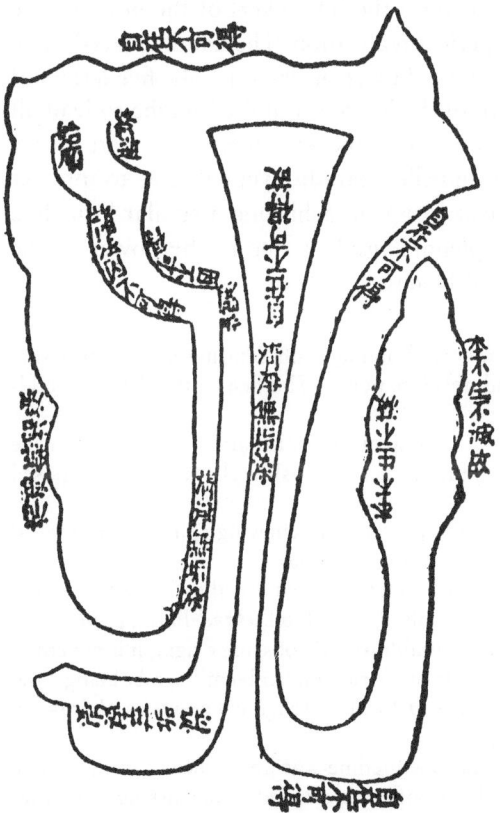

Figure 24. Enlarged detail of the schematic drawing of the flow of incense smoke of the "[Incense] Seal of Avalokitésvara Bodhisattva" from *The Tripitaka in Chinese.*

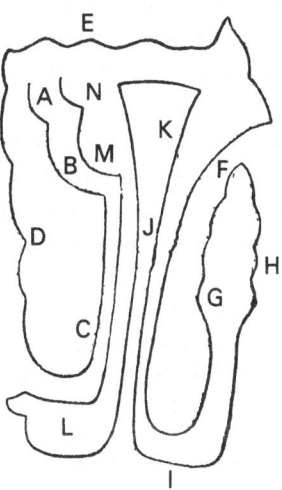

Figure 25 Translations of the inscriptions on the schematic drawing of the flow of incense smoke, Fig. 24:

A. Beginning
B. The origin of all being is ungraspably empty
C. Undefiled purity
D. Undefiled purity
E. Ungraspable freedom
F. Ungraspable freedom
G. Innately unborn and undying
H. Because innately unborn and undying
I. Ungraspable freedom
J. Undefiled purity
K. Ungraspable freedom
L. Undefiled purity
M. The origin of all being is ungraspably empty
N. Ending

This miraculous pattern is known as the "Great Compassion Which Allays Suffering." When incense is burned, this smoke pattern manifests the True Principle; when the incense smoke is consumed it represents the arising and passing of all things within Emptiness.

The following procedure should also be observed: the five letters *oṃ*, *va*, *jra*, *dha*, *rma* each contain unlimited creativity. Each one transforms itself into the body of all the Buddhas and Bodhisattvas and manifests itself in the relative world in order to save sentient beings. These letters bestow upon the practitioner unlimited good fortune and contain the complete wisdom of all the Buddhas. The letters protect the practitioner from all misfortune and confusion in the same manner as the miraculous Lotus Flower [of the True Law]. The letters are to be cherished, for they allow rebirth within a lotus in the highest Paradise. They possess saving wisdom and beneficence. Buddha can be seen in their forms.

Master this mantra[17] [and you enter the] undefiled world. Miraculous fragrance will be emitted from this very body, perfuming 10,000 lands and countless sentient beings. Realize all these principles without delay and proclaim the unlimited virtue of this teaching. [The five virtuous letters] should be inscribed on the lid [cover] of the burner. The center of the lid symbolizes the aspiration for perfect realization. The handle placed there should be *vajra*-shaped[18] with an eight-petaled lotus flower on the top. The five letters of the mantra revolve around the center, a symbol of the intention of the Buddha to lead all to realization. Gaze on this form, compose the mind, and enlighten the innate Lotus nature. In such a manner, one is reborn in an undefiled paradise yet still able to interact with the world [of *samsara*][19] and assist sentient beings in achieving a similar lotus-like [state], free of contamination or stain. This splendid result is due to the power of the original vows made by past Buddhas to save all beings.

[15] Also Vairotchana, the highest of the Trikaya, corresponding with Dharma in the Triratana, and the personification of essential *bodhi* and absolute purity. The first of the Dhyani Buddhas. Eitel, *Hand-Book*, pp. 192–93.

  Mudra literally means "the seal of the law." A system of gesticulation, derived from the Yogacara school, with the fingers distorted to imitate ancient Sanskrit characters of supposed magical efficiency. Eitel, *Hand-Book*, p. 101.

[16] The ancient practice of ecstatic meditation for the purpose of achieving spiritual or magical power, revived by the Yogacara school. Eitel, *Hand-Book*, p. 208.

[17] The *mantra* is a form of short magical sentence first adopted by the followers of the Mahayana school, then popularized in China by Vadjrabodhi. Eitel, *Hand-Book*, p. 96.

[18] The *vajra* is one of the most important symbols in Buddhism. Of obscure origin, it is generally interpreted as "diamond," and "a destroying but indestructible emblem" symbolizing "the ultimate reality." Its magical nature gave birth to a form of religion known as Vajrayana Buddhism. Liebert, *Iconographic Dictionary*, p. 318.

[19] *Samsara* or transmigration. Buddha claimed that one lifetime was not sufficient for adequate appreciation of the Buddhist truth of the *dukkha* or suffering entailed in all existence, and that it required viewing the unending chain of rebirth and the sum of misery therein entailed. Ling, *Dictionary of Buddhism*, p. 167.

  The passage may also be translated as follows: "In such a manner, one is reborn in an undefiled Paradise yet still able to interact with the world [of *samsara*] and assist sentient beings in achieving a similar lotus-like [state], free from contamination or stain."

A practitioner should concentrate his thoughts on these symbolic forms and grasp their meanings. The *hrīḥ* pattern of the incense smoke symbolizes the aspiration for perfect enlightenment. That pattern is the sacred form [of an enlightened being], simultaneously revealing the original vow [to enlighten all] and the fruit [of obtaining Buddhahood in this very body]. Each is the symbolic principle of aspiration for perfect enlightenment. When incense is burned one venerates the sacred form and smoke activates the *mantra*. Thus, there is total attainment. The lid should look like this (Fig. 26).

Through this circular pattern one arrives at unsurpassed enlightenment. Those who are still troubled and confused and wish to continually utilize this *mantra* should rely on this miraculous pattern. Use sandalwood, lotus, or other fine incense in such a burner, and every day the Diamond Realm [of Tantric realization] will shine brighter and brighter. All this is accomplished through the mantric principles described above. The incense burner pattern truly manifests [the Tantric teaching]. Perform the various *mudra* and recite the *mantra*: *oṃ va jra dharma hrīḥ*! [Hail to the Adamantine Dharma (Wisdom) of Thousand-Armed Kannon!]

If one utilizes this *mantra* all calamity and sickness will be avoided; one will be reborn in the highest paradise after death and be a facilitator of the liberation of all. Rely on this teaching, practice it, and perfect enlightenment will quickly be attained.[20]

The text is followed by a colophon which noted that it had been

Compiled by the monk Mutō of Myōn-ji Temple in the summer of 1735 based on a manuscript prepared by Priest Jōgon. Use this propitious text to kindle incense [properly] and endlessly billow forth clouds of [holy] smoke in the shape of the letter *hrīḥ*, the true image of reality.[21]

The Shingon monk Jōgon (1639–1702), whose family name was Ueda, was a noted scholar and gifted calligrapher. Well informed on Chinese classical writings and Buddhist literature, he undertook the study of Siddham at an early age and lectured on the subject in the temples of Kōyasan. Eventually he settled in Edo [Tokyo] and in 1691 founded the Reiun-ji temple in that city. He was favored by the Tokugawa Shogun, and highly respected in the community of Japanese sinologues. In his later years he devoted himself to the study and editing of the Mantrayanic *sutra*.[22]

Inasmuch as the foregoing is a Tantric Buddhist text, it presupposes familiarity with esoteric symbolism and additional oral (secret) instruction. It is to be noted that the incense to be burned was specified to be made of lotus or sandalwood. Sandalwood (*Santalum album*) is an aromatic wood known in India from the

---

[20] This Tantric scripture was translated and brought to the author's attention by Professor John W. Stevens of Tohoku Fukushi University at Sendai, Japan.

[21] *The Tripidaka in Chinese*, no. 1042. Translated by Professor John W. Stevens.

[22] Van Gulik, *Siddham*, p. 133.

most ancient times, and exported from there to China in the early centuries of the Christian era.

The design of the cover specifies that it is to be divided into five sections, each featuring a syllable in Siddham script of the *mantra-dharani reading: oṃ va jra dha rma hṛíḥ*. This is the *mantra-dharani* for Avalokitésvara, [the Japanese Kannon], "the Fearless Bodhisattva of great compassion" and the incarnation of the Amida Buddha, "Lord of Light and Universal Compassion." Of the numerous manifestations in which the Amida Buddha is represented, that of the one-thousand-armed deity, symbolizing the Bodhisattva's infinite ability to assist sentient beings, is the most popular.[23]

The character *hṛíḥ*, which forms the design of the template for the incense trail, is the symbol of the enlightened Amitabha, who achieved enlightenment to the level of Avalokitésvara. It is used also for other Bodhisattvas. It is a combination of the four characters, *ha, ra, í,* and *aḥ.* When offering the incense from this burner it is vital to know the phases of total awareness of the universe, the meaning of which is indicated by the four characters. This *mantra-dharani* is included in the Tantric text *Sadhanamala*, confirming that its use in an incense seal relates to Tantric Buddhism as derived from lamaistic influence. Such *mantra* are difficult to translate into any language, and for that reason in China, Japan and Korea, they have remained in Siddham script (Figure 27).[24]

The inscription on the cover, *oṃ, vajra dha rma [hṛíḥ]*, rendered in the Tantric interpretation, can be translated as follows:

Hail to the Diamond-like wisdom of the True Law!

and on a popular level would be understood as

Hail to the Kuan Yin, the Thousand-Armed Bodhisattva!

The characters perforated on the cover combined with the seed-syllable of the template form an esoteric Buddhist *mantra*. The character *oṃ* is the sacred seed-syllable used in Hindu and Buddhist esoteric teachings to symbolize, among other

---

[23] Amida (Amitabha in Sanskrit) is represented by the great image Daibutsu at Kamakura. Originally this deity was viewed as an abstraction, the ideal of boundless light. As this Bodhisattva was about to enter Nirvana, there was great reaction from everyone, lamenting his imminent demise. Accordingly, he renounced final release until everyone else had entered Nirvana. Having perfect freedom and not bound in any way, he was a master of inordinate skills which he used for bringing people to salvation, assuming many forms. He responded to any request for salvation, appearing as Buddha, a king, or any other human even of humble station. Avalokitésvara is portrayed in both male and female forms. Stevens, *Sacred Calligraphy*, pp. 48, 54; Chamberlain and Mason, *Handbook*, p. 45.

[24] Bhattacharyya, *Sadhanamala*; *Bukkyō daijiten*, pp. 316–17.

things, the ultimate act of worship and salutation. The meaning of *vajra* is "diamond," representing the indestructibility of enlightened wisdom. Dharma is "the Truth of the Universe" as revealed in Buddha's all-embracing teaching. The text explains only the mystical significance of the pattern forming the incense seal by means of Siddham letters. There is no suggestion in the text that the device could be used also for measuring segments of time of the Buddhist ritual, nor does it provide specific details relating to materials and construction.[25]

No example of a seal of Avalokiteśvara of Indian origin has come to light, nor has evidence been found of the use of incense seals for time measurement in India as part of the Buddhist ritual or in other forms. It seems likely, therefore, that the timetelling function of the incense seal was incorporated by the Chinese at some date after Amogha's translation of the Tantric scripture had become available. It appears that the timetelling function would have been a simple addition made to the incense seal of Avalokiteśvara of the form prescribed in the Tantric text without requiring modification.

An example of the incense seal of Avalokiteśvara, featuring the *hríḥ* seed-syllable, that appears to be of Chinese origin, possibly of the seventeenth or early eighteenth century, has come to notice which was recently acquired in Hong Kong from a dealer who had purchased it from a government warehouse in mainland China. It is a smaller version of the incense seal of Avalokiteśvara, made for time measurement, as indicated by the nature of its accompanying utensils (Fig. 28).

The seal is circular in form, made entirely of heavy brass, the outer surface entirely plain, the only decoration being a narrow copper beading on the rims of each section. It consists of three tiers, the uppermost of which contains the incense tray, and the others accommodating the utensils and a supply of incense.

The cast domed cover, originally terminating in a finial in the form of an eight-petaled lotus on a decorative stem, as prescribed, is perforated with the five syllables in Siddham script of the *mantra-dharani* reading *oṃ va jra dha rma hríḥ*. The consonant character for *rma* is malformed, however, possibly due to difficulties of casting or carving the inscription. The utensils, also of brass, consist of a

---

[25] Communications from Professor John Stevens, August 5 and September 30, 1987. The Shin or True Sects of the Pure Land are extremely popular in Japan. Their central object of worship is Amida, the Buddha of Measureless Light and Life. Widely accepted in China between the fourth and seventh centuries, the sect was transmitted to Japan and developed by the priest Hōnen Shōnin. The Shingon sect is one of eleven leading Buddhist sects of Japan, founded in about 806 by the Buddhist priest Kūkai (774–836), later known as Kōbō Daishi. Its oldest and principal temple is the Kongōbu-ji, which serves as the sect's headquarters. It was built in 816 on Mt. Kōya on the Kii peninsula south of Ōsaka. The present building is of much later date. Armstrong, *Buddhism*, pp. 79–94.

tamper, a spreader of the incense, and a template perforated in the shape of the seed-syllable *hrīḥ* (Fig. 29).[26]

Undoubtedly other early examples of Siddham incense seals have survived which have not yet come to notice. That they were in use in China during at least the ninth century is suggested in several lines by the poet Tuan Ch'eng-shih (d. 893). A scholar and official of the T'ang Dynasty who rose to the position of sub-director of the Court of Sacrificial Worship, his writings were about the notable sights of the ancient capital of Lo-yang.[27] He wrote also of the incense seal. In a couplet he spoke of *gatha* – metrical narratives or hymns having a moral purpose – and Siddham inscriptions of seal-characters used in Buddhist rites:

> Translated and clarified are the *gatha* from under Western skies;
> Burned is the balance of incense in Sanskrit characters.[28]

In Japan the incense seal of Avalokitésvara Bodhisattva is known as the "*kirikuji* incense burner" (*kirikuji kōro*), and may have been introduced by Kūkai in the ninth century with the transmission of the Siddham script (*bonji*). The *kirikuji kōro* is described and illustrated not only in the *Tripitaka* but also in the great dictionary of Buddhism, the *Bukkyō daijiten*. The Siddham seed syllable *hrīḥ* symbolizes Amida Buddha (Amitabha) and is known as "a seed of Amida" (*kiriku*) (Fig. 30).

Traditionally the *kirikuji kōro* is used only in temples in which an image of Amida Buddha or Kannon Bosatsu is worshiped, particularly Senju Kannon Bosatsu, Avalokitésvara Bodhisattva of the thousand arms formally known as Sahasvabhujaryavalokitésvara. Examples are to be found in a number of Buddhist temples of the esoteric Shingon and Tendai sects, founded by Kūkai and Saichō. These examples range from the seventeenth century to modern times, and are often inscribed in Chinese with the name of the donor, the chief priest then presiding in the temple, and the date.[29]

Older examples of the *kirikuji kōro* are generally made of cast bronze and are gilded, although those produced in modern times are of gilt cast brass. They are made in the shape of a covered chalice representing a lotus blossom supported

---

[26] Communications from Dr. Akira Yuyama, April 8, 1983 and from Dr. Yutakia Ojihara, November 30, 1982. This incense seal was acquired from a dealer in antiquities in Hong Kong in 1978–79, having recently been brought from mainland China.

[27] Tuan Ch'eng-shih, "Tseng chu shang jen lien chu." Quoted in Schafer, *Golden Peaches*, p. 161; Giles, *Dictionary*, p. 788.

[28] A *gatha* is a metrical hymn or chant usually consisting of 4, 5 or 7 words, sung in verse in praise of an adored object, often recurring in sutras. Soothill and Hodus, *Dictionary* p. 342.

[29] Information from Rev. Exsei Kawaguchi, Chief Priest, Tōzen-ji Temple, Yokohama, who also provided descriptions and photographs of several other examples surviving in temples in Japan.

upon a pedestal base. Lotus leaves, incised or sculptured in relief, surround the sides of the bowl and the round base is likewise encircled with lotus leaves in relief. The rounded cover is smooth, with a finial at its center, terminating in a lotus blossom supported upon an elongated stem and having eight petals representing the three phases of meditation of Avalokitésvara. At equal intervals the surface of the cover is perforated with the five Siddham characters, *oṃ va jra dha rma [hríḥ]* for the emission of smoke and fragrance. The body of the chalice is filled with finely powdered wood ash to a level just below the rim. Upon this bed an incense trail is formed by means of two utensils, the first of which is a "tamper" to compress the bed of ash and to impress the pattern upon it. The tamper's upper surface, which is equipped with a handle, is inscribed with the single Chinese character *shang*, meaning "above," "top," or "first," indicating that this tool is to be used before applying the second. Both utensils measure 30 cm in diameter, the flanges which form the character *hríḥ* project 1.6 cm, indicating that the bed of ashes must have a substantial depth and that the resulting incense trail would be wide and deep and consequently would continue to burn for a long period of time (Figs. 31–32).

The second utensil is the template, a heavy plate perforated with the seed-syllable *hríḥ*. It would be pressed upon the bed of wood ash, and powdered incense sifted through its perforations and tamped firmly to form the incense trail in the shape of the Siddham script character *hríḥ*. A spout on one side is used to remove the excess incense. These utensils form what appears to have been the immediate predecessor of the earliest form of the incense seal, used for the purpose described in a Buddhist temple dedicated to the Bodhisattva, "the Goddess of Compassion and Unlimited Skillful Means" (Fig. 33).

The *kirikuji kōro* vary in size from 18 cm, 24 cm to 30 cm in diameter, and even larger examples may exist. A set of these utensils in the temple of Hōshōji at Minami-ku, Yokohama, one of the historical temples of the Shingon sect, measure 15.5 cm in diameter. Other examples are to be found in the Kenchō-ji Rinzai-shū temple at Kamakura, and in the Hōju-in and the Chōgen-ji temples in Yokohama. The *kirikuji kōro* has also been used in ceremonies in the Shingon Kongōbu-ji temple on Mt. Kōya. Among the older examples is the *kirikuji kōro*, stored in the Kongō-in temple in Kyoto Prefecture, which has been declared a national treasure and is maintained by the head priest, Jinyō saku. Despite the survival of the *kirikuji kōro* in a number of Japanese temples, its existence and purpose appears to be not always well-known in the Buddhist priesthood.

Early examples of the *kirikuji kōro* in bronze are quite rare. It is to be noted that *kirikuji kōro* are being made at the present time of gilt brass, incorporating all traditional features, and offered for sale by several firms specializing in altar furnishings for Buddhist temples. The proprietors of these firms report that because sales of these items are extremely infrequent, the *kōro* are not kept in stock; rather, they are made to order and require several months for delivery.[30]

The physical features of the exterior of the incense seal of Avalokitésvara immediately bring to mind its similarity to the exterior appearance of the venerable incense seal in the Shōsōin at Nara. The multi-layered lotus leaves of the exterior and the pedestal base of the latter suggest a derivative relationship, although the Shōsōin *kōinza* does not include an incense trail in the form of the Siddham *hríḥ* character (Fig. 34).

A possibility that must not be overlooked, of course, is that the incense seal (*hsiang yin* or *hsiang chuan*) may have existed in a primitive form in China prior to the eighth century and that it pre-dated the incense seal of Avalokitésvara, but if so, however, no record of it has been found in the literature. It is much more likely that a means of measuring longer periods of time with greater precision than was possible with the simple incense stick and the incense coil evolved from the use of Avalokitésvara's incense seal in Buddhist Tantric ritual. Based upon the same principle as the template for that seal, it was a simple matter to devise templates having other specially designed seals or "chops" formed with a pre-determined width and depth and length in which ignited compressed powdered incense was consumed during a specified period of time, ranging from one day of twelve Chinese hours to as long as 720 hours or a full month. The incense seal may be compared to the common "chop," of which the path's edges represented the "chop's" vermilion ink. Extensive experimentation would be required to establish the appropriate width and depth of the path in relation to its length, the trail's dimensions carefully calculated by trial and error from the speed with which incense burned. It was eventually discovered that dampened pulverized wood ash provided the optimum base. Powdered incense was introduced through the openings of the template, firmly compressed with a tamper, and the template removed to leave an unbroken trail of the same dimensions. The end of the trail was ignited and the incense was consumed at a constant rate as the glowing head progressed through the labyrinth of the seal character along the part of the seal or "chop" that was not inked.

[30] Information from Rev. Exsei Kawaguchi and Professor John W. Stevens.

6

In Chinese Buddhist rites

At some time after the translation into Chinese of the Tantric text of "The [Incense] Seal of Avalokitésvara," the Chinese cleverly conceived that the incense seal could serve the function of timetelling as well as of prayer. This could be easily effected without modifying the seal, but merely by introducing time indicators along the incense trail. Whether the incense seal made with the surviving utensils described in the previous chapter served this additional function cannot be determined with certainty, but appears to be most likely.

Since Buddhist services were required to be performed six times during the course of the day – early morning, midday, sunset, early evening, midnight and past-midnight – a means of dividing the time was essential. It is probable that each of the temples was provided with an incense seal for this purpose.[1]

One can only speculate on the nature of the base on which the earliest seal-characters of incense were burned. It seems likely that in an early period use may have been made of the "incense basin" (*hsiang p'an*), a broad low vessel of Buddhist origin. It was derived from an earlier bronze vessel of the Chou period known as a *p'an*, which was shaped in the form of a hollow basin and used to contain water in ceremonial rites (Fig. 35).[2]

Early incense burners or censers consisting of decorative flat pans on which incense was burned are noted in the frescoes in the Ch'ien-fo-tung caves at Tun-huang. One of the long-handled basins is illustrated in the encyclopedic dictionary of Morohashi. Modern examples copied from the old forms are being sold by Japanese firms specializing in altar furnishings.[3]

The "incense basin" (*hsiang p'an*) may have been combined with the famous *Po-shan hsiang lu* of the Han period, eventually to become the receptacle of the incense seal. It is to be noted in this connection that later incense burners, merely known as *hsiang lu*, were modifications derived from the earlier *ting*, a bronze sacrificial three-legged vessel.

[1] Takigawa Seijirō, "Kawachi," pp. 9–27.
[2] *Zenrin zōkisen*, cited in Takigawa; Bushell, *Chinese Art*, vol. 1, pp. 75–76; Willetts, *Chinese Art*, vol. 1, pp. 154–55.
[3] Morohashi, *Dai Kanwa jiten*, vol. 8, p. 8962 d, vol. 9, p. 967; Needham, *SCC*, vol. 3, p. 129.

The date of the adaptation of the incense seal for time measurement is not known. Literary evidence suggests that it occurred in the eighth century, probably shortly after the incense seal of Avalokitésvara was introduced into China. Unfortunately, clepsydra manuals produced prior to this period have not survived, although authors of some of these works are known. Among them were Ho Jung (c. 102 A.D.), Chu Shih (c. 563 A.D.) and the Astronomer-Royal, Sung Ching. A more comprehensive treatise was produced later by Huang-fu Hung-tse of the Sui or early T'ang period.[4]

The earliest incense seals probably were made of stone or wood and were relatively crude. The seal-character either was formed by sifting loose powdered incense used in temple rites upon the surface of a flat panel of stone or a hard-wood such as mountain pear or cedar, or sifted into grooves carved directly into the surface. It is estimated that the seal platform was approximately 30 cm in diameter and 2.5 cm in thickness.

Eventually the temple priests realized that the particles of varying size contained in the incense produced erratic burning, and finely pulverized incense was substituted. The length of the groove was determined by testing the period required for the incense to burn a specified length of the trail, generally twenty-four hours, and the segments estimated therefrom. The trail was marked at these points with indicators. These were merely small markers made of wood or bamboo, each bearing the name of one of the twelve double-hours and possibly attached to a pin, and inserted along the incense trail at the end of each of the pre-determined segments.

During the development of the incense seal for time measurement as well as for the ritual of Buddhist temples, it acquired several characteristics. Its major feature, of course, was the use of a seal-character to form the path of burning incense, and it is from this element that the device derived its name, "incense seal" (*hsiang yin*). Technically, the terms "incense seal" (*hsiang yin*) and "incense seal-character" (*hsiang chuan*) refer only to the design of the incense trail when formed in the shape of a seal-character. The encyclopedia *Chung wen ta tz'u tien* provided two definitions for *hsiang chuan*; namely, "To form incense into shapes of *chuan*-style characters, and light it to measure time;" and "When the smoke from lighted incense coils and twists about to form *chuan*-style character shapes."[5]

The name of the complete incense time measurer, including the tray upon which the seal-character is formed and the container in which it is maintained, is more precisely rendered "incense seal burner" (*hsiang yin lu* or *hsiang chuan lu*).

---

[4] Needham *et al.*, *Heavenly Clockwork*, pp. 89–94; Needham, *SCC*, vol. 5, p. 147; Maspero, "Les Instruments," pp. 185–217.

[5] From the *Chung wen ta tz'u tien*, cited in Crump, *Songs*, pp. 196–97.

However, in common usage it became identified by either of the phrases *hsiang yin* or *hsiang chuan*. The design of the incense trail in the form of a seal-character formed by means of the template is the *hsiang chuan* or "incense seal-character."

Hsiang yin may also be interpreted as "stamped incense" or "imprinted incense," indicative of a tablet or pastille of incense formed from hard paste and imprinted or engraved with the characters of the time divisions in the traditional Chinese characters used for engraving seals. These were marked at equal intervals with the hours so that the passage of time was indicated with the consumption of the stick or tablet.

Consequently, the terms "Aromatic [incense] seal," or "Aromatic [incense] seal-character" were used to signify incense "clocks." In the same manner, the word *k'o* [notch] has also been commonly used to mean a clock, so that "aromatic [incense] notch" also became synonymous with incense clock.

Although the phrases *hsiang yin* and *hsiang chuan* are used interchangeably to identify the device commonly known as the "incense seal," and both words *yin* and *chuan* are translated in modern usage as "seal," the meaning is not exactly the same for both. A *yin* is a "seal" in the sense of a personal stamp or "chop" employed to impress the outline of a calligraphic character upon a piece of writing or other surface. The word *chuan*, however, signifies a "seal-character," forming part of a specific calligraphic style used before the writing reform of the third and second centuries B.C. But when applied to incense timepieces, the meaning of both terms is the same, namely, a highly stylized pattern resembling "the seal form of the character [*chuan*]," or "the characters that are used on seals [*yin*]."

Early in the evolution of the Chinese language, after its beginnings in the Chou period, three new types of characters were added to the original pictograms of which the language consisted. It was at this time that the form of writing known as the "seal script" was introduced.

Two versions existed of the seal script (*chuan shu*). The Great Seal script (*ta chuan*) was invented by Chou, a recorder at the court of Emperor Hsüan in the Chou Dynasty. This style was adopted and used with slight variation until the Ch'in Dynasty. It was at this time, when China's several feudal states were unified, that Prime Minister Li Ssu determined to consolidate also the several scripts in use, and devised the style known as the "Small Seal script" or "Lesser Seal" (*hsiao chuan*).

The *ta chuan* style differed little from the several earlier styles of ancient script (*ku wen*) found inscribed on bone, stone and bronze artifacts, and it was in effect a synthesis of these, whereas the "Lesser Seal" script provided a standard character for each object and action.

The standardization brought about by the "Lesser Seal" script constituted the last step in the evolution of Chinese archaic scripts. It was the predecessor of the official or "clerical" script (*Li*), which in turn was succeeded by the cursive, running and standard scripts. The clerical script was derived from the "Lesser Seal" script, reaching its height in the Eastern Han period. The "Lesser Seal" and "clerical" scripts have survived to modern times primarily among calligraphers.[6]

Originally the seal scripts were styles of writing not connected with the engraving of seals, a characteristic form of Chinese art, but eventually they came into common use for that purpose. These seals continue today to serve as the personal stamp of an artist added to his work, on stamps or chops used for personal identification, and they also appear as Chinese porcelain marks.[7]

What are believed to be the earliest known surviving examples of the Chinese incense seal are preserved in the Shōsōin, the famous Imperial Treasury near Nara, Japan. This repository, unique in the history of fine arts, was established in A.D. 749 by the widowed Empress Kōmyō as a memorial to her late husband, Emperor Shōmu. In it she assembled all the imperial treasures and personal belongings of the deceased, to which were added gifts from the mother of the Empress, other royal relatives, ministers and high ranking court officials. It was dedicated to the Vairoshana Buddha of the Tōdai-ji, the imperial temple at Nara.

Originally the treasures were stored in two buildings of the Tōdai-ji, identified as "a pair of warehouses" (Sōsō), under the supervision of a high priest. Because the treasures dedicated to the late Emperor were too numerous for the space available, it became necessary to construct another building between the two warehouses, and all three were joined under a single roof. The three buildings – the North Building (Hoku-sō), the Middle Building (Chū-sō), and the South Building (Nan-sō) – were thereafter referred to collectively as the "Repository in the Enclosure" (Shōsōin).

The entire structure is made of wood, without the use of metal nails or earthen walls. Each building must be entered separately and there is no communication between buildings from within. The treasures were installed in A.D. 756, and despite the intervening twelve hundred years and more, the buildings have remained unchanged. They have been under the Imperial Seal of each succeeding emperor, and although the Todai-ji was burned twice during civil warfare, the Shōsōin itself remained undisturbed. It was opened only on rare occasions, often at intervals of a hundred years and more, until the Meiji era.

In 1872, the fifth year of the new era, imperial sanction permitted the opening of the Shōsōin for the first time, every autumn between November 1 and November 14 for a period of airing (bakuryō). At first the treasures were stored in *kara-hitsu*, chests of the Tenpyō period, and it was not until 1892 that glass cases

---

[6] Shen C. Y. T. Fu *et al.*, *Traces*, pp. 40–55.

[7] Chiang Yee, *Calligraphy*, pp. 41–58, 217–20; Shen C. Y. T. Fu *et al.*, *Traces*, pp. 43–60; Kwo Da-Wei, *Brushwork*, pp. 9–30.

were provided. During the bakuryō, it was the practice of the imperial household to issue a limited number of permits to visit the Shōsōin.

Emperor Shōmu (724–748), the forty-fifth emperor of Japan to whom the Shōsōin was dedicated, ascended the throne at the age of twenty-five. He was determined to rival the splendor of China's T'ang Dynasty. Much of that civilization already had been imported into Japan during the previous century and had in fact almost reached a saturation point.

During Shōmu's rule, which was called the Tenpyō period meaning "Heavenly Peace" and which lasted until A.D. 748, he attempted to promote the cultural movement initiated by Prince Shōtoku Taishi. This was a tranquil interlude, without social unrest and with excellent cooperation between the clans, which enabled the imperial house to achieve a golden age that actually rivaled China's T'ang Dynasty. Buddhism flourished and numerous temples and pagodas rose throughout the land, with Buddhism's great center developed at Nara, Shōmu's capital city.

Preserved in the Shōsōin's South Building (Nan-sō), the incense seals are zealously guarded together with other treasures of past emperors. They consist of five pieces, including a pair of bases called "incense-mark stands" (kōinza). The word literally means "a seat or stand for a censer or fragrant [incense] board." These are circular blocks of white marble, bowl-like in shape, which served as bases for incense seals having Sanskrit or Chinese seal-characters forming the incense pattern. One of the kōinza measures 56 cm in diameter and 17 cm in total height. The second example is 55.6 cm in diameter and 18.5 cm in overall height (Fig. 36).

According to the Shōsōin Office, "seal-like marks are drawn with powdered incense on their flat tops; the marks are ignited at one end and caused to burn slowly towards the other end, as a lasting offering of incense to the Buddha." The Shōsōin authorities attribute the kōinza to be of Chinese origin of the eighth century A.D.

In use, each kōinza is placed upon a wooden seat or pedestal called a "Lotus Flower Pedestal" (ganza). Each ganza is built upon a wooden base carved to represent rock which is painted brown and ultramarine. Inscribed on the underside of each in black ink are the words "stand for kōin" (kōinza), and "a figured plate lacquered and painted gold" (urushi-kinpaku-eban). The latter phrase may have been intended to describe the shitsuban, another component of the incense seal (Fig. 37).

The two ganza are made alike, differing only in the decorative motifs of the petals. Four copper plates serve as a matrix for the leaves. Each is cut into a sunbeam shape having eight radiations, and a lotus petal carved of thin wood is affixed to each of the eight terminals. The four copper cut-outs are placed one upon another in such a manner that the lotus petals form alternating tiers, and the whole assembly is riveted upon the rock-shaped base. The result is a pedestal in the form of a lotus flower with an open center. The "pip" is made of wood and lacquered in black.

The pedestal's surface is concave so that it will accommodate a *kōinza*. Its sides are covered with gold leaf, and pistils painted in cinnabar are added. The lotus petals are painted on both sides in patterns with red, vermilion, dark blue, blue, green, white, yellow, black, and gold leaf. The obverse sides of the eight petals of the top tier are covered with gold leaf, four decorated with *Hossōge* flowers, two with mandarin ducks and two with birds bearing flowers in their beaks in the central circles, surrounded by floral and bird motifs and scrolls. They are bordered with bands of arrow-feathered patterns in *ungen* colors.

The petals of the second tier are painted blue with the edges finished in gold leaf. Four have *Hossōge* flowers, a mandarin duck, a bird bearing a flower in its beak, a dragon-headed beast, and a strange bird also bearing a flower in its beak (Fig. 38).

The petals in the third tier are covered with gold leaf and painted alternately with four-lobed and round frames surrounded by floral scrolls. Two enclose "man-birds," and one each a lion, a floral spray, a bird with a ribbon in its beak, paired birds, mandarin ducks, and a *Hossōge* flower.

The petals of the bottom tier are painted vermilion and edged with gold leaf. The petals are alternately painted with *Hossōge* flowers, and two versions of flower-and-bird motifs (Fig. 39).

The pedestals have been said to be undoubtedly the most beautiful of the imperial treasures, and remarkably they have retained their original brilliance through the centuries. Represented are "Chinese flowers" (*toka*), "flowers of treasure" (*hō so-ka*), "birds feeding on flowers" (*hana-kui-dori*), "birds of honor" (*ganju-dori*), "Chinese phoenix" (*hōō-dori*), "the mandarin duck" (*oshi-dori*) carrying floral pendants in their bills, and "the lion" (*shishi*).

The most famous is the creature called the Kalavinka, a half-human creature with the lower body of a bird, which is famous in Buddhist scripture for its sweet voice. It is represented on a middle petal between two petals of the bottom, but actually belonging to the second tier. The creature wears a crown and a necklace, and also bears a flower in its beak. The legend claims that it is chanting a mass for a better life in the future, and seeking to escape from its beastly form by means of admirable conduct. It may possibly be associated with the Garuda, a half-man and half-bird in Indian mythology with a golden body and red wings on which the god Vishnu rides through the air.[8]

---

[8] Matsuoka, *Sacred Treasures*, pp. 1–12; Yone Noguchi, *Emperor Shōmu*, vol. 1, pp. 3–6, 21–28, 109–16, vol. 2, pl. 30–31.

At the right of the Kalavinka is a representation of the *Hossōge* flower blooming with a smile because it believes it will achieve the final blessed state of Buddha. The flower reflects the promise of rebirth into a better life under the laws of Karma which is promised in Buddhism, and its attempt to solve all problems of the world with one word of humanity.[9]

The "Record of imperial treasures" notes that the two *ganza* originally were accompanied by mats having borders and cushions of *nishiki* silk, and padded coverlets of *ashiginu* silk, now lacking. One of the petals of the bottom tier of one of the *ganza* has become detached but is preserved separately.[10]

The *kōinza* are recorded in the *Amida kōkaryō shizaichō*, an inventory of property compiled "on August 30th in the first year of Keiun" (A.D. 704) by a Japanese Buddhist priest named Bunzon. This document is preserved in the Tōdai-ji at Nara.[11]

Related to the Shōsōin's *kōinza* is a "wooden lacquered plate" (*kōinban*, *kōinoukeban*, or *shitsuban*), finished in black lacquer. Measuring 20.5 cm in diameter and 4.1 cm in height, the "seat" itself is 19 cm in diameter and 1 cm in height. The only information available from the Shōsōin authorities about this piece is that it is believed to have served as an "incense press" for impressing the incense trail on the *kōinza*, because the groove formed by the pattern on the bottom of the plate begins and ends at the same point. In fact, the sculptured impression appears to be the traditional Chinese flower pattern (*karahana*), a variety of the *karakusa moyō*, confirming its Chinese origin (Fig. 40).[12]

There seems to be no doubt that the *kōinza*, *ganza* and *shitsuban* were used during the reign of Emperor Shōmu, as an early version of the incense seal which was developed as the *jikōban*. The *kōinza* and related equipment in the Shōsōin may have been brought to Nara from Ch'ang-an, the Chinese capital in the T'ang Dynasty, by the Buddhist priest Chien Chen. This famous Chinese scholar was invited to Japan by the emperor in A.D. 753 to officiate at the inauguration of the Daibutsu image.

This immense seated statue of the Birushana (Vairocana) Buddha, commonly known as the Daibutsu, was completed in A.D. 749 by the noted sculptor

---

[9] C. A. S. Williams, *Encyclopaedia*, pp. 392–93.

[10] Yone Noguchi, *Emperor Shōmu*, vol. 1, pp. 115–16, vol. 2, pl. 30–31; *Treasures of the Shōsōin*, pp. 26, 31–32.

[11] *Treasures of the Shōsōin*, pp. 31–33.

[12] Copies of the document were provided by Dr. Shunichi Sekine, associate curator of the Nara National Museum and by Dr. Kosei Morimoto, chief librarian of the Tōdai-ji Temple Library. The inventory also lists three incense burners made of "nickel," presumably paktong, in height 16 inches, 8 inches and 7 inches respectively.

The *kōinza* and *shitsuban* are described and illustrated in the "Central Storehouse section" of the *Shōsōin Gobutsubyō betsumokuroku* and are illustrated also in the *Shōsōin gyobutsu zuroku* 11, vol. 11, plate 22, 27, and in the *Shōsōin hōbutsu* (Tokyo, 1960).

Kimimaro. It weighs approximately five hundred metric tons and is approximately 16 m in height. The possibility suggests itself that the *kōinza* and accompanying equipment may have been used in the ceremonies consecrating the opening of the eyes of the statue.[13]

Another *kōinza* or *kōinban* is reported to be preserved in the Mikumari-jinja, the Japanese Shintō shrine in the Nara Prefecture. It bears an inscription with the date 1490.[14]

The Chinese continued to improve the incense seal and developed it from its first crude form to a sophisticated time measurer. In one discoidal form, Professor Joseph Needham suggested, the incense trail was arranged in such a manner as to be Yin during one double-hour – narrowing and centripetal – and Yang during the next – expanding and centrifugal.[15]

It was discovered early on that the rate of combustion of incense could be much more accurately regularized if the incense trail was made of uniform dimensions throughout its length and if it was laid upon or into a bed of finely sifted dampened wood ash. This was required so that oxygen from the air had equally restricted access along the entire length. The ash is incombustible and permits air to pass between its particles and serves as a thermal insulator; thus each side of the trail could be expected to burn as fast as the others. All sides of the trail had to be unobstructed, or equally obstructed, not only so that oxygen had equal access to all sides, but also so that the cooling by radiation and convection was the same on both sides as well as the bottom of the incense trail. If these conditions were not met, then the sides that burned hotter could be expected to burn faster. In such a case the burning on the hotter sides would meet the markers, made of bamboo or metal, more quickly than on the cooler sides and the end of the time interval would not be clearly identified.

If the trail were to be laid directly on metal, then the side that touched metal would be cooled by heat being conducted away, and the access of oxygen would be obstructed. Therefore this side would burn especially slowly.

If the trail were to be laid on wood, then the wood could be expected to char, burning like the incense itself, driven by the oxygenated smolder of the incense. The effect would be similar to the charring of a table top by an unattended cigarette stub, and the underside of the trail would burn more slowly than the other sides.

---

[13] Asahina, *Tokei*, pp. 576 ff.; Watanabe, *Tōyō monyō shi*, p. 581; Nishimura, *Karakusa: Nihon no monyō*, vol. 9, p. 211. A sentence in *Baien nikki* appears to support the opinion that the *kōinza* and *shitsuban* were used for time measurement with incense.

[14] Shigeru Aoyama, *Nara*, pp. 8–9; communication from Dr. Shunichi Sekine, associate curator of the Nara National Museum, August 15, 1987.

[15] Needham, *SCC*, vol. 5, part 2, p. 147.

It was discovered, furthermore, that the furrow into which the incense was laid could be either rectangular or triangular in cross section, but it had to be of a specified uniform thickness. The powdered incense had to be packed tightly into the depression forming the trail of the early examples made of wood or stone. Later the Chinese made the incense seals from a variety of metals, utilizing a perforated template to form the incense pattern. Powdered incense prepared in accordance with one of many recipes created for the purpose was spread over the template with a small shovel, pressed into place through its openings by means of a tamper-like utensil, and the excess swept or blown off. When the template was removed, a continuous trail of incense reproduced the seal-character. A gap in the incense trail was required, so that ignition at a specified point would not proceed in both directions.[16]

The device required an enclosure, which was provided in the form of an open box, which at first was made of wood and later of metal. A latticed cover was added to permit the smoke and heat to dissipate and yet prevent disturbance of the burning trail from movement of air caused by drafts or breezes which might accelerate combustion or extinguish it.

The pattern of the seal-character had been carefully calculated to burn for a specified period of time, which varied from twelve or twenty-four hours to thirty days. Small markers of bamboo or metal with the characters for the hours were placed alongside the trail at intervals to mark the double-hours so that the passage of time indicated by the burned trail could be readily determined. One of the additional segments of the incense seal was used for storing the utensils, and when there was a third, it was used for storing a supply of incense (Figs. 41–42).

The early use of the incense seal in Buddhist temples was confirmed from time to time in Chinese poetry. It was mentioned by the T'ang court official and poet Wang Chien (768–833) of Ying-chou in Anhwei,[17] and in the poetry of Fang Kan (fl. A.D. 860), for example. The latter wrote:

> The incense seal [*hsiang yin*] finished
> [its trail] when [the monk] came out of [finished] his
> meditations.[18]

Several lines in a work by the tenth-century poet, T'ao Ku (902–970), attest to the use of a wooden base for the incense trail in the early form of the incense seal. He suggested

---

[16] Communication from Dr. Robert M. Organ, formerly director of the Conservation-Analytical Laboratory, Smithsonian Institution, March 29, 1987.

[17] Wang Chien graduated as *chin shih* and served as governor of Shan-chou in Honan for eight years. His career ended when he criticized a relative of the emperor. Giles, *Dictionary*, p. 814.

[18] *P'ei won-wen yün fu*; Needham, *SCC*, vol. 4, part 2, p. 147 d.

> If, when you are using a wooden mold, with incense
> fragments spread in a seal-character text, you
> quickly invert it, that makes "winding river incense."[19]

"Winding river incense" was not a variety of incense, but rather described the appearance of the incense trail which, after it had been formed into the shape of a seal-character, was quickly flipped out onto a flat surface. The trail would appear in reverse but would no longer sufficiently retain the image of the seal-character to be legible; instead, it would spread formlessly like the waters of a "winding river."

Reference to the common use of incense seals in the early Buddhist environment of China in the T'ang period is to be found also in the Tun-huang manuscripts, a treasure trove of Chinese Buddhist history spanning the period from the early fifth to the early eleventh century. One manuscript, compiled by a certain Tz'u En in approximately A.D. 911, contains the inventory of the belongings of a Buddhist nun called Shan Sheng. Noted among her possessions were "One square and one round incense seal." Although the character for "round" is incorrectly written, its meaning is nevertheless clear.[20]

The *Tsun sheng pa chien* (Eight Notes on an Orderly Life), a work of the Ming Dynasty written in about 1570 by Kao Lien, provides a particularly enlightening account of the incense seal in addition to a *Yin hsiang kung fo fang* (Recipe for Incense Seal for Use in Buddhist Worship):[21]

> Studios and prayer rooms require the daily burning of incense. One of the methods used employs an incense seal in Buddhist worship. The seal template is die stamped. There are models [of such incense seals] which burn one day and others which burn for six [Chinese] hours. The ingredients in recipes of incense for incense seals are discretionary, but the weight proportions will allow for only minor adjustments from the recipes, such as the old recipe at left [the text of which follows]:

Meng Chüeh An miao kao hsiang fang

Sublime Incense Recipe from the Meng Chüeh Temple[22]
[These 24 aromatics provide 24 "airs" (scents) for use in Buddhist worship]:

	Grams[23]
Ch'en su [roots of the Aloeswood tree, *Aquilaria agallocha*, Roxb.]	200

---

[19] T'ao Ku, *Ch'ing i lu*, b, p. 59a, *Hsi yin hsüan ts'ung shu*; *Hsiang pu hui k'ao*, fol. 43; also noted in Schafer, *Golden Peaches*, p. 168.

[20] Tun-huang manuscripts, Pelliot 3638. Courtesy of Dr. Carole Morgan.

[21] Kao Lien, *Tsun sheng pa chien*, c. 1570, in *Hsiang pu hui k'ao*.

[22] Meng Chüeh Temple has not been identified.

[23] The gram weight does not exist in Chinese, but the measures in the text have been equated to

Huang t'an ["yellow sandalwood," *Santalum album*]	200
Chiang hsiang [Lakawood stems, *Acronychia laurifolia Bl.*]	200
Mu hsiang [Inula]	200
Ting hsiang [cloves, *Caryophyllus jambosa* Ndz.]	300
Ju hsiang [frankincense, *Olibanum*]	200
Chien yün hsiang [may be rue or *Rutaceae* product]	300
Kuan kuei hsiang [cassia, *Cinnamomun cassia*, Presl.]	400
Kan sung [*Valeriana officinalis*, L.]	400
San lai [*Kaempferia Galanga*, L.]	400
Chiang huang [*Curcuma longa*]	300
Yüan shen [*Scrophularia oldham, oliv.*]	300
Tan p'i [root bark of tree peony, *Salvia miltiorrhiza* (?)]	300
Ting p'i [may be bark of clove tree]	300
Hsin i hua [magnolia flowers, *Magnolia kobus*, D.C.]	300
Ta huang [rhubarb, *Rheum officinalis*, Baill.]	400
Kao pen [*Nothosmyrnium japonicum*, Miq.]	400
Tu huo [*Archangelica gmelini*, D.C.]	400
Huo hsiang [*Lophanthus rugosus*, Fisch. et Mey]	400
Mao hsiang [*Hierochloe borealis*, R. et S.]	400
Pai chih [*Angelica anomala*, Pall.]	300
Li chih [skin of litchi fruit, *Litchi chinensis/Nephelium Litchi*, Camb.]	400
Ma t'i hsiang [aloeswood root joints]	400
T'ieh mien ma ya hsiang [unknown]	500
Huai Ch'an Mo Hsiang [unknown]	500
Ch'ao hsiao [potassium nitrate]	5

The last two above-named ingredients aid combustion and promote continuous burning. This recipe is for use in large or small incense seal braziers.

The templates for the four incense seals shown above have "ears" or flanges cast on the sides to facilitate inserting and removing them [from the brazier]. First, spread ash level in the bottom of the incense seal receptacle. Then shovel powdered incense through the tracery openings of the template upon the ash, so that the incense forms a pattern on top of the ash. Carefully fill all sections of the negative tracery image. Carefully remove the seal template, revealing a raised incense seal-character upon the ash. The large seals burn all day, the smaller seals burn for one to two [Chinese] hours. Essential for reading sutras and in Buddhist worship.[24]

the metric weights. To convert the quantities in the recipe back to the terms used in the original: 1 *ch'ien*=5 grams; 1 *liang*=10 *ch'ien*; 1 *chin*=10 *liang*; 1 *fen*=0.5 grams.

[24] Kao Lien, *Tsun sheng pa chien*, in *Hsiang p'u hui k'ao*.

Another recipe described in the same source is not associated with a particular temple, and is listed merely as

Yin hsiang fang
[Recipe for Incense Seal]

	Grams
Huang shu hsiang, aloeswood roots	2500
Su hsiang, [unknown]	500
Hsiang fu tzu [rhizome of flatsedge, *Cyperus rotundus*, L.]	50
Hei hsiang [unknown]	50
Huo hsiang [*Lophantus rugosus*, Fisch. et Mey]	50
Ling ling hsiang [*Coumarouna odorata*, Aubl.].	50
T'an hsiang [sandalwood, *Santalum album*, L.].	50
Pai chih [*Angelica anomala*, Pall.]	50
Po hsiang [*Arbor vitae, Thuja*]	1000
Yün hsiang [rue, *Ruta graveolens*, L.]	50
Kan sung [*Valeriana officinalis*]	400
Ju hsiang [frankincense, *olibanum*]	50
Ch'en hsiang ["heavier than water," black heartwood of aloes wood, *Aquilaria agallocha*, Roxb.]	100
Ting hsiang [clove, *Caryophyllus jambosa*, Ndz.]	50
Ch'en hsiang [aloes wood, trunk wood]	200
Sheng hsiang [unidentified]	200
Yen hsiao [potassium nitrate]	0.5

Pulverize all ingredients, form into incense seal and ignite.[25]

[25] *Ibid.*, fol. 30.

7

In Chinese civil life

The incense seal had already been long established as part of the Buddhist ritual before it was adopted for time measurement in civil and social life in China. Its earliest use in community life occurred during the Northern Sung period.

The incense seal for communal use was designed to supplement and sometimes replace the public clepsydra, and consequently functioned for a much longer period than incense seals used in Buddhist temples. The first incense seal for community use was unquestionably based on those already developed in Buddhist temples. However, the first description of it, which appeared in the writings of Shen Li (c. 1010–1075/86), stated that it was an original invention. A Chinese prefect with a considerable interest in technology, whose *tzu* or "courtesy name" was Li-chih, Shen Li achieved fame as a hydraulic engineer whose work in the control of the waters of the Yellow River was particularly noteworthy. He died at the age of seventy-one, in the reign of Shen Tsung, known also as Chao Hsü (A.D. 1068–1085).[1]

According to Shen Li's account in his modest publication entitled *Hsiang p'u*, (Catalogue of Aromatics), the invention, called the *Po k'o hsiang yin* (The Hundred Graduations Incense Seal) was conceived in 1073 by an official with the courtesy name Mei-ch'i for the purpose of establishing a regular time for sunrise and sunset (Figs. 43–44).[2]

Wu Cheng-chung, a craftsman of Kuang-te (in the present province of Chekiang), studied the description and specifications of the invention in Shen Li's work and proceeded to construct a practical model of it, adding improvements of his own. He also concocted a recipe for a special incense to be used with it. These he presented to Shen Li in 1074. After conducting some trials with the timekeeping device, Shen Li became greatly impressed with its workmanship and accuracy. So great was his admiration for Wu's talents, as demonstrated by his improvements upon Mei-ch'i's original design, that Shen Li caused an account of Wu's

[1] *Chung-kuo jen ming ta tz'u tien*, p. 487, c-d.

[2] Mei-ch'i; is the courtesy name of an otherwise unidentified individual. The same courtesy name was used by two scholars in the twelfth and thirteenth centuries, but no person so named could be found in the eleventh century.

work to be engraved on stone at Hsüan-chou so that he could "pass it on to all those having fond interest in affairs."

Although no copy of Shen Li's *Hsiang p'u* has survived, fragments of his writings were preserved by incorporation into two works on aromatics and incense published in the thirteenth century. These fragments also were reprinted in the second chapter (*chüan* II) of a later work entitled *Hsin tsuan hsiang p'u* (Newly Compiled Book of Aromatics [Incense]), as well as in another rare work entitled *Hsiang ch'eng*.[3]

It is to be noted that repeated inconsistencies occur in the titles of the incense seals described by Shen Li when quoted in the later works. "The Five Watch Aromatic Notch timekeeper," which occurs later in the text, may in fact have been simplified to "The Five Watch Incense Seal."

The account began with a brief summary of the clepsydrae employed in that period:

[Time keeping instruments] with a hollow pitcher such as clepsydrae and a piece of floating wood as an indicator-rod, have been esteemed since the time of the Yu-hsiung Clansman.[4] That usage has been followed from the three periods of antiquity, through the two Han Dynasties,[5] down to the present day [eleventh century]. Although the construction of these [clepsydrae] has at times been clever, and at other times clumsy, there has been no reason to change from this system. At the beginning of our Dynasty [the Sung period], we acquired the water steelyard [clepsydra] of the T'ang Court.[6] Its functioning has been delicate and ingenious. It corresponds very closely to the steelyard clepsydrae of Tu Mu[7] at Hsüan-chou and Jun-chou. Later in our Dynasty, when Yen Su, of the Dragon Chart Library,[8] was Magistrate of Tzu-chou, he made the Lotus Flower Clepsydra, and presented it to His Highness.[9] Recently, in addition, Wu Seng-jui has newly created the steelyard clepsydrae of Hang-chou and Hu-chou, but these examples are both inaccurate and imprecise. At the beginning of the Wu-tzu Year of the Ch'ing Li Reign [A.D. 1048], it was desired to set the Court ceremonials in new order. In the Twelfth Month, the Court Recorders withdrew and made an announcement permitting all the officials to examine

---

[3] Ch'en Ching, *Hsin tsuan hsiang p'u* or *Ho-nan Ch'en shih hsiang p'u*, and *Hsiang ch'eng*.

[4] "Yu-hsiung Clansman" is another name for Huang-ti, a mythical emperor who was claimed to have invented the clepsydra in about 1082 B.C.

[5] The two Han Dynasties, called Western and Eastern, are also known as the Former and Later Han Dynasties, dating from 206 B.C. to A.D. 9, and from A.D. 25 to A.D. 220 respectively. They were separated by the Wang Mang Interregnum

[6] The names of the various clepsydrae follow those in Needham, *SCC*, vol. 3, pp. 313–29.

[7] Tu Mu (A.D. 803–852), born in Lo-yang, rose to the position of Secretary of the Grand Council. He achieved distinction as a poet and was spoken of as Tu the Younger to distinguish him from Tu Fu. Giles, *Dictionary*, p. 783.

[8] "The Dragon Chart Library" was a repository for the storage of imperial writings and book collections.

[9] Wu Seng-jui has not been identified.

the new steelyard clepsydra in the Court Hall. Consequently, they obtained a detailed look at it, and committed it silently to memory. It was only then that they knew that neither the ancient nor the modern [T'ang] systems had been minutely thought out or probed to their deepest. Presumably, if you make the water case for the second steelyard too small, it will cause either slowness or celerity in the dripping [of the water] from the clepsydra. Have not the difficulties of all times past been grandly corrected in the research of our Dynasty? Once I brought forward my own stupid shortcomings and presumed to develop a method in imitation [of the above discovery], which I applied in the Drum and Cornet Towers of Wu-chou and Mu-chou.[10]

This record of eleventh-century clepsydrae is of considerable interest, supplementing our knowledge of the subject derived from other sources. Shen Li then continued with the description of *Pai k'o hsiang yin* (The Hundred Graduations Incense Seal). He wrote:

In the Kuei-ch'ou Year of the Hsi-ning reign [A.D. 1073], there was a great drought. In the summer and autumn, [there were only] unseasonable rains, and the wells and springs were dried up and exhausted. The people were hard pressed for drinking water. It was at this time that Mei-ch'i, an official awaiting promotion, first made the Hundred Notch [sic] Incense Seal, in order to regulate sunset and sunrise. In addition, he set up the Five Night [sic] Aromatic Notch as follows:

The last sentence is perplexing because instead of the description of the "Five Watch Aromatic Notch Timekeeper," the reprinted Shen Li continued with a description of

"The Hundred Graduations Incense Seal"

The Hundred Graduations Incense Seal is made of hard wood. Mountain Pear[11] is best, with camphor[12] and [wood of] the *nan mu* tree[13] next, in that order.

It [the Seal] is 3.04 cm [*nan my*] thick.[14] The outer diameter is 33.02 cm. The diameter of the center part is 2.54 cm and no more. Take the part with the pattern and divide it into 12 sectors of zig-zag lines. The pattern is perpendicular [to the sector radii], with 21 paths lying over [each other]. All paths are .38 cm in breadth. Bring the top part to a point. The depth is likewise like this [0.38 cm]. Each graduation [*k'o* or "notch"] is 6.09 cm in length. One hundred graduations has a total length of 609.6 cm. Each HOUR [*shih*] is almost 60.96 cm in length, and the total [of the twelve HOURS] is 609.6 cm.[15] [Each hour

---

[10] The "Drum and Cornet Towers" were probably watch towers from which alarms were sounded throughout the city in the event of trouble.

[11] *Pirus serotina*, var. *culta*.

[12] *Cinnamomum camphora*.   [13] *Machilus nanmu*.

[14] The inch [*ts'un*] is one-tenth of a foot [*ch'ih*].

[15] The statement "almost two feet" is puzzling, because $2.4 \times 8\frac{1}{3} = 20$ and 20 inches=2 feet. If 2.4 is multiplied by the decimal 8.333, however, a slight shortage occurs. The total given appears to be in error and should be 731.52.

consists of] eight and one-third *k'o* [graduations].[16] In the narrow part close to the center, six of the ring [segments] connect with the adjoining [ring segments, making six pairs]: PIG and RAT, OX and TIGER, RABBIT and DRAGON, SNAKE and HORSE, SHEEP and MONKEY, and COCK and DOG. *Yin* exhausts itself to reach *Yang*.[17]

When [the flame] is at the end of DOG, it enters PIG, because the six long ring [segments] each connect with each other on the outer rim. There are six *Yang* HOURS. They all go forward, from the small into the large, from the imperceptible to the manifest.[18]

At the long parts facing the outer rim, six of the ring [segments] likewise connect with the adjoining [ring segments, making six other pairs:] RAT and OX, TIGER and RABBIT, DRAGON and SNAKE, HORSE and SHEEP, TIGER and RABBIT, DRAGON and SNAKE, HORSE and SHEEP, MONKEY and COCK, and DOG and PIG. The *Yang* culminates in order to enter *Yin*.[19]

When [the flame] is at the end of PIG, it reaches RAT, because the six [short] ring [segments] each connect with each other in the center. There are six *Yin* HOURS. They all go backward, from the large into the small. *Yin* presides over the extinction [of the flame].[20]

There is no stopping point at all. It is similar to a ring, which has no end. Each time one starts the fire, one goes by the [appropriate] HOUR. Generally, one starts at the Top o' the HORSE [noon].[21]

[16] Twelve Chinese hours (twenty-four Western hours) will then consist of exactly 100 *k'o* or "notches." In this text 8⅓ *k'o*=1 *shih*. Concerning the *k'o* used for clepsydrae, the Japanese encyclopedic dictionary, *Dai Kanwa jiten*, states: "In ancient times a day and a night were divided into 100 *k'o*. At the Vernal and Autumnal Equinoxes, there were 50 *k'o* each for the day and for the night. At the Winter Solstice, there were 40 *k'o* for the day and 60 *k'o* for the night, while at the Summer Solstice it was vice versa." This ancient value of 100 *k'o* for a twenty-four hour period is applicable to the incense seal.

[17] The *yin* and *yang* hours are explained further in the text. The *yin* hours are those for which the incense trail reduces from wide to narrow, or from the outside inward. This is, of course, in complete harmony with the general concept of *yin* – negative, female, weak, waning, decreasing. *Yin* never disappears entirely; when at the end of its process it becomes *yang* and begins the opposite course.

[18] Indented separated segments are notes by the author of the original text and appear in the form shown. Explanatory words or phrases within brackets are additions by myself.

[19] Referring to the diagram, the pairs that join on the outer periphery are indicated by the encircled "2." The *yang* hours are those for which the incense trail enlarges from narrow to wide, from the inside outward. *Yang* is positive, male, strong, waxing, increasing. Odd numbers are considered to be *yang* and here the odd numbered hours are the *yang* hours, first, third, fifth, seventh, ninth, and eleventh. Another meaning for *yang* hours might be "sunlight hours" and for *yin* hours "night time hours" respectively, although such meanings do not relate to this passage.

[20] In traditional Chinese thought, even numbers are considered *yin*, and the *yin* hours on this incense seal are the second, fourth, sixth, eighth, tenth, and twelfth.

[21] The Chinese *wu-cheng* refers to the middle of the HORSE HOUR. "Top o' the HORSE" may be a somewhat fanciful rendition but it nonetheless provides a clear explanation. The *wu* or "HORSE HOUR" lasts from 11:00 A.M. to 1:00 P.M. and consequently "HORSE upright" indicates 12:00 o'clock noon.

The third path close to the center is the correct [place]. Some begin [there] at sunrise. [For example], one looks at the calendar [and finds that] sunrise is in RABBIT, [so he starts the fire] at Top o' the RABBIT [or] several graduations [to one side or the other as appropriate].
We set neither an end point nor a place for starting the fire (Fig. 45).[22]

The foregoing description is derived primarily from the *Hsin tsuan hsiang p'u*, and supplemented by comparison with related passages in the *Hsiang ch'eng*. Once again, Shen Li's original account appears to have suffered in the later renderings, a confusion that may have been due to the nature or condition of the surviving fragments of Shen Li's work.

As might have been expected from a trained engineer, Shen Li's instructions for the construction and operation of the *Po k'o hsiang yin* were precise, and no detail was overlooked. In the translation, the Chinese word *shih* has been rendered as the word "HOUR," capitalized to denote the Chinese hour of one hundred and twenty minutes, which is equivalent to two Western hours. Twelve *shih* elapse in the course of a twenty-four-hour day. Each *shih* is named for one of the characters in the list of "Twelve Branches" and is symbolized by one of the twelve animals. The names of the animals have been used to translate these cyclical signs and have been written in upper case to indicate that each represents an HOUR instead of a Western hour. Although the Chinese word *k'o* technically means "notch," it is used in the translation to mean "graduation." Each *k'o* represents fourteen and four-tenths minutes in Western time, and each *shih* or double hour includes eight and one-third *k'o*.

"The Hundred Graduations Incense Seal" illustrated in the *Hsin tsuan hsiang p'u* was undoubtedly reproduced from the stone engraving Shen Li had ordered to be made. The "part with the pattern" to be divided into twelve sectors of zig-zag lines is understood to mean the area of the total circle exclusive of the center ring.

In the line rendered as "the pattern is perpendicular," the original text stated "the pattern is criss-cross" with respect to the twelve sectors, which when drawn, define twelve areas crossing back and forth across them.

The rendition of Shen Li's account continued with a description of *Wu yeh hsiang k'o* (The Five Watch Aromatic Notch Timekeeper).[23] It was made with

---

[22] This note is somewhat puzzling, because if one wishes to ignite the incense seal at 12:00 noon, it would have to be done in the middle of the incense trail that traverses the HORSE sector. This suggests, however, that the seal should be ignited in the third path from the center (see overlay), which would appear to make the seal "slow." Possibly the correct interpretation is "the third path approaching the center [from the outside]." However, this would require that the path be ignited on the path numbered 19 on the overlay, which would appear to make the seal "fast."

[23] The two titles, "Five Watch Seal Notch" and "Five Night Seal-Character Aromatic" refer to the same incense seal. Another name used is "Five Night Aromatic Notch."

three different lengths, for the long, medium and short nights of the different seasons:

The Upper Seal lasts the longest period of time.[24] It is used alone from after the "Slight Snow" [November 22–December 7], through the "Heavy Snow" [December 21–January 6], until after the "Slight Chill" [January 6–January 21]. Following it are four Seals A, B, C and D (*chia, i, ping,* and *ting*).[25] Both notches are used simultaneously.[26]

The Middle Seal is the most average [in length of burning]. It is used alone from after the "Excited Insects" [March 5–March 20] until after the "Vernal Equinox" [March 20–April 5]. The autumnal division is the same. Before and after it there are the seals *wu* and *chi* [E and F].[27] Each one is used alone.[28]

The Last Seal is the shortest. It is used from [after] the "Grain in Ear" [June 6–June 21][29] until after the "Summer Solstice" [June 21–July 7] and the "Slight Heat" [July 7–July 23]. Preceding it are the four Seals G, H, I and J (*keng, hsin, jen,* and *kuei*).[30] Both notches are used simultaneously.

The First Seal is the longest because it consists of the maximum number of *k'o* or graduations, a total of 60. It was used only during the period of the year when the nights were approximately 60 *k'o* long. These were the nights following the festival day of "Slight Snow" which occurs on November 22, until after the night of the festival day of "Slight Chill," or January 6. In actuality these dates are only approximate, as clarified in the "Table of Unified Data for the Five Watch [Aromatic Notch Timekeeper]" (Fig. 46).

The four Seals A, B, C and D have a double use in that they can be used twice during the year. As an example, Seal A can be used before and after the winter solstice. Seals A, B, C and D also have two sets of lengths. Seal A can be either 59 or 58 *k'o*, for instance. The text is not explicit concerning how this variability is achieved, but from the diagrams it is apparent that Seals A, B, C and D each have

[24] This is a condensed description made by the editor of the *Hsin tsuan hsiang p'u* of the *Wu yeh hsiang k'o* ("The Five Watch Aromatic Notch Timekeeper") which also had been engraved on stone by Shen Li.

[25] *Chia, i, ping,* and *ting* are the first four of the "ten celestial stems." When the ten celestial stems are combined in regular order with the "Twelve branches" they form the Sexagenary Cycle.

[26] The last sentence is puzzling, for the Chinese word is *ping*, meaning "side by side" or "of equal rank." Presumably the intention is "side by side," both notches (graduations) being used.

[27] *Wu* and *chi* are the fifth and sixth characters in the list of the "ten celestial stems."

[28] Apparently the Middle Seal is used also during the period in which the autumnal equinox occurs, when the length of the days and nights are similar to the spring.

[29] "Grain in Ear" is the name for the period during which the beard, or awn, appears on growing grain.

[30] *Keng, hsin, jen,* and *kuei* are the last four of the "ten celestial stems."

a spiral terminal at their center into which the burning track presumably continues to allow for the greater number of *k'o*, such as 59 in the case of Seal A. On the other hand, if the spiral is closed off, it allows for the lesser number, or 58 *k'o*.

The Middle Seal is the mean or average Seal because the total number of *k'o*, which is 50, is the mean between 40 and 60 *k'o*, the totals of the First and Last Seals respectively. Its single use is during the periods when the nights are approximately 50 *k'o* in duration, or just before and after the two equinoxes.

Seals E and F can be used twice during the year, as indicated in the "Table." Unlike Seals A, B, C and D, Seals E and F each have only a single *k'o* total – Seal E has 51 *k'o* and Seal F has 49. Neither has a spiral terminal in the center, bolstering the supposition concerning the function of the spirals which form part of Seals A, B, C and D.

The Last Seal is the shortest because it has the least number of *k'o*, totalling 40. Its single period of use is during the period when the nights are approximately 40 *k'o* in length (Figure 47).

Seals G, H, I and J function similarly to Seals A, B, C and D.

In the text following the brief summation of the functioning of "The Five Watch Aromatic Notch Timekeeper" is a series of drawings with captions explaining the use of this form, reproduced from the *Hsin tsuan hsiang p'u*, in which they appear without captions. The same designs are illustrated in *chüan* 22 of the *Hsiang ch'eng* with descriptive captions, but the drawings are of much poorer quality.[31] These are the first four of the "ten celestial stems," namely, *chia*, *i*, *ping*, and *ting*. *Wu* and *chi* are the fifth and sixth stems, and the last four are *keng*, *hsin*, *jen*, and *kuei*.

The instructions embodied in the captions have been compiled and arranged in the "Table of Unified Data" for facility of comparison. The Seals are listed by name as they appear in the text, followed by the number of *k'o* for each as designated. The diameter of the character is given in each instance, in addition to a measurement of the length of the incense trail, followed by the dates when each is to be used (Fig. 48).

In nine instances the dates have been found to be in conflict with other data in the table. Some of these conflicts are due to an obvious corruption in the text, and others to a disparity between the approximate dates provided in Appendix A of the *Dictionary of Chinese Dates* by R. H. Mathews,[32] and the dates derived by reckoning the specifications of the text. In each instance of conflict, the dates have been amended to fit the scheme of the Seals. The dates as amended are tabulated in the final column and these are shown to be consistent with each other.

[31] The edition of the *Hsin tsuan hsiang p'u* used for this study is the version included in the *Shih yüan ts'ung shu* published by Chang Chün-heng in 1914.
[32] Mathews, *Dictionary*, p. 179.

The duration of use for the various Seals has been plotted against an approximately elliptical curve illustrating the length of the night as it varies throughout the year. The dates for which the Chinese dates are in conflict but which have been amended in the "Table of Unified Data" are shown with dotted lines.

Two additional seal-characters for incense seals are illustrated without explanation or description in the *Hsiang ch'eng* but are not included in the *Hsin tsuan hsiang p'u*. The first utilizes the Chinese seal-character for "long life" (*shou*) in an abbreviated and stylized form. The second is the greatly stylized character for "blessings" (*fu*) (Fig. 49).[33]

Also surviving is Shen Li's description of Wu Cheng-chung's recipe for the blend of incense required for the incense seal with one hundred graduations:

> In the case of The Hundred Graduations Incense seal-character, if you use ordinary aromatics [incense], it will not [burn] evenly. Now we use the two fragrances "Wild Thyme" and "Pine Ball." Mix them together and have them placed evenly in a new earthenware vessel, and then [they may be] applied and used. The "Wild Thyme" is the leaf of the *jen* Tree.[34] Pick them before the middle of autumn, lay them out in the sun to dry, and then make them into powder. Ten ounces for each batch. The "Pine Ball" is the withered pine flower. Gather them at the end of autumn, selecting the ones that have fallen by themselves. Lay them out in the sun to dry, cut away the center, and make it into a powder. For each batch use eight ounces.[35]
>
> Formerly, when I compiled the preface for the Aromatic Catalogue, the account of the Hundred Graduations Incense Seal was not greatly detailed. Wu Cheng-chung of Kuang-te (in Chekiang) fashioned his Seal-character Notch as well as the prescription for the aromatic.[36] On testing it, I have found that it is quite minutely thought out and considered. Were it not for his refined talents and subtle thought, how could we have been able to attain this? Consequently I have engraved it on stone so as to pass it on to all those with fond interests in affairs.[37]

> The Second Day of the Second Month of
> Spring, in the Chia-yin year of the Hsi-ning
> Reign [A.D. 1074]. Right Censor, Grandee, and
> Magistrate of Hsüan-ch'eng Commandery,
> Shen Li.

The greater part of the remainder of the *Hsin tsuan hsiang p'u* consists of recipes for aromatics for general use as well as incense to be used specifically for

---

[33] Also included in the *Hsiang ch'eng, chüan* 22 and in the *Shih-yüan ts'ung-shu*.

[34] The "Jen Tree' is *Perilla ocimoides (frutescens)*. The powder of the dried leaf was included in incense to promote slow and even burning.

[35] It is to be noted that it is the flower of the pine that is meant, not the cone.

[36] "Wu Cheng-chung" appears to be a *tzu* or courtesy name, but efforts to identify it with a personality of that period have been unsuccessful.

[37] This paragraph may have been engraved on stone, from which Ch'en Ching may have copied the text for insertion in his work.

incense seals. Generally each incense seal recipe is identified as "The Seal Aromatic Used in the Public Storehouse of" and then the name of a city. An example follows:

"The Seal Aromatic [Used in] the Public
Storehouse of Ting-chou"

Chien hsiang [Aromatic]	1 ounce
T'an hsiang [Sandalwood][38]	1 ounce
Ling-ling hsiang [Misty Tumulus Aromatic][39]	1 ounce
Huo hsiang [Betony][40]	1 ounce
Kan sung [Sweet Pine, Spikenard][41]	1 ounce
Ta huang [Big Yellow, Rhubarb][42]	1 ounce
Mao hsiang [Reed Aromatic][43]	½ ounce

Very finely powdered. By soaking it in water, drying it in the sun, and then roasting it with a slow fire, the color becomes yellow.

Take the above ingredients and pestle them through a sieve, making powder, and then use it according to the usual method. Whenever one makes Seals or Seal-characters, you must mix the powder with a little almond powder; the aromatic then will not raise dust. Everyone follows this method after they have had some aromatic blow away on them.

Following the section on recipes in the *Hsin tsuan hsiang p'u* is a woodcut entitled "Illustration of the Greatly Elaborated Seal-Character," which represents a later improvement of the incense seal (Fig. 50). It is included also in the *Hsiang ch'eng*, and is captioned as follows:

Whenever aromatic powders for Seals or Seal-characters are combined, do not use *Chien* [aromatic], or "Nipple Aromatic" [probably frankincense], (*Pistacia khinjuk*) or *Chiang chen hsiang* ["Aromatic that Calls Down the True Ones"]; [or] *Rutaceae Acronychia laurifolia*, for they are oily, liquid and bubbly, and often cause the fire not to burn.

The text continued:

Tsou Hsiang-hun was presented with this illustration. Hsiang-t'an, whose adult name was Chi-lung, and whose courtesy name was Shao-nan, was a native of Yü-chang [in Kiangsi Province]. His post was in Tz'u-li [Prefecture] of Yü-li. He loved the old "Ele-

---

[38] *Santalum album.*
[39] This is either *Lamiaceae ocium basilicum* or *Melilotus arvensis.* "Misty Tumuli" is the name of a range of mountains in southern Hunan.
[40] *Stachys officinalis.*
[41] *Valerianaceae Nardostachys jatamansi.*
[42] *Rheum spec.*
[43] *Hierochloe borealis.*

gancies;"[44] he was particularly excellent on the *Book of Changes*. He was held in high esteem by many worthy gentlemen and grandees.

First Day of the Good Month [October] in the
2nd year of the T'ien-li Reign [A.D. 1329].
Written by the Retired Gentleman of the
Central Studio.[45]

The *Hsin tsuan hsiang p'u* (Newly Compiled Handbook of Aromatics [Incense]) in which Shen Li's description was included, was compiled by Ch'en Ching of Honan, probably between 1275 and 1322, although it may have been completed as early as the middle or latter part of the twelfth century. The work was known also by another title *Ho-nan Ch'en shih hsiang p'u* (The Handbook of Incense by [a Member of] the Ch'en family of Honan).

The clue to the date of publication of Ch'en Ching's compilation may lie in the presence of one of his sources, the work entitled *Ch'ien Chai hsiang p'u shih i* (Incense Catalogue supplement by Ch'ien Chai), which appears to date as late as the end of the thirteenth century. Ch'ien Chai was a pen name used by two writers living at the end of the Sung and beginning of the Yüan Dynasties. It is also possible that Ch'ien Chai was the pseudonym of a yet earlier writer.

Ch'en Ching collected into a single work all the writings that he could find by earlier writers on the subject, noting the existence of eleven monographs of earlier dates. He included the available fragments of the work of Shen Li, and he presumably had access to the inscription on stone of 1074. Of the previous works, only the *Hsiang p'u* by Hung Ch'u, written in about 1115, has survived in its entirety.[46]

Among the other writings included, of which only the titles have survived, were the *Yen Ch'ih-yüeh hsiang shih* (History of Incense by Yen Ch'ih-yüeh); the *Hsiang lu* (Record of Incense written in 1151 by a Sung writer named Yeh T'ing-kuei; the *Hsiang p'u* produced in about 1322 by Hsiung P'eng-lai, a provincial professor of the Yüan period. Treatises on blending of perfumes and incense had been produced as early as the first and second centuries A.D. but were already lost by Ch'en Ching's time. One which has not survived but which would have been particularly interesting was entitled *Lung-shu p'u-sa ho hsiang fang* (Incense Blends of the Bodhisattva Nagarjuna).

The *Hsin tsuan hsiang p'u* has several prefaces. The first was written Hsiung P'eng-lai (1322). Hsiung may possibly be identified with the "Retired Gentleman of the Central Studio" since he derives from Yü-chang, the same area as the one he praised in the caption.

---

[44] Name of an early work of glosses on ancient literature.
[45] Not identified.
[46] Cited in Needham, *SCC*, vol. 5, part 2, pp. 134–35.

The second preface is taken from the *Hsiang p'u* of Hung Ch'u, and the third is the preface to the *Hsiang shih* by Yen Ch'ih-yüeh and the fourth the preface to the *Hsiang p'u* by Yeh T'ing-kuei, dated 1151. Yeh T'ing-kuei was a customs inspector at the southern seaport of Ch'üan-chou and undoubtedly was familiar with aromatics being imported into China from southeast Asia. A later preface, dated 1322, was included in the *Shih yüan ts'ung shu*. None of the prefaces relates to incense seals.

Ch'en Ching's *Hsin tsuan hsiang p'u* was subsequently included as part of two later works, the collectanea Shih yüan ts'ung shu, and of the more elaborate compilation, the *Hsiang ch'eng* (Comprehensive Account of Aromatics [Incense]) published by Chou Chia-chou between 1618 and 1641.

Notable among later works on the subject were the *Hsiang chien* (Incense Notes), a short work by T'u Lung (d. A.D. 1577), and the *Hsiang kuo* by Mao Chin, produced in the early seventeenth century.

A century after it was first recorded, "The Hundred Graduations Incense Seal" was again described by Hung Ch'u, a writer of the Northern Sung period, in his book *Hsiang p'u*:

"The Hundred Graduations Incense Seal"
[In recent times], persons who esteem oddities have incense seals [*hsiang chuan*] made for them, the texts of which conform to the twelve double-hours; they are divided into one hundred graduations [*k'o*], and burn all through the day and night.[47]

Incense Seals
Engraved into wood, patterns in the form of seal script characters are revealed when the incense in them is burned at drinking parties or before images of Buddha. These incense seals [*hsiang yin*] frequently measure up to two or three [Chinese] feet in diameter.[48]

The date of the invention recorded by Shen Li was noted again in the Ming Dynasty. In his *Hsiang ch'eng*, Chou Chia-chou wrote:

In the sixth year of Hsi Ning ... was first made the incense seal of a hundred notches and ... the five night incense seal was added.[49]

The sixth year of the Hsi-ning era, during the reign of Emperor Shen Tsung of the Sung Dynasty, corresponded to the fifth year of Enkyū in the reign of the Emperor Shirakawa in Japan, or the year A.D. 1073.

By the advent of the (Southern) Sung Dynasty (1127–1279), the incense seal

---

[47] Hung Ch'u, *Hsiang p'u*, translation communicated by the late Professor Edward H. Schafer, November 3 1959.

[48] From Tso Kuei, *Pai ch'uan hsüeh hai*. One Chinese foot is equal to approximately one-third meter.

[49] Chou Chia-chou, *Hsiang ch'eng*.

had become a common feature in urban life. It is described in the *Hsiao hsüeh kan chu*, an early Chinese encyclopedia issued in ten *chüan* or parts, and compiled in about 1270 by a prolific writer named Wang Ying-lin. It was not published until 1299, a few years after his death.

In the first *chüan* of his compendium, Wang Ying-lin quoted Hsüeh Chi-hsüan, a writer of the Northern Sung period, who stated

In these times [twelfth century] timekeeping devices [*kuei lou*] are of four different types. There are the bronze vessels [clepsydrae, *t'ung hu*], the [burning] incense seal-characters [*hsiang chuan*], the sundial [*kuei piao*], and the revolving and snapping springs [steelyard inflow clepsydrae].[50]

The modern Japanese writer, Yabuuchi Kiyoshi, in commenting on the same passage, added

[The] incense seal-character [*hsiang chuan*] is a clock which measures time by means of a joss stick. This clock was called a smoke seal clock [*yen chuan*] in a work entitled *Lou k'o ching* by an unknown author, which was published probably in the fifth to the sixth centuries, and it is clear that such a clock or timekeeper was used since that time. These clocks were employed for private or personal use in order to measure time at night, but they were too inaccurate to be used for official or public purposes.[51]

In interpreting the phrase incense seal-character (*hsiang chuan*) to mean the graduated incense stick (*hsiang pang*), Yabuuchi appears to have been confused, but not in identifying the incense seal with the *yen chuan*.

The mid-twelfth-century Sung scholar, Hsüeh Chi-hsüan, also noted that time was measured in that period not only with sundials and clepsydrae, but also by means of the incense seal-character (*hsiang chuan*). The incense seal was commented upon also by another Sung writer, Hung Ch'u, in his work, *Hsiang p'u*.

After Shen Li's time, the incense seal continued to be made of wood and possibly also of stone, for at least a century, as indicated in the account by Hung Ch'u.

As the incense seal was improved, it was discovered that there were considerable advantages in making it of metal. The earlier metal examples were probably of cast bronze, a metal generally used for vessels intended for ceremonial use.

Although no record exists of the first use of metal for making incense seals, the transition from wood or stone probably occurred early in the Sung period. The components of metal incense seals consisted of a container made in several segments or tiers, often with a supporting base. At first they were probably of common shapes, square, rectangular or round, although in the course of time

---

[50] Wang Ying-lin (A.D. 1223–1296), *Hsiao hsüeh kan chu*.
[51] Yabuuchi, "Chūgoku no tokei," pp. 19–25.

other special shapes were devised, such as the melon, gourd, *ju-i* scepter, bell, and leaf. The equipment included a template, a tamper, and a shovel.

The use of metal also made it possible to produce incense seals that were portable in compact form of smaller sizes, a great advantage for seals intended for use in private and social life. For such a purpose the pattern was designed with varying path lengths comparable to the lengths of the unequal night watches which varied with the seasons, as well as for the twelve double-hours and the hundred quarters standard throughout the year and requiring no change in pattern. Incense seals intended for public use – in palaces, government offices and for the community – were made sufficiently large to measure time ranging from the half day to a full month.[52]

During the Sung Dynasty incense seals appear to have achieved considerable popularity for public and private use and were being commercially produced in substantial numbers for Buddhist temples as well as for the public. Furthermore, the shops which dealt in them operated under strict government regulations. Confirmation of this development was noted in Meng Yüan-lao's famous work, *Tung ching meng hua lu*:

[As regards] incense seal makers who supply seals on a daily basis, everything from the size of their shops and the boards on which they advertise their trade [signboards], to the time when certain Buddhist figures are to be stamped on the seals, is officially regulated.[53]

The same passage is quoted in *Meng liang lu*, a later work written in c. A.D. 1270 by Wu Tzu-mu, who added,

They supply the seals daily and submit their bills at the end of the month.[54]

Seal-characters were used by the Chinese from an early period to signify auguries of good will on the numerous occasions that lent themselves to ceremonious formalities, such as betrothals, weddings, anniversaries, birthdays and other memorable events that required acknowledgment. Seal-characters particularly favored for conveying congratulatory wishes in salutations and inscriptions were those representing longevity, happiness, good luck, and double or conjugal happiness. The character for longevity was used not only as an augury for the individual, but also for his family line and to wish prosperity for his descendants. These seal-characters were carved or painted upon gifts to be presented on appropriate occasions, and skillfully incorporated in decorative designs in an infinite number of ways.

In time, the intellectual and wealthy classes in China realized the desirability of

---

[52] That incense seals of longer duration were produced for public purposes is verified in the examples provided by Shen Li.

[53] Courtesy of Dr. Carole Morgan.   [54] Courtesy of Dr. Carole Morgan.

giving incense seals as gifts. The devices were eminently suitable for acknowledging important auspicious events and for housewarming gifts for use on the Buddhist altar in the home. As their popularity for this purpose increased, they were made of the several metals in common use, including bronze, copper, pewter, and paktong. These portable metal incense seals probably ranged from simple forms made with minimum elaboration at modest expense to extremely costly and excessively decorated examples to suit the particular occasion.

The auguries of good will rendered in seal-characters proved to be eminently suited not only for decorating incense seals intended to be given as gifts or acquired for personal use, but also for the designs of the templates and the perforated covers.

An example having unusual characteristics is a square seal, made of bronze in two tiers, footed, with the cover perforated with the inscription "Great wealth and honor and a long life." In smaller characters on either side of the larger inscription is the date, "The first month of Spring, of the *chia-tzu* year." Inasmuch as only the cyclical date is provided, it is impossible to determine the year precisely, but it appears to be of the eighteenth century or earlier. The sides are inlaid in silver with poetic quotations from the *I ho ming* (Inscription to a Departed Crane) and several other popular stelae. The hallmark, which is cast into the bases of each of the two sections, consists of the phrase *Yüeh-heng*, which may be the maker's name. The first character is "Mountain peak" and the second "constant" or "persevering." Consequently it is not revealing of the identity of the maker with certainty, or the place (Figs. 51–52).

Incense seals given as gifts were generally supplemented with elaborate carved bases and containers consisting of boxes made of fine woods fitted with brass hardware and bearing decorative inscriptions, such as *Yen yün kung yang* (Misty Offering).

Brass for making incense seals appears to have been used infrequently. It may have been reserved for use in the temples, such as the example with Siddham inscriptions. If used also for smaller portable seals, examples have not come to notice. However, the latticework grids of the covers of pewter seals were invariably made of brass because it was more resistant to heat than the softer pewter metal. They were perforated with decorative designs featuring a seal-character of augury. Occasionally brass was also used to make the segment containing the incense tray in an incense seal made of pewter, again for the same reason. Brass (*t'ou shih* or *huang t'ung*), the most common alloy, had been used in China and India since the first millennium. This alloy was produced in China to some degree, and it also was imported in quantity from Persia and India.[55]

[55] Needham, *SCC*, vol. 5, part 2, pp. 195–211.

The greater number of incense seals produced in the eighteenth and nineteenth centuries were made of paktong, and it is probable that this metal had been used for the same purpose in earlier periods. The literal meaning of "paktong," derived from the Cantonese pronunciation of the name (*pai t'ung*), is "white copper," descriptive of its appearance. Paktong closely resembles both silver and the Western alloy known as German silver.[56]

The alloy is an extremely hard and tough but nonetheless malleable metal which can be cast, hammered and polished. It is relatively heavy, sonorous when struck, having a lustrous sheen which can be highly polished. It is less liable than other metals to corrosion or tarnish from the atmosphere, and less likely to acquire verdigris than other metals. However, many examples of incense seals obtained in the 1970s and 1980s from the mainland of China (after it was opened to Western trade) and sold in Hong Kong were heavily coated on all surfaces with a thick and stubborn tarnish. This was the result of neglect and storage under damp conditions for a long period of time.

Paktong is an alloy consisting of approximately four parts copper, two parts zinc and one part nickel. There is considerable variance, however, depending on the source of the nickel ores. As a consequence, the appearance of the alloy can range from white and silvery to a light shade of yellow approaching the appearance of brass.

Paktong was used extensively in China for at least 2,000 years, before nickel was isolated as a separate metal. The Chinese considered it a precious metal, less valuable than gold and silver, to be sure, but nonetheless reserved for special uses. During the Sui period, in about A.D. 585, the imperial coinage was struck of paktong, and during the T'ang period paktong was noted as a metal of which valuable ornaments were made.

The processing of the alloy was a closely guarded secret. It is believed that the metal was obtained by the reduction of a metallic ore found in China and containing the ingredients of which paktong is composed. As they did with brass and copper, the Chinese cast it into sheets, which were then cut into suitable strips to form the parts desired, and then skillfully soldered together. The remarkable technique of "invisible" soldering, which the Chinese had perfected over the centuries, was unknown in the Western world, and it was never better demonstrated than in the Chinese metal incense seals, particularly those made of paktong. A raised beading of red copper was frequently added to the edges of the major sections of paktong incense seals for decorative emphasis. Incense seals were also made of cast paktong of several thicknesses.[57]

[56] *Ibid.*, pp. 225–32.
[57] Fyfe, "Analysis," vol. 7, pp. 69–71; "Analysis," *London Journal* (1823), pp. 18–19; Needham, *SCC*, vol. 5, part 2, pp. 190–91, 225–30; Tiffany, *Canton*, p. 97.

There is relatively little mention of paktong in the literature of the Ming and Ch'ing periods, although the *T'ien kung k'ai wu*, published in about 1637, noted that when arsenious trioxide (*p'i shuang*) and other arsenical chemicals are mixed and combined with zinc carbonate, or calamine (*lu kan shih*), or with zinc metal (*wo ch'ien*), the derivative is paktong (*pai t'ung*), namely, "white bronze" or "white copper." Arsenious trioxide and arsenolite had been exploited in China since early times. It was mentioned in the *Kuang ya* of the third century as "Tan-yang copper," which is another name for paktong.[58]

The sale of paktong was observed by the Spanish missionary, Fray Manrique, during his travels in the Far East. While in Arracan (Arakan State, Burma) he noticed the sale of gems, gold and silver in plates and bars, incense and "Calain and Tutunaga." The last two were metals similar to tin, but considered to be much superior in the opinion of skilled chemists. Calain was the name used for "tin copper vessels" sold in India, and "Tutunaga" was the name Manrique used for paktong.[59]

Exportation of paktong from China in the form of the crude alloy or as minted coins was forbidden, and any attempt to smuggle it out of the country resulted in instant execution. The trade was highly lucrative, however, and smugglers were undeterred, for the metal was always in short supply and commanded high prices.

The first mention of the alloy in the Western world appeared in the *De Natura Metallorum*, a work published in 1597 by Andreas Libau (d. 1616). Known as Libavius, the German naturalist, physician, and chemist described the alloy as a white bronze. In English it was first described in 1736 in a translation of P. du Halde's *History of China*. After mentioning the output of the copper mines in Yunnan and Kweichow, he wrote:

> But the most singular, is the [*Pe Tong*] White Copper; it is naturally of a white Colour, and still more so inwardly than outwardly. Several Experiments have been made at Peking, to try if it owes its Whiteness to any Mixture, by which it was found that it did not; on the contrary, all Mixtures, except Silver, diminish its Beauty. When polish'd, it is exactly like Silver, but what detracts from its Value, is, its being more brittle than any Copper.[60]

Halde's text was later quoted by the English chemist, Richard Watson (1737–1816), Bishop of Llandaff, in his *Chemical Essays*, which appeared in 1781. Watson also described the importation of tutenag and paktong to England.

The earliest examples of paktong brought to England appear to have arrived in the first several decades of the eighteenth century, transported overland from

---

[58] Needham, *SCC*, vol. 5, part 2, pp. 225–42.
[59] Manrique, *Travels*, vol. 1: Arakan, p. 380.
[60] Du Halde, *General History*, vol. 1, pp. 16, 23, 257.

Canton to India and then by sailing ships to London. Inasmuch as the Chinese never were willing to export the unwrought alloy, the examples which arrived in England were probably in the form of small boxes and other objects purchased in China and brought out illicitly by shipmasters on ships of the East India Company. The first analysis of the alloy was made in Sweden in 1776 by the chemist Gustave v. Engestrom.[61]

The earliest use made of the alloy in England was by the celebrated architect and interior designer Robert Adam (1728–1792) after his return from a sojourn in Italy. There he may have been first introduced to paktong, where it was to be found in the villas and country homes of the wealthy, probably imported by the Portuguese East India Company. It is not known how Adam obtained his supply, but he had elegant fire-grates and fenders made of paktong to his designs, and it was also used when available by London silversmiths for making candlesticks.[62]

Paktong has always been considered somewhat of a mysterious metal, and relatively little has been written about it in the Western world. Writers have often confused it with tutenag, the refined commercial metal derived from zinc carbonate, a mineral ore known as calamine.

Tutenag is commercial zinc or spelter and not an alloy. Its Chinese name is *pai-yüan* or *pai ch'ien*, while the Indian name is *tutenag*, probably derived from the Sanskrit *tuttha*. It has been known and used in eastern India and China for more than a thousand years. In China tutenag was mined chiefly in Hunan Province from depths ranging from 121 to 152 m. At one time it was used for making coinage, and its export as raw materials and in minted form was severely limited and controlled.

With the development of trade with Europe, tutenag was exported in large quantities chiefly from India to England and the Netherlands. The tutenag imported into England was sold as spelter or zinc and was not identified by its Indian name until the late eighteenth century. In the course of time tutenag became confused with paktong; this confusion has persisted, despite the fact that tutenag is one of the three components of paktong.[63]

Another metal of which incense seals were made commercially was pewter, called *pai la*, an alloy consisting chiefly of tin and lead. In China pewter has been used largely for altar furniture in temples and the home. It is not certain when this metal was first utilized for making incense seals, but it seems probable that it was not before the late eighteenth or the nineteenth century. Its history in China is not

---

[61] Watson, *Columbia*, vol. 4, pp. 28, 108 ff., vol. 5, p. 251; John and Coombes, *Paktong*, pp. 11–14; Engestrom, "Pak-tong, a White Chinese Metal," quoted in John and Coombes.

[62] John and Coombes, *Paktong*, pp. vii–ix, 2–18; Bonin, *Tutenag*, pp. 18–51.

[63] Bonin, *Tutenag*, pp. 3–17.

certain, but it is believed that the metal was not produced prior to the Ming Dynasty. The major centers for its manufacture have been Wen-chou and the region around Swatow, a port at the mouth of the Han River in the province of Kwangtung. Once a fishing village, by the nineteenth century it became an important trading center and emigration port, and was opened to foreign trade in 1869.[64]

Almost all the incense seals made of pewter that have been examined bear hallmarks which identify the place of origin as Swatow, suggesting that this region was an important if not unique center for their production.[65]

[64] Hommel, *China At Work*, pp. 346–48, 354–56; Needham, *SCC*, vol. 5, part 2, p. 218 n.
[65] Needham, *SCC*, vol. 5, part 2, p. 217; Hommel, *China At Work*, pp. 354–56.

8

In Chinese social life

By the time incense seals could be made of metal instead of wood and stone, and in compact and decorative forms, their production assumed a commercial dimension. There is no evidence of mass production, to be sure, and every example appears to have been individually hand-crafted, but in time the incense seal became a product for multiple production. No longer were they limited to use in the temple or for the community.

The burning of incense had always served as inspiration for Chinese scholars and writers, and the brazier or censer was invariably to be found close at hand. That the presence of incense was a requisite for inspiration is confirmed by U. A. Casal:

The censer is as necessary an article in a scholar's study as the writing utensils and the books themselves. The subtle fragrance aids exalted thoughts and contributes to discussions in an appreciative nod. As an old Chinese sage has it: "The use of incense gives manifold benefits. When retired scholars, detached from the world, are sitting together discussing *Tao* and its application, they burn incense to purify their hearts and rejoice their spirits. At the dead of night, when the morning moon is in the sky, artistic and sad poetical folk burn incense, and their hearts are elated and they whistle carelessly. By the bright window copying old famous scrolls, or leisurely humming, fly-whisk in hand, or when reading at night under the lamp, incense is burned to drive away the demon of sleepiness. Therefore incense may be called the Old Companion of the Moon."

Melancholy, Casal noted, is a characteristic of most exalted Far Eastern poetry, and sadness seems to prevail as a reflection of older native feelings. Undoubtedly the concept of the vanity of things, inherent in Buddhism, also played a role. Cherished preoccupations were the practices of making copies of famous manuscripts and executing paintings in ink, for not only did these activities provide excellent training for the hand and eye, but also exerted the greatest moral influence.[1]

The decorative portable metal incense seal proved to be particularly desirable for scholars and writers: its small size more convenient for the writing desk than

[1] Casal, "Incense," pp. 56–58.

the larger brazier or censer. The trailing smoke emanating from the burning seal
character was assumed to provide inspiration while engaged in thinking or writing at the same time that the burning trail measured the passing hours.

The designs with which the portable incense seals were decorated were selected
with particular care. Elevating thoughts expressed in literary style and lines of
poetry selected from the classics inscribed in archaic scripts were esteemed to
reflect scholarship and were especially popular decorations for incense seals given
to or used by scholars and writers.

Frequently featured were the eight precious objects (*pa pao*); the eight Buddhist
emblems of happy augury (*pa chi hsiang*); the symbols of ancient Chinese lore,
such as the eight trigrams of divination (*pa kua*); the dualistic *yin yang* symbol;
the symbols of the Eight Immortals of Taoism such as the fan with which Chung-li
Ch'üan revived the dead; the magic pilgrim's gourd of Li T'ieh-kuai; the bamboo
tube and rods of Chang Kuo-lao; the peach considered as "the fruit of life"; and
the sacred fungus (*Polyporus lucidus*), in addition to many others. Favored were
representations of flowers and other plants, such as the prunus, pawlonia blossoms and grasses.

Over the course of the past two centuries, incense seals for this market were
produced in a great variety of shapes representing classic Chinese motifs.
Although the larger proportion of the seals were round, square or rectangular in
shape, many were made in a variety of symbolic shapes, such as the leaf, the stone
chime, the lute, the melon, the fan, and other exotic forms.

The smaller seals are often footed, and consist of two to five sections,
sometimes with a separate base. The upper section accommodates the incense
tray, the second section serves as a compartment for storage of tools, and a third
for storage of incense. A latticework cover protects the burning incense from
movement of air which might extinguish it. The seals are equipped with the
necessary utensils, including a template, a tamper to compact the ash and incense
in place, and a miniature shovel to spread the incense.

Many of the seals bear hallmarks stamped on the base of the bottom section
and occasionally on the underside of all individual parts. These marks generally
include the name of the maker and a place name in seal-characters, and are
frequently without indication of date.

Originally, many of the incense seals were equipped with a skillfully carved
decorative base of hardwood, some of which survive. Those intended as special
gifts were housed in elaborate presentation cases of polished wood inscribed with
a suitable augury (Fig. 53).

The covers with delicate latticework to emit the heat and smoke generally
feature a seal-character of good wishes at the center. Bats, a crane, leaves of iris or

prunus or other symbols of happiness are cleverly incorporated into the perforated design.

A favored form in which incense seals were made was that of the *ju-i* scepter, a symbol derived from the shape of the sacred fungus or "plant of long life" called *ling chih* (*Ganoderma lucidus* or *Polyporus lucidus*). One of the Taoist emblems of longevity, the literal meaning of *ling chih* is "the divine fungus." It may have been introduced into China from India, perhaps as a reflection of the Soma, the magic mushroom in the *sutra* of RgVeda, in the Ch'in Dynasty. It was first mentioned during the reign of Emperor Shih Huang.

For thousands of years the Chinese have considered the sacred fungus to be a symbol of life among the immortals. The Taoists adopted the concept of the sacred fungus in an early period, and exploited it in their writings so extensively that its original meaning was buried in numerous imaginary divagations. It began to appear in every form of Chinese art as well as in literature from the late thirteenth century and assumed considerable importance. The concept spread also to Korea and Japan. In Japan the mushroom is known as *reishi* or *mannentake*, "the ten thousand year mushroom."[2]

Another interpretation of the *ju-i* scepter represented it as a form of a short blunt sword made of iron. It was said to be used for "pointing the way" and for "guarding against the unexpected." The scepter often appears as the symbol of Lao-tzu, the reputed founder of Taoism. This legendary figure, who is said to have been born in 604 B.C., left China late in his career for the western regions. Before doing so, it is said that he recited his teachings to the keeper of the pass. These teachings were recorded and became the famous *Tao-te ching*, the basic Taoist text. The carved panel depicts Lao-tzu on his journey, prominently carrying his *ju-i* scepter. In Buddhism the *ju-i* scepter represented the mystic lotus, which is generally featured on the scepter's superior end (Fig. 54).[3]

Incense seals in the *ju-i* form were produced in all of the metals commonly used for making incense timepieces. Particularly finely executed examples were equipped with elaborate bases of contrasting metal. The template for the incense trail invariably consisted of a stylized version of a seal-character, and is identical in a great number of seals that have survived. Much favored was the single character *shou*, meaning "long life," or the two characters *yen nien*, signifying "prolonged years."

The seal-characters featured on the cover and the template of incense seals given as gifts were carefully selected to reflect the occasion. Designs worked into

---

[2] Wasson, *Soma*, pp. 77–93; Imazeki, "Studies," pp. 29–51; Imazeki, "Mannentake," pp. 11–15.　　　　　　　　　　　　　[3] C. A. S. Williams, *Encyclopedia*, pp. 249–50.

the cover included many of the auspicious seal-characters, as well as pictorial features.

The *shou* character appears to be most popular among the Chinese and has been used as a decorative device from ancient times. It is formed into more than one hundred variations, some of which have become basic forms of Chinese design and ornamentation. Occasionally the seal-character was doubled, as in the ideogram *yen nien*. Another doubled character was *shuang hsi* meaning "double happiness," "double felicity," or "conjugal happiness," which were particularly suitable as gifts for a newly-wedded couple, for example.

In addition to those already mentioned, favored decorations were pictorial representations of the phoenix, double fish, leaf, butterfly, *yin* and *yang*, the eight trigrams, prunus blossoms, bamboo, five auspicious clouds, pawlonia blossoms, and inscriptions in several archaic scripts. Later examples included delicately perforated fretwork in a variety of the traditional Chinese lattice designs, often copied from examples in temples. Various forms of symmetrical ice-rays, and bamboo and prunus designs were among the favorites for bordering the incense seal covers (Figs. 55–72).[4]

One category of the scholar's incense seal, of which numerous examples survive, is made almost entirely of pewter, including the utensils. Generally these seals are square in shape and consist of three tiers in addition to the cover. The latter invariably has a perforated grid made of brass, its design usually in the form of the symbol for "longevity" or "double happiness." Often the upper rim of each tier is decorated with brass beading. Many of the pewter seals are hallmarked on the underside of the lowermost tier, or of all the tiers (Figs. 73–74).

The outer surface of the panels of each tier on all four sides of pewter seals was either left entirely blank or engraved with decorative designs and inscriptions. The motifs included bits of landscape scenes, figures of seated scholars, plants and grasses. The inscriptions consisted of wise sayings or poetic excerpts executed in different scripts (Figs. 75–76). For example, one incense seal is inscribed with separate sayings, here freely translated:

> Cleanse the mind and smell the wondrous incense;
> Official rules are impartial;
> Treasure this incense burner through the generations;
> The lasting fragrance of the incense resembles spring;
> The fragrance of the incense goes through my red sleeves as I
>   read the book at night;
> Use with care this auspicious burner.

---

[4] Dye, *Lattice Designs*.

Each tier of this particular incense seal, except for the cover, bears an inscription stamped on the underside, which reads

> The Chou State arose brilliantly. She is
> now strong enough to think of launching a war.[5]

Another pewter example in the same form includes the following inscriptions (Figs. 77–79):

> Good fortune and use it as treasure;
> Appreciation of extraordinary prose;
> Excellent scenery faces the Hall;
> Duke Chou makes vessel for King Wen, made by the Duke of Lu
>   [son of the Duke of Chou];
> Increase years and add longevity;
> Good luck and wishes come true;
> Conversations with helpful friends;
> Clear breeze enters window;
> As if the deities were above.
> [Garbled inscriptions from an archaic Chou *ting*];

The hallmark of this incense seal has been rendered as

> *Ch'ao-yang* [district]. The shop is in Swatow –
> "The Genuine Old Shop" – Genuine material tin,[6]

A similar example, in the collection of the Science Museum, London, made with two tiers, bears the hallmark "Original Inscription by Pioneer City [Shop?]" (*Chao ch'eng ming hsin*).

An exception to the usual size of pewter incense seal is an excessively large example, measuring 26.3 cm square and 29.2 cm high, and made with three tiers. Because of its size, it may have served a temple or some public purpose (Figs. 77–79).

The production, apparently in the same region in the same period of almost identical pewter incense seals but with differing hallmarks, suggests that artisans copied from one another, or that they worked from standard forms or models. Another possibility, far less likely, is that they were all produced by the same several makers or establishments and that the hallmarks were added later by vendors.

The range in quality of the material of incense seals made of paktong is noteworthy. Several thicknesses of plate were used, of which at least three have been examined. The thinnest plate appears to be found in examples in which the cover

---

[5] In the William Barclay Stephens Collection, Bernice P. Bishop Museum, Hawaii.
[6] In the collection of the author.

is executed in latticework designs of extreme delicacy. The bodies of these examples generally are engraved with appropriate decorative designs, such as leaves or bats, in simplified form. A thicker plate is to be found in seals of square, round and rectangular form without decoration, in which the cover design appears to have been cut in a bolder form. Finally, there are seals made of unusually thick plate, some of which have been cast, and others soldered. The engraving on the sides and the execution of the cover design are of the finest quality and are examples to be particularly prized.

The introduction into China of Western mechanical clocks and later of European and American-made watches between the seventeenth and late nineteenth centuries apparently had no impact on the production and use of scholars' incense seals. This was undoubtedly due to the fact that the latter were not primarily being used for timekeeping, but for inspirational use as an adjunct of the scholar's table.

Few if any Chinese incense seals made or used prior to the late Ming period are known. Some may survive in Buddhist temples, palaces, government offices or in the private collections of old families in China, but if so they are not public knowledge.

Early examples of incense seals were made of gold and silver for the court aristocracy and most of them undoubtedly have suffered the fate of many early treasures, and were subsequently sold to antiquarians by the government of the People's Republic of China. In recent years a few silver incense seals have appeared on the market in Hong Kong, presumably obtained from mainland China. Of considerable rarity, they have been described as being of the finest workmanship and decoration.[7]

Others, made of more common metals and categorized under applied arts, were not considered to be *objets d'art* and were of lesser value, suffering the fate of many personal possessions of the Chinese as a consequence of political upheavals.

Another factor affecting the scarcity of incense seals is that with the passage of time their original function was forgotten by subsequent generations, and they were used merely to burn incense or else were discarded. Examples illustrating such use are undoubtedly to be found in regions to which families migrated from China. In Hawaii, for example, one *haole* (white woman) owned a Chinese pewter incense seal made in Swatow, which probably was originally used on the family Buddhist altar. It had been brought from China and remained in the family for several generations. This woman's earliest childhood recollection of it was that it was used by her family for burning insect repellents.[8]

---

[7] Not seen by the author; reported by Mr. Joseph C. Y. Tse, Hong Kong.
[8] Communication from Dorothy Rainwater, December 3 1959.

An important chapter in the later history of Chinese incense seals was contributed by a reticent nineteenth-century Buddhist scholar named Ting Yün (c. 1800–1879), who assumed the courtesy name of "Moon Lake" (*Ting Yüeh-hu*). Evidence suggests that he was personally responsible for what may have constituted a revival in China of the incense seal in the second half of the nineteenth century. In a memorial volume published by his friends after his death an autobiographical account by Ting Yün noted that his residence was Mai Yü Wan ("Fish-monger's Bay"), on Ch'ung Ming. This is a small, lozenge-shaped island at the mouth of the Yangtze River, just north of Shanghai. Among his intimates was Shih Yün-sheng, whose courtesy name was Lan-pin, who lived at Shih Kang ("Stone Harbor"), which appears to have been in the same region. Shih Kang is at approximately 32′ 14″ N, 121′ 00″ E, almost directly north of the district capital, Nan-t'ung, and directly across the Yangtze, north of urban Shanghai. She stated that she was Ting Yün's cousin and the eldest daughter of Ting Yün's father's oldest brother.[9]

She identified Moon Lake's home as being "some twenty or thirty miles north of Lang Shan (Wolf Mountain) at a place called Yü Wan (Fish Bay)." The mountain is the site of a major Buddhist temple, and the possibility exists that Ting Yün was in some manner connected with it.

"Mai Yü Wan" has not been identified as a geographical location and may in fact have been not a place name but a name Ting Yün used to identify the premises or building in which he lived. The naming of buildings was a common practice among Chinese literati.

No biographical account of Ting Yün has appeared, and what is known about him is derived principally from a collection of dedicatory prefaces written by his friends and colleagues for publication in a memorial volume, *Hsiang lu t'u p'u* (see Appendix C).

In all probability Ting Yün was related to one of the several notable contemporary Ting families of the Ch'ing Dynasty, members of which distinguished themselves as public officials, men of property, and as literary scholars. Several contemporary members of these families were possibly relatives.

For example, Ting Jih-ch'ang (1823–1882) of Feng-shun in Kwangtung was a magistrate, army leader, and celebrated collector of rare books. While serving as a Tao-t'ai at Shanghai, he founded an academy which later became a normal school.

Another, Ting Pao-chen (1820–1886) from Niu-chang, was a public official like his father and grandfather before him. He led a private army against outlaws on several occasions, became Governor of Shantung, captured and executed the

---

[9] The details of Ting Yün's life and work which follow have been gleaned from his own writings and the dedicatory prefaces prepared by his friends for the memorial volume. See Appendix C.

eunuch An Te-hai, became Governor-general of Szechwan, and was one of the most highly respected men of this time.

Ting Ping (1832–1899) was a noted bibliophile, publisher and philanthropist and inherited an important family library. His elder brother, Ting Shen, shared his literary pursuits. Late in life Ting Ping built three two-storey buildings in Hang-chow to house his library. A younger brother, Ting Wu, was a writer. A son named Ting Li-chung and a nephew named Ting Li-ch'eng were also bibliophiles.

Another potential relative was Ting Yen (1794–1875), a scholar of Shan-yang in Kiangsu, who emerged as a military leader and important public official. He was the director of several academies and the author of more than fifty scholarly titles. His two sons, Ting Shou-ch'ang and Ting Shou-ch'i, also were public officials and writers.[10]

Ting Yün provided no clues concerning his ancestry. As a youth he traveled extensively about China observing the country's great mountains and rivers. Whether he did so on official business was not indicated. He had many friends and acquaintances among the nobility, the gentry and the learned, from whom he often received gifts during his visits to government posts south of the Yangtze River.

The recluse of Fish Bay wrote that he had a passionate love for "things ancient," and developed a consuming interest in *I ching* (The Book of Changes) and in Taoist writings. He was said to be skilled as a writer of both prose and poetic meter. Among his lesser talents was a demonstrated competence in the ancient drinking game of *t'ou hu*. He was an accomplished painter and a skilled calligrapher, and had mastered not only the *chi chiu chang* or *chang ts'ao* style, a cursive form standardized during the reign of Han Dynasty Emperor Yüan Ti who ruled from 47 to 33 B.C., but also the *mo yin* script, a seal form based on one used in the Ch'in Dynasty.[11]

Of a humble nature, Ting Yün had no political or commercial ambitions. He lived a retired life in his remote refuge, a bamboo farmhouse, devoting himself solely to the pursuit of the scholarly activities which had preoccupied him since youth. He cultivated a small garden which included banana trees. He whiled away his leisure playing the lute, composing poetry, and "sitting in meditation before burning incense" of one of his incense seals.

Over the course of the years he developed a particular preoccupation with the incense seal. He sought out and studied surviving examples, copying their designs and inscriptions. He also researched the literature of the history of incense and its uses, and of incense burners. In the spring of 1876 he set himself the task of

---

[10] Hummel, *Eminent Chinese*, vol. 2, pp. 721–28.
[11] Kwo Da-Wei, *Brushwork*, pp. 30–33; Chiang, Yee, *Calligraphy*, pp. 67, 80, 93.

recording his designs of covers and templates for the devices, based on the examples of archaic form which he had observed. He constructed examples of the seals he designed, experimenting with various materials, including earthenware pottery and metal. He forged, cast, hammered and cut bronze, copper, brass and paktong, and he may have experimented also with pewter. In his designs of incense seals he incorporated symbolism derived from the *I ching* and inscriptions copied from ancient bronzes. For the most part his products were made in portable form suitable for the home altar and the writer's desk, designed to burn for six to twelve Chinese hours, or twelve to twenty-four Western hours.

Turning from his workshop to his drawing board, Ting Yün began to render the designs of his incense seals into drawings, for the purpose of recording them in published form. By the winter of 1878 he had completed many of the drawings and had written a preface for the proposed work. Before the year ended he had these cut in wood block and printed by Ai Wu-lu, a printer who presumably lived nearby. In his colophon he indicated that work on the cutting of the wood blocks began in 1878, "the fourth Year of the Kuang-hsü period."

Ting Yün then distributed prints of these designs and copies of his preface to fifteen of his friends and colleagues and requested their comments. By the end of the seventh month of 1879 a number of them had responded with dedicatory prefaces lauding his talents.

The designs were much admired by all who saw them, and his friends encouraged him to proceed with his publication. He permitted Lan-pin to make the necessary arrangements, and there is a suggestion that he also planned to include a study of aromatics or incense in the work as well.

In the midst of this ambitious project, Ting Yün died unexpectedly. His death occurred in 1879, "in the eighth lunar month," before the completion of the proposed work. Those of his friends who had not yet prepared prefaces submitted them later, rendered as memorials, and they were added to the work by Lan-pin. The study on aromatics was never completed, if indeed it had ever been begun.

The memorial work was published in two volumes near the end of 1881 or in 1882 by the printer Ai Wu-lu. The work was probably produced in an extremely limited edition, with perhaps only enough copies for distribution to the contributors. Surviving copies of the work are extremely rare; only three are known in the United States, and there are none in the major libraries of France or England.[12]

A detailed comparison of several copies reveals minor differences, indicating that more than one edition had been produced of at least some of the pages. A

---

[12] Copies in The Library of Congress, the M. V. Starr East Asian Library, Columbia University, and in the collection of the author.

likely explanation is that after the original edition was issued, some of the wood blocks were destroyed. Later requests for copies required that new wood blocks be made for some of the pages.

The most informative of the writings in the work is the "Author's Preface," which the scholar had completed before his death. In it Ting Yün described the pattern, sculpture and style of the covers and templates of the incense seals, which he called *hsiang lu* (incense burners). He noted that the designs could include birds, insects, dragons, snakes, corals, plants and flowers. The shapes, which could be square or round, could contain the meaning of Nature, symbolize harmony, utilizing the great skills of those making them.

Since the form of the incense smoke varied continuously, Ting Yün noted that it was difficult to understand the meaning of the shapes it formed. The smoke seemed empty, yet without color; it seemed close yet far away; it was not a plant yet seemed to have leaves. The emptiness of the smoke inspired one to think about life and its complications, but life did not vanish, inasmuch as the lamp in front of the Buddha was never extinguished. Hence the hearts and minds of men should learn to be open, to overlook physical possessions, in order to achieve peaceful lives.

The incense seals had the carved patterns of many animals in order to derive happiness from the incense, he wrote, and to perpetuate the history of men, and for this reason Ting Yün carved these words to reveal its meaning. He also noted that incense seals were used in earlier times to tell fortunes, and accordingly he designed seals which would reveal the *Yang* and *Yin* of the *I ching* (such as in Figs. 64–65). He stated:[13]

> Patterning metals for [articulation] of smoke,
> Grinding incense into powder,
> Making a pattern like that of seal ink or a drawing in sand,
> But without using any tools, such as awls or reeds.[14]

> But rather like the crow's [?]
> What pigeon sitting on a wall could be more stupid than I?
> My heart is on course, but my head is full of weeds
> And the movement of my hand is thorny and difficult.[15]

---

[13] In the following translation and the translation of the dedicatory prefaces in Appendix C, translated by Kirby R. Vining, question marks within brackets have been added in instances when the translation is uncertain, and passages that have defied translation are denoted unintelligible.

[14] The last line relates to the end product in which the incense is patterned by means of the device, not by hand.

[15] The last line suggests that Ting Yün may have been afflicted with arthritis.

Seal script copies are done like this,
The words rising up in bas-relief,
Cast in silver [or other precious metals],
The mold is drawn in steel.[16]

Like bookworms eating paths through books,
Like birds drawing near [?]
Like ants passing through pearls,
Like scorpions spinning, running.[17]

Along the perimeter, in a spiral, in a circle,
Burning evenly,
Burning smoothly,
Making the paths closer still,
Bending in and out, finally unfolding.

Moving in accord with heaven's design,
As duplicated by human artifice,
To reveal the heart's flower
As if from within a cloud.[18]

Butterflies appear as if in a dream,[19]
Twisting and reeling about like dragons,
Like birds, like the phoenix,
Like worms in spring, like snakes in the fall.

Twisted and tangled,
Like coral, like jade [locust?] trees,
Like branches of trees[?]
Like a jade tablet, like a string of pearls.

---

[16] The form of calligraphy which most interested Ting Yün was seal script, and it was incorporated on many of the seals illustrated in his work. The phrase "cast in silver" relates to the form in which early examples of seals were produced, few surviving examples of which are known.

[17] The lines are descriptive of the patterns themselves.

[18] The foregoing is quite obscure, vaguely referring to Taoist cosmology, but includes the first reference to images visible in clouds.

[19] In a famous story, the ancient sage Chuang-tzu alludes to the Taoist dichotomy of reality – "Is what I dream real and that which I see waking false, or vice versa?" – in which he makes use of the butterfly imagery, which may be the reference in this verse, as well as to the images Ting Yün sees in the clouds of smoke.

Joined auspiciously like the sun is linked with the moon,
In the end all is perfumed,
Only the intestines are restless and hot,
Cyclically spiraling back and forth.[20]

Steadying the heart is most important
Passing through quickly, but not to subdue,
To understand the path of the pattern,
[Is to] understand that emptiness is the void.[21]

The head is the path [or Tao],
Unmoving, yet not static,
Fine as [three words illegible],
As surely as a banana tree has leaves.

First displaying the shoots, then unfurling its leaves,
The different layers appear, but are not distinct,
[Unintelligible] the plough comes [unintelligible],
[Unintelligible line].

I have collected examples of the hundred varieties,
And all varieties from every source,
The ancients drew divination symbols
In fragments and sections.

Today I combine these with the making of incense,
Two models, four images,
All based on
The T'ai Chi.[22]

---

[20] The "end" to which Ting Yün refers may have been the end of life – his own. The intestines are the seat of consciousness in certain traditional Chinese belief systems, and also an excellent metaphor for the spiral pattern of the incense form. The use of the word "hot," if referring to one's own intestines, is a double reference to the *yang* and *yin* of Chinese medicine – hot is not necessarily fever, merely an extreme of condition, and the consequent confused emotions experienced by one with such an intestinal imbalance – anxiety, physical and emotional.

[21] "Emptiness is the void, the void is emptiness" is a parable at the very core of Buddhist thought. Emptiness is the Buddhist reference to life on earth, its pleasures and sorrows, which are one and the same in Buddhist philosophical teachings.

[22] The content is confusing. The *T'ai chi* or "Supreme Ultimate" is possibly the Chinese equivalent of "the music of the spheres," the celestial blueprints of all existence, the understanding of which is one of the goals of Taoists.

Unraveling the obscure until circumstances are clear,
Without help from another,
The Tao is understood
And fathomed.

Pass through jade to the soft bamboo-like core,
Tongues of green lotus lapping everywhere,
Without hindrance or obstructions,
The stillness of Ch'an.[23]

Unlike a lamp lit before an image of Buddha,
This lamp [understanding or enlightenment] does not extinguish.
Like a colt caught in a crevice,
Ceaselessly struggling to be free.

Spouting fog, shooting out clouds,
Not slowly, not quickly,
Remembering when I first cut the images,
It seems forever twilight.

A companion to me as I write,
Arranged on several tables,
Or as I chant or sing,
As dignified as it is warm.[24]

My heart draws deep of this profound fragrance,
Emitted from this wondrous man-made mechanism,
Enlightening me, enabling me to converse with the spirits,
Sending my own soul soaring aloft.

Reciting the ancient truths [chanting the sutras?]
Ten thousand concerns are cleansed away,
Suddenly, I am beyond this world,
Beyond things – free.

Shih, my friend Lan-pin [Orchid Guest]
Has troubled herself over three matters for me, [for which I
     would like to thank her],
She assembled and arranged this book of images
And took it to the printers for engraving.[25]

---

[23] *Ch'an* is the Chinese word for Zen (Buddhism), and parent of the Japanese Zen.
[24] The references seem to be to Ting Yün's own use of incense seals, and their convenience as hand warmers.
[25] Lan-pin, a female cousin and close associate of Ting Yün.

I told her that I thought it was not worthy,
I had a sort of a mother's anxiety about it,
Her masterful rhetoric won me over, though,
I still wonder whom my worthless writings would interest.[26]

What it the need for this rubbish,
Will it feed the hungry?
Shih would have none of it,
And urged me to heed her words.

How many years is a life?
As ephemeral as a goose's footprint in snow,
Or as timeless as a stone stele,
Or a commemorative tablet.

Today [unintelligible] two [unintelligible]
[unintelligible] different [unintelligible]
     ancient times,
Refined and excellent,
His will is indeed equally strong.

I hear this kind of talk,
I feel humiliated;
With less shame than a flock of pigeons,
It is done and Mr. Hu will engrave it immediately.[27]

> Winter, 1878.
> [Signed and Sealed] Ting Yün
> [personal name] Yüeh-hu [at] Yü Wan.

Fourteen of the one hundred designs for incense seals produced by Ting Yün and included in the *Hsiang lu t'u shih* incorporated the signature *Yüeh-hu* (Moon Lake) cleverly combined into the archaic form characters. Of these, seven also include a date. On several the signature is elaborated, such as "Moon Lake, imitating the ancients," or "Imitated by Moon Lake." A few others included the inscription "Made by Moon Lake." The inscribed dates range from merely the year, which was either 1878 or 1879, to include also the season, such as Spring,

---

[26] The poetical phrase, "masterful rhetoric," means literally the carving of dragons and is one of the highest compliments made to a person's skill in writing. "Worthless writings" literally means "paper useful only for covering condiment jars," and is extremely self-deprecatory.

[27] The foregoing may be intended as further compliments to Lan-pin. In the last line "Mr. Hu" may be "Mr. Crane," a metaphor for eternity, inasmuch as printing would make this writing immortal. The meaning of the phrase, "I hear this kind of talk," is not established.

Early Summer, Autumn and Winter of the years 1878 and 1879; the latter were undoubtedly among the last ones he designed.

Most of the surviving incense seals bearing Ting Yün's signature are cast in bronze or copper, and others are made of paktong. No examples made of pewter or brass are known. In addition to those which he may have personally constructed, others were made from Ting Yün's designs by one or more metalworking craftsmen, possibly under his direction. Presumably they were made for commercial consumption in the Shanghai region, which, as a major center of Chinese commerce, was particularly wealthy and provided a market for the product.

Confirmation of the use of other craftsmen by Ting Yün is to be found on at least one example, made in the form of a *ju-i* scepter. Cast on the underside of the base is an elaborate hallmark which is particularly revealing. It identified "Moon Lake" as the designer and a craftsman named Li Hsüeh-lu of Shih Kang as the maker:

> *Yüeh-hu fang-ku*
> *Shih-kang, Li Hsüeh-lu tsao*
> "Moon Lake, copying the antique"
> Made by Li Hsüeh-lu of/in Shih-kang."[28]

Although the standard Chinese encyclopedia of place names lists two places named Shih Kang (Stone Harbor), in the provinces of Kiangsu and Kiangsi, it was undoubtedly in the province of Kiangsu that Li Hsüeh-lu lived and worked, probably in the environs of Shanghai. The designs of the cover and template of this seal form part of Ting Yün's memorial volume (Figs. 80–83).

Ting Yün apparently has a particular fondness for the *ju-i* scepter form, for several variants were included among his designs. One version of the *ju-i* scepter bearing his "Moon Lake" signature in the archaic inscriptions on the cover, exists in a number of exact copies in private collections. It was made in several sizes, either entirely of copper or of paktong with a copper cover (Figs. 84–85).

The inscription of the cover is to be read counterclockwise around the head of the scepter and then along the stem:

Reading a chapter of *Changes* [*I ching*], played a tune on the lute, and sitting [in meditation?] for a while, cleanses the heart of envy. Made by Moon Lake.

Of all the known surviving incense seals in private and public collections that have been surveyed, a substantial number proved to have been made from Ting Yün's designs. Many of these appear to have been brought out of mainland China by dealers in antiquities in Hong Kong and elsewhere during the past several decades. The number of these suggest that Ting Yün not only produced them for

---

[28] In the author's collection, obtained in Hong Kong.

his friends, as he related, but that they were commercially marketed as well, either by the scholar himself or by craftsmen who produced them for him. It is also conceivable although hardly likely that others not associated with Ting Yün produced incense seals based on his designs after his death, in the last several decades of the nineteenth century.

A study of known examples suggests that Ting Yün did not begin making incense seals in 1876, as implied in the dedicatory prefaces. It seems quite certain that in fact he had been producing them for some years previously, and that it was in 1876, with the encouragement of his friends, that be began to record the designs he had been using based on existing earlier examples.

As Ting Yün admitted in his preface, many if not all of his designs for incense seals were copied in part or in whole from surviving earlier examples. The phrase "imitating the ancient" which appears on several of Ting Yün's incense seals, could refer to the form of the incense seals, the scripts, calligraphy, or all three aspects. From an examination of several surviving incense seals made in earlier periods and which were illustrated among Ting Yün's designs, compared with examples made either by Ting Yün or from his designs, it is apparent that in such instances his designs were copied from earlier forms, literally without change or additions except in choice of materials and size. Consequently the wood block prints of his designs constitute a veritable catalogue – an invaluable and possibly unique record – of the forms and designs of a wide range of the Chinese incense seals produced during the past several centuries prior to Ting Yün's time (Figs. 84–88).

Particularly interesting is a comparison of three versions of an incense seal made in the shape of the melon, a magical symbol which assumes special meaning in Chinese decoration. It is the symbol of Li T'ieh-kuai, one of the legendary Taoist Eight Immortals (*pa hsien*) who devoted himself to the practice of magic and necromancy (Figs. 89–90).

Of the three almost identical examples studied – and there may be others – the first considered is tentatively dated to the eighteenth century, not made by Ting Yün. It bears a hallmark cast into the base in reverse writing, which may be interpreted as *Tsung cheng chai tsao* (Made by the Studio of Ancestral Rectitude), or as the name of an individual named Tsung Cheng-chai who made it.[29]

The exterior parts of the incense seal are of cast bronze or copper with the bases of the compartments made of brass. It is shaped in the form of a melon, and the template for the trail pattern incorporates the character *mien* meaning "melon," doubled in Siamese fashion, in reverse writing. Unlike the majority of incense seals, the cover is not made of latticework or having a geometric pattern for the

---

[29] Examples in the collection of the writer and of Lydia B. Voorhees of Maitland, Florida.

emission of smoke, but is pierced with the figures of butterflies, symbolizing summer and representing conjugal felicity, in addition to melons and the characters of an inscription. The latter consists of the words "The melon vine is long and extended" (*kua tieh mien mien*) with the last two words in reverse writing (Figs. 91–93).[30]

The phrase is an echo of the first line of a poem in the classical anthology, *Shih ching* (The Book of Songs). The poem is an ancestral hymn of the Chou Dynasty, which contains the line *Mien mien kua tieh*. The traditional interpretation is that just as the melon close to the base of the vine is invariably quite small, the one at the end of the vine is generally very large; in the same manner the founders of the Chou Dynasty, although having very obscure origins, eventually produced a great country.

Another interpretation of the line rendered by Arthur Waley, contrasts substantially with the one traditionally accepted in China. He has interpreted it as a reflection of the widespread belief prevalent among many Asian and African peoples that mankind has its origin in melon seeds.[31]

The *Shih ching*, a collection of 311 poems, most of which were produced during the Middle Chou period, was among the most important of Chinese literary classics. The contents include sacrificial songs which originally were probably sung to music during ritual dances, and poems dealing with political and legendary topics. For the most part they are purely lyrical, folk songs relating to contemporary life.

A second melon incense seal, also made of copper, is almost identical but of nineteenth-century vintage. It is slightly smaller and appears to have been made by Ting Yün or copied from his design[32] (Figs. 94–95).

The third example is made of paktong, identical in design to the second and of the same period. Slightly smaller in size than the others, it has perforated decorative feet, and does not bear a hallmark (Fig. 96).[33]

Another incense seal incorporating the "Moon Lake" signature in the inscription in archaic characters on the cover is made of copper in the form of a lute. It is included among Ting Yün's designs. An earlier example of exactly the same design is not presently known, but one that may be contemporary, made of paktong, varies substantially in the design of the cover (Figs. 97–99).

A particularly attractive design of which examples in thin paktong were produced, is in the form of a leaf. This particular leaf undoubtedly had some traditional

[30] Translated by Dr. Gari K. Ledyard, 1960.
[31] Karlgren, *Book of Documents*, pp. 190–91; Waley, *Book of Songs*, pp. 246–49; Pound, *Classic Anthology*, pp. 151–52.
[32] In the collection of Lydia B. Voorhees, Maitland, Florida.
[33] In the collection of James M. Kallison, San Antonio, Texas.

significance, but the tree or plant of which it forms a part has not been positively identified (Figs. 100–1).[34]

A number of other incense seals of bronze, copper and paktong have been identified with designs in Ting Yün's collection. In some instances it can be determined that they are the work of Ting Yün or others who may have made the incense seals for him. Yet other examples, the designs of which were recorded by Ting Yün, appear to be by makers of other periods (Figs. 102–6).

[34] Dr. Stephen A. Spongberg, Research Taxonomist and Editor of the Arnold Arboretum of Harvard University suggests that the leaf represents a member of the *Aroid* family or *Araceae*, or less likely, of a species of the genus *Sterculia*, perhaps *S. nobilis*, sometimes used in southern China for wrapping rice.

9

In Chinese and foreign writings

The prevalent use of incense seals in public and private life as a means of time measurement resulted in their mention from time to time in the works of Chinese poets and scholars, particularly during the T'ang Dynasty, when literature and exotic tastes in particular reached their highest form.

Such frequent mention in poems of the mid and late T'ang Dynasty suggests that a strong vogue for the device existed during that period. Many of these references are contained in the *P'ei-wen yün-fu*, a comprehensive dictionary, compiled in the seventeenth century at the request of Emperor K'ang-hsi (1654–1722), of two-character (although sometimes three- and four-character) poetic phrases. The main entry provides seven poetic lines or couplets all including the phrase "incense seal" with a variation in the addenda.

A phrase that occurs much more frequently in Chinese poetry than even "incense seal" (*hsiang yin*) is "incense seal-character" (*hsiang chuan*), or "seal calligraphy in incense." Two meanings for the phrase are given in Morohashi's *Dai Kanwa jiten*, the authoritative dictionary of classical Chinese phrases explicated in Japanese. The first is a synonym for incense seal (*hsiang yin*). The second describes how the smoke wafted from an ordinary incense stick appears to be writing seal-script calligraphic characters in the air as it rises, and it is the latter meaning that is generally intended in poetic usage.[1]

In the *Chung-wen ta t'zu-tien*, two other definitions are given for *hsiang chuan*. The first is "To form incense into shapes of *chuan*-style characters [seal script], and light it to measure time." The other is "When the smoke from lighted incense coils and twists about to form *chuan*-style character shapes."[2]

Among the earliest to mention the incense seal in his work was the famous T'ang poet, Li Po (701–762), a writer familiar even to those not particularly informed about Chinese literature. It is said that he wrote poetry from the age of ten, loved adventure and became a wanderer. He was brought to favorable imperial attention by his writings, but after joining the court began a life of wild

---

[1] Morohashi, *Dai Kanwa jiten*, courtesy of Professor Jonathan Chaves.
[2] *Chung wen ta tz'u tien*, quoted in Crump, p. 197.

dissipation. Having made enemies of influential courtiers, he was forced to leave the imperial court, and joined the service of Prince Lin of Yung (b. 678).[3]

Again misfortune followed him, and after almost losing his head, he escaped to seek refuge with a relative. Legend has it that while making his way by boat, he leaned too far over the vessel's edge in a drunken effort to embrace the moon, and drowned.[4]

In a quatrain he described a lonely young lord awaiting the dawn in his room while burning an incense seal of aloes wood:

> Curling, swirling – the smoke of "Sinks-in-water,"
> A crow cries out – the spectacle of a worn night,
> A winding pond – the ripples among the lotuses,
> The waist-girding white jades are cold.[5.]

"Sinks in water" (*ch'en hsiang*) is incense made from the heart and joints of aloes wood (*Aquilaria*) which sink when placed in water.

Wang Chien (768–833), a soldier, poet, and official of the court who later became a recluse, described his own lonely vigil while in seclusion.[6]

He watched the incense seal as it burned to its end while he awaited the coming of the light of dawn, which enabled him to read the letters on a stone tablet in his garden:

> The Incense seal
> I sit at peace – burning an incense seal,
> Which fills the room with scent of pine and cedar.
> When all the burning stops, a clear image is seen,
> Of the green moss upon the epigraph's carved words.[7]

One of China's greatest poets, and a high-ranking T'ang official, Po Chü-i (772–846), writing in the same period commented how

> The incense seal's morning smoke thins out;
> The gauze lantern's evening flame burns bright.[8]

---

[3] Giles, *Dictionary*, p. 487.

[4] *Ibid.*, pp. 455–56.

[5] Schafer, *Golden Peaches*, p. 164.

[6] A native of Ying-chou in Anhwei, he served as governor of Shen-chou from 827 to 835. His official career ended abruptly when he voiced criticism of an imperial relative or supporter. Giles, *Dictionary*, p. 814.

[7] From *Ch'üan T'ang shih*, p. 3421. The last line appears to be a metaphor for the appearance of the incense ash. Also in *Ch'in ting t'u shu chi ch'eng, Hsiang p'u hui k'ao* (Section on Fragrances), Booklet No. 556, fol. 37. Also quoted in Schafer, *Golden Peaches*, p. 160.

[8] From the *P'ei-wen yün-fu*. Po Chü-i was a child prodigy, who graduated as *chin shih* and entered into a public career at seventeen. He became a member of the Han-lin Academy but was suddenly banished to become Magistrate at Chiang-chou. He abandoned his public

and Liu Yü-hsi (772–842 A.D.) mentioned

> Four characters – incense writing seal.[9]

Among the earliest references to the incense seal is a line by the T'ang poet and official Yüan Chen (779–831) of Ho-nan Fu. A brilliant individual, he wrote poetry at nine years of age and had already attained an appointment to an official position at fifteen. He served in various capacities from Supervising Censor to secretary in the Imperial Banqueting Court.[10] He wrote:

> In the incense seal, white ashes fade.[11]

Chou Ho (fl. c 821) had a practical comment:

> The incense seal – rain dampens its ash,[12]

while his contemporary, Hsü Hun, wrote:

> The classics in their cases, moistened by dew, text now turning
> dark;
> The incense seal, blown by the wind, words half swept away.[13]

Somewhat later, the poet Lu Kuei-meng (d. c. 881), called "the wanderer among rivers and lakes" because he delighted in roaming alone in a small boat accompanied only by his books, fishing tackle and tea service, wrote contemplatively:

> With clear wine in cup beneath the trees I watch the incense seal;
> in window framing distant peaks I hang my alms-bowl bag.[14]

career and built a retreat at Hsiang-shan, where he devoted himself to contemplation and poetry. This and the following seven extracts from the *P'ei wen yün fu* were kindly furnished by Professor Jonathan Chaves of the Department of Far Eastern Languages and Literature, The George Washington University.

[9] From the *P'ei wen yün fu*. Liu Yü-hsi was born in Chih-li and became a censor after graduating as *chin shih*. Because of his close association with Wang Shu-wen, he was banished to a post in Yunnan when the latter fell from power. Returning later to court, he was appointed Secretary of the Board of Rites and again he was dismissed to provincial posts. His poetry was recognized, and just prior to his death he was promoted to president of the Board of Rites. Giles, *Dictionary*, pp. 530–31.

[10] From the *P'ei wen yün fu*. Yüan Chen became Supervising Censor and later was appointed Secretary of the Imperial Banqueting Court. He developed a form of poetry known as the "Yüan Ho style," which became popular in court circles. Although having risen to the highest state offices, he died in failure and disgrace while governor of Wu-ch'ang in Hupeh. Giles, *Dictionary*, p. 964.

[11] From the *P'ei wen yün fu*.   [12] *Ibid.*   [13] Hsü Hun, from the *P'ei wen yün fu*.

[14] From the *P'ei wen yün fu*. Lu Kuei-meng abstained from food brought to the market, and also from participation in the annual festivals and ceremonies of mourning and burial. Giles, *Dictionary*, p. 546.

Cheng Ku (d. c. 896), like Po Chü-i, mentioned the

> Incense seal and gauze lantern, just like long ago![15]

Another poet of the T'ang period, Fang Kan, who also was quoted in the *P'ei wen yün fu*, as previously noted, wrote:

> The incense seal [*hsiang yin*] finished [its trail]
> when [the monk] came out of [finished] meditation.[16]

In a line from an unidentified poem, the Sung poetess Chu Shu-chen (960?–1126?), described how

> A seal-character burns its brazier [under?] my afternoon pillow,

referring to the occasional practice of burning incense or opium under or in the pillow, presumably to induce sound and more pleasant sleep. Called the "censer pillow," it was often a hollow structure made of rattan or wood, and of ceramic for the summer months. In Japan the "censer pillow" was known as the *kōro makura* and used by the *daimyō*.[17]

Kuan-hsiu (832–912), a Ch'an (Zen) monk of the T'ang period, also described the incense seal. Greatly admired in his time as much for his poetry as for his painting, he was a scholar of the classical texts of Confucianism and Taoism as well as of Buddhism. After his studies were interrupted by the collapse of the T'ang Dynasty, he traveled incessantly during the second half of the ninth century. He finally settled at Ch'eng-tu, where he was greatly honored with the title "Grand Master of the Ch'an Moon." Celebrated as a visionary artist, calligrapher and teacher, he was honored in his lifetime also as a poet of great sensitivity. His work was characterized as of "lonely and profound expression."[18]

It was in this mood that in his poem, "An idle stay by the T'ung River," he described how he took asafetida (*a wei*) with tea while burning an incense seal and

---

[15] From the *P'ei wen yün fu*. Born in I-ch'un in Kiangsi, Cheng Ku graduated as *chin shih* and later distinguished himself as a poet. He was often called "Partridge Cheng" because he said that no one should sing his "Song of the Partridge" in the presence of individuals from south China since it would render them homesick. Giles, *Dictionary*, p. 114.

[16] From the *P'ei wen yün fu*; Needham, *SCC* vol. 4, part 2, p. 147 d.

[17] *Hsiang p'u hui k'ao*. In Far Eastern countries the pillow was generally made of solid wood, and, for use in the summer, of ceramics, covered with a small bag containing tiny seeds, etc. The wooden pillow often contained a drawer or opening in which amulets were kept. In Japan during the Tokugawa period, the *daimyō*'s pillow was often a lacquered box having a slightly curved upper surface perforated with ornamental designs. Through these was emitted smoke of burning incense stored within, to ensure redolent slumber for the *daimyō*. Casal, "Incense," p. 58.

[18] Wu Chi-yu, "Le Sejour," vol. 2, 1969, pp. 158–78.

reading *sutra*. Chih-tun, to whom he referred, was a hermit monk of the fourth century, and a great admirer of horses:

> In the quiet room I burn a sandal-seal;
> In the deep brazier I heat an iron flask,
> The tea, blended with *a wei* is warming,
> The fire, sown with cypress roots, is fragrant.
> Some few single cranes have come flying,
> A good heap of *sutra*s is read through;
> What hinders me from stealing away like Chih-tun –
> And riding a horse up into the blue darkness?[19]

This poem has marked similarity to another by Ch'a Te-ch'ing, a poet of the Yüan Dynasty:

> *Ch'un-k'un*
> [Spring Langour]
>
> My shutters fast,
> The busy voices cease –
> The first of twilight time is past.
> The incense burner's warm to touch
> Though its ash no longer glows.
> Behind closed doors,
> On brocade couch
> My vision slowly goes –
> Half the time my eyelids flutter,
> Half the time I doze.[20]

Ou-yang Hsiu (1007–1072), one of the most famous Sung poets, mentioned the incense seal in an untitled poem extolling the pleasures of the ideal retired life of the scholar. In his poem the incense seal was intended as a symbol of elegant retirement from government service:[21]

---

[19] From *Ch'üan T'ang shih*, han 12, ts'e 3, ch. 5, p. 5b, translated by Schafer, *Golden Peaches*, p. 183. Chih-tun was a famous monk who lived in the Chin period, dying in approximately 367. He was known as a lover of horses and other animals. *A-wei* is *Ferula asafoetida*. The use of plant smoke to disinfect against invertebrate pests is still being practiced in southern China and Southeast Asia in the form of the popular "mosquito incense" coils. Needham, *SCC*, vol. 6, part 1, pp. 271, 473.

[20] Crump, *Songs*, p. 34.

[21] Ou-yang was born in Lu-ling in Kiangsi. After having learned to write on reeds, he achieved a reputation as a writer by the age of fifteen. Upon the completion of a commissioned history of the T'ang Dynasty in 1060, he was appointed Vice President of the Board of Rites and became a chancellor in the Han-lin Academy. He rose to the position of President of the Board of War. He produced the earliest work on ancient Chinese inscriptions, and an exposition on the Book of Odes, among other works. Giles, *Dictionary*, pp. 606–7.

Lute [*ch'in*] and books worth a thousand gold pieces,
Day by day and month by month the hundred-graduations
     incense seal
Burns away the idle hours.[22]

In the poem, "To the Air of *P'u-sa man*," Li Ch'ing-chao, the Sung poetess, wrote:

My home village
     – which way
     does it lie?
Forget?
     Only one way
     – drink
Guru incense [aloes wood, *ch'en hsiang*] lighted when I went to
    bed
Incense
        burnt out
        The wine cloud remains . . . [23]

One of the most interesting mentions of the incense seal occurs in a poem of Su Shih (1036–1101), sometimes known as Su Tung-po, a Chinese statesman and one of the most distinguished poets of the Sung Dynasty. Su Shih bore the same family name as Su Sung and was his contemporary. They had studied under the same master, the great scholar Ou-yang Hsiu, and both were affiliated with the conservative party. An acknowledged scholar from his youth, Su Shih obtained an appointment as Magistrate, but his opposition to important political figures of his time resulted in his transfer from post to post and finally brought about his dismissal and banishment. Eventually he was recalled, rising to the position of President of the Board of Rites. His political enemies again brought about his banishment, however, first to the barbarous regions of Hui-chou in Kwangtung and later to the island of Hainan.

Situated off the coast of Viet Nam in the South China Sea, Hainan long served as a place of exile for political figures and disfavored court officials, and was considered as the worst possible punishment.

His poem speaks of the exiled Su Shih's brooding as he watched the passing of the endless boring hours, measured by the incense seal on his desk, one of the few cherished objects which he was allowed to take with him into an exile from which he did not believe he would ever return. Nonetheless he was hoping for a change of heart at the capital which would allow him to see his home again.

[22] *Hsiang pu hui k'ao, Hsiang pu hsüan chü.*
[23] Kwock and McHugh, *Friends*, p. 48.

Su Shih composed the poem in about A.D. 1100. According to its title, it was "Written on the Lantern Festival, Fifteenth day of the first lunar month, in Tan-chou, Hainan Island, while my son, Kuo, attended a party given by the local prefect":

> The prefect laid out a table of wine that we might all be drawn
>   closer together,
> But I thought there'd be no harm in staying home instead,
>   keeping an eye on the house.
> I lie down to watch the moon from the window of my study and
>   noticed a small lizard on the window sill,
> Quietly waiting for the tasty mites to fall down off the curtains.
> As the flame of the candle in the *teng hua* lantern smoldered out
>   I lapsed into daydreaming –
> My son will be returning when the incense seal has burned
>   through its course.
> Thinking back over the sorrows and sadness of my last ten years
>   [here in exile on Hainan],
> I see that I've traded my court robes for these tropical oranges.[24]

Su Shih cultivated the acquaintance of the few literate individuals on the island, and it was not unusual for him to have allowed his son to attend a party given by the Prefect, who appears to have been the senior administrator on the island. The occasion was an important one, the traditional Chinese holiday of the Lantern Festival.

Su Shih's mention of the exchange of his court robes for oranges was a negative allusion to all things southern, and which were abhorred by the northern Chinese. Oranges served as a symbol of exile far from home to a place which no northern Chinese would otherwise visit.[25]

Another year or two passed after the poem was written before Su Shih received his pardon. He was restored to honor once more by Emperor Hui Tsung, and was traveling from the capital to his home when he died.[26]

Mention of the incense seal in the form of the "seal-word" (*chuan yen*) occurred occasionally also in the *san ch'ü*, the song poetry of the thirteenth and fourteenth centuries. Included in the *Ch'üan Yüan san ch'ü* (Complete Yüan Dynasty *san ch'ü*) are two poems by Ch'iao Chi (d. 1345) about the "Hsian p'an" (Incense burner), their metrical structure derived from *P'ing lan jen*. In the first short *ch'ü*, the poet compared the sparse tendrils of rising incense smoke forming the archaic curvilinear forms of Chinese seal-characters:

[24] *Complete Works of Su Tung-p'o*, vol. 37, p. 16, translated by Kirby R. Vining.
[25] Schafer, *Shore of Pearls*, pp. 100–1.
[26] Giles, *Dictionary*, pp. 680–82.

> Coiling dragon shed its chrysalis of heat;
> Bones of fragrant dust lie at its feet.
> Like billowing silks, sinuous, cloud-tipped
> Smoke has written ancient script,
> From the last of the incense ash to burn.
> There lingered warmth in my precious urn,
> While moonlight had already died
> In the garden pool outside.[27]

In a *hsiao-ling* with that same title, the same poet wrote:

> Heart's fire twisting burns
> As do The Nine Turns
> Of Viscera. Meditation, reverence
> Paid for *samadhi*'s[28] sweet incense.
> Ashes of *kalpas*[29] written time and again –
> But still does an inch of time remain.[30]

It is to be noted that the ashes are "written" in the seal-character form of the incense seal type of burner, and that the last line consists of a *double entendre* of both space and time.[31]

In the same period a secular song by Ch'iao Chi presented a vivid view of the incense-seal, character:

> *Hsiang chuan*
> [The incense seal-character]
>
> Carven brazier's ember, glowing, small –
> A firefly survived the fall?
> Silk-skeins of smoke from incense trace
> Little melancholy clouds
> Above her vanity-case
> Alas, before its ardent course is run
> This burning heart leaves ashes in its place.[32]

---

[27] *Ch'üan Yüan san ch'ü*, No. 594.5; Crump, *Songs*, p. 196.

[28] "Samadhi" is literally translated as "fixity" or defined as "perfect tranquility," "meditative abstraction" or moral self-deliverance from passion and vice. Eitel, *Hand-Book*, p. 140.

[29] *Kalpas* are long periods during which a physical universe is created and destroyed. Eitel, *Hand-Book*, p. 68.

[30] *Ch'üan Yüan san ch'ü*, No. 595.1; Crump, *Songs*, p. 199.

[31] Communication from Professor James I. Crump, January 2, 1979. He explains that the word "meditation" (*pi kuan*) in this context literally means to focus one's inner eye on the back of the bridge of one's nose to induce undisturbed meditation.

[32] *Ch'üan Yüan san ch'ü*, No. 594.1; Crump, *Songs*, p. 199.

Similar references occur in several of Ch'iao Chi's other poems. The "brazen beasts" which were mentioned were the early incense burners made in various zoomorphic forms, and the "*chuan*-character smoke" left no doubt that the burners were incense seals:

> *Chuan*-character [incense seal character] smoke and burnt
> incense
> Have cooled in the brazen beasts.[33]

In yet another quatrain the same poet again referred to incense seals and also to the drum of the night-watch and the community clepsydra:

> Precious *chuan* characters
> Burn out in the brazen beast.
> The Painted Drum of the watch
> Hastens the clepsydra on.[34]

A verse by Shang Tao (fl. c. 1311) writing in the same period confirmed that the *hsiang chuan* referred to time:

> In the painted pavilion
> The watch clepsydra is chilled;
> In the bronze burner
> The seal-character incense is burned out.[35]

Hsieh Chin (1260–1368), another poet of the Yüan Dynasty, also referred to time recorded by an incense seal:

> Smoke from an incense seal marks the passing
> Of a fragrant afternoon.[36]

Later, in the early fifteenth century, Chü Yu, an official of education and poet of the Ming Dynasty, wrote imaginatively of the

> *Hsiang-Yin* [Incense Seal]
>
> Incense powder fills the left-spiraling maze which was so
> skillfully carved.
> Like a firefly, the red ember's glow traces the lines of the ancient
> seal-character in its tray.
> Like a smoke-belching Phoenix traveling through the linked-
> words' tracery.[37]

[33] *Ch'üan Yüan san ch'ü*, No. 450.3, quoted in Crump, *Songs*, p. 196. No. 454.2 has similar content.
[34] *Ch'üan Yüan san ch'ü*, No. 881.4, quoted in Crump, *Songs*, p. 196.
[35] *Ch'üan Yüan san ch'ü*, No. 25.1, in Crump, *Songs*, p. 196.
[36] *Hsiang pu hui k'ao, Hsiang pu hsüan chü.*
[37] *Hsiang pu hui k'ao*, Booklet No. 556, fol. 30, in *Ch'in ting t'u chi ch'eng.*

Another poem of the Ming period entitled *Hsiang yin* (Incense Seal) defies a satisfactory interpretation:

> Smoke, sinuous like a coiled dragon,
> A flower blossoms on a twig,
> Fire leads ice to the cocoon's thread [?]
> Burning up a poem in elegant calligraphy,
> It seems a time of heartbreak for a man.[38]

An old Chinese poem known as a "short stop" describes a waiting wife complaining of the leaden feet of Time:

> It seems that the Clepsydra has been filled up with the Sea
> To make the long, long nights appear an endless time to me.
>
> The incense stick is burned to ash the water clock is stilled
> The midnight breeze blows sharply by and all around is chilled.[39]

That the incense seal continued to be a familiar household item in China in the eighteenth and nineteenth centuries is attested to by the many references made to it in the literature, not only in poetry, but occasionally in prose writings as well. It was featured, for example, in *Shih t'ou chi* ("The Story of the Stone," also known as *Hung lou meng* "The Dream of the Red Chamber") by Ts'ao Hsüeh-ch'in. Although the novel was written in about 1760, it did not appear in published form until 1792. In the intervening years manuscript copies were circulated in various forms and with various titles, the best known of which was "The Dream of the Red Chamber." Often described as the most popular book in all of Chinese literature, this work is considered to be China's great novel of manners.

The author, Ts'ao Hsüeh-ch'in, was born in Nanking into a family of great wealth and prestige, members of which had held the office of Commissioner of Imperial Textiles for three generations. The importance of the family can be judged by the fact that it had entertained the Emperor K'ang-hsi on four occasions.

Eventually the family fell upon unfortunate times. Their properties were confiscated and they were reduced to extreme poverty. Ts'ao Hsüeuh-ch'in was living without adequate means near Peking when he wrote the novel. In a narrative purported to parallel closely the rise and fall of his own family, he described how an eminent family fell into decline, reduced from the highest position in the

---

[38] *Ch'üan Yüan san ch'ü*, No. 1046.2. Communication from Professor Crump, August 10, 1987.
[39] Earle, *Sun-Dials*, p. 54.

Manchu Dynasty to a state of abject poverty in which the author lived and died. The work is notable for the account of considerable detail and poetic elements that were present in the daily Chinese life of the period, interspersed with a strong element of humor.

In the novel, the children of the household are described during a riddle party. Each of the children wrote out a poem incorporating the clues to the riddle, and these were hung on a lantern made in the form of a three-leaved screen. Other forms of lanterns also were used. Poems of various forms were submitted, ranging from the couplet and quatrain to longer verses, and prizes called "Lantern Festival presents" were prepared for the winners.

Tai-yü, one of the principal characters, proposed a riddle:

> At court levee my smoke is in your sleeve:
> Music and beds to other sorts I leave.
> With me, at dawn you need no watchman's cry,
> At night no maid to bring a fresh supply.
> My head burns through the night and through the day,
> And year by year my heart consumes away.
> The precious moments I would have you spare:
> But come fair, foul or fine, I do not care.
>   A useful object.[40]

The use of the romantic form in the poem is a strategy employed by one of the enamored hero's female friends to gain more of his attention. The correct solution to the riddle is "incense seal" or "incense clock."

The Chinese incense seal is featured also in modern fiction, in two short stories by the late Dutch diplomat-writer Robert H. Van Gulik (1910–1967) about the adventures of the internationally famous Judge Dee. Van Gulik, a distinguished connoisseur of classical Chinese art, music and literature, borrowed the figure of a T'ang Dynasty judicial official Ti Jen-chieh (630–700). Ti Jen-chieh later rose to prominence as minister of state. The Judge's successes in crime-solving were described in a sixteenth-century Chinese novel entitled *Ti kung an* (Three Murder Cases Solved by Judge Dee), translated into English by Van Gulik in 1949 and first published in Tokyo.[41]

During the next two years, Van Gulik conceived and produced the first of his series of original stories, written in Chinese-style, about Judge Dee and his four lieutenants, entitled *The Chinese Bell Murders*. Subsequently he wrote a total of twenty-two Chinese-style novels and short stories about Judge Dee, each of which immediately became popular in a number of languages.

---

[40] Ts'ao Hsüch-ch'in, *Story of the Stone*, vol. 1, *The Golden Days*, p. 449.
[41] Roggendorf, "In memoriam," pp. vi–vii; Van Gulik, *Remarks*, pp. 1–7.

In a short story entitled "Five Auspicious Clouds," contained in the volume *Judge Dee At Work*, Van Gulik featured an incense seal as the clue to the identity of a murderer.[42] In the story, the incense seal was in the form of the "five auspicious clouds," made of brass [sic], and measured approximately 30 cm in diameter. It was the interrupted burning of the seal's incense trail that provided a clue to the solution of the murder. The murderer had spilled tea from a teapot upon the last segment of the incense trail to suggest that the crime had been committed at that time, although in fact it had occurred several hours earlier. At the trial, Judge Dee explained to the coroner,

The pattern excised in the cover is that of the Five Auspicious Clouds, each cloud being represented by one spiral. If one lights the incense at the beginning of the design, it'll slowly burn on along the spirals of the pattern as if it were a fuse. Look, the tea spilling from the spout of the teapot moistened the center of the third spiral, extinguishing the incense about halfway through the part of the design. If we could find out when exactly the incense-clock was lit, and how long it took the fire to reach the center of the third spiral, we would be able to establish the approximate time of the suicide.[43]

In a colophon, Van Gulik explained that he had copied the design of the incense clock from a collection of designs published in 1878 [sic] entitled *Hsiang yin t'u k'ao*. This was, in fact, the collection of Ting Yün's designs, *Hsiang lu t'u p'u*, although Van Gulik did not identify the author.[44]

"Five Auspicious Clouds" is a reference to the Chinese belief that clouds can appear in as many as five colors. The traditional Chinese view is that there are five primary colors instead of three. These primary colors are green/blue, white, red, black and yellow. Each of the cloud colors has a meaning used in fortune telling and astrology, as follows:

Green/Blue	. . . insects, worms, and other pests
White	. . . mourning
Red	. . . military disaster
Black	. . . water
Yellow	. . . abundant harvest

It was the practice of Chinese families on the morning of a child's birth to observe the colors of the clouds and make a prediction concerning the child's future. This appears to have been more common in farming communities than among urban dwellers.

The colors of the five clouds encompassed all possible fates, and a literary reference to the five clouds may have been intended as a reminder that men's lives

---

[42] Van Gulik, *Judge Dee At Work*, pp. 1–19.
[43] *Ibid.*, pp. 6, 174.   [44] *Ibid.*, p. 174.

are controlled by greater forces. In his "New songs of Empress Wu's reign" the poet Sung Chih-wen wrote

> While among the stones there still remain
> Immortal coaches of five-colored clouds.[45]

Certain highly placed state officials were said to communicate with the palace on a type of paper which invariably had swirls of colors stained into it, denoting the writer's status. Various anecdotes relating to imperial correspondence concerning an individual's influence with the reigning power used the term "Five Clouds."

In his work, Ting Yün included the design of an incense seal in the form of the "Five Auspicious Clouds," with two templates. An example of this form, probably of nineteenth-century vintage but apparently not the work of Ting Yün, has come to notice (Figs. 107–9).

In a short novel also featuring Judge Dee entitled *The Chinese Maze Murders*, Van Gulik again utilized the pattern of an incense seal. It formed the design of an outdoor maze in the garden of Governor Yoo, a feature which played a role in the solution of the murders. This pattern also was borrowed by Van Gulik from Ting Yün's work, in which it was labeled "The Tower of the Void" or "Emptiness"[46] (Fig. 110).

European missionaries and emissaries who visited China in the sixteenth and seventeenth centuries occasionally confirmed in their writings observations of the use of incense for time measurement. The first of these occurred in the memoirs of Father Matteo Ricci, S. J. (1552–1610), founder of the Jesuit mission in China. In a journal entry written probably after 1601, while he was still residing in Peking, Ricci wrote

> As for their [the Chinese] clocks, there are some which use water, and others [which use] the fire of certain perfumed fibres made all of the same size; besides this they make others with wheels which are moved by sand – but all of them are very imperfect ... [47]

Several transcriptions and translations were made of Ricci's journals by later writers, each amplifying the original account and rendering it somewhat differently. The earliest was included by the Reverend Nicolas Trigault, S. J. (1577–1628), in his account of the Jesuit missions in China, which was published in 1615. Trigault's reference to the timekeepers mentioned by Ricci was as follows:

> *Horis metiendis vix habent instrumenta, quae habent, vel acqua vel igne mensurantur. Aquea sunt velut ingentes clepsydrae; Ignea ex odorifero cinere confecta, tormentorum nostrorum fomites imitantur ...* [48]

---

[45] Owen, *Early T'ang*, p. 305.
[47] D'Elia, *Fonte*, vol. I, p. 33.
[46] Van Gulik, *Chinese Maze Murders*.
[48] Trigault, *De Christiana Expeditione*, p. 22.

Trigault's Latin passage has been translated in several published editions, each with a somewhat different interpretation. The Reverend Louis J. Gallagher, S. J. rendered it as follows:

This land possesses few instruments for measuring time and in those instruments which they have, it is measured either by water or by fire. The instruments run by water are fashioned like huge waterpots. In those which are operated by fire, time is measured by an odoriferous ash, somewhat in imitation of our reversible grates through which ashes are filtered. A few instruments are made with wheels and are operated by a kind of bucket-wheel in which sand is employed instead of water, but all of which fall short of the perfection of our instruments, are subject to many errors, and are inaccurate in the measurement of time . . .[49]

In a more recent work, another version of the same passage is proposed:

They [the Chinese] have very few instruments for measuring time, and those which they do have measure it either by water or by fire. Those which use water are like big clepsydrae, and those which use fire are made of a certain odoriferous ash very like that tinder which is used for torture among us . . .[50]

The second version is based on the premise that Trigault referred to *moxa* or medical cauterization, inasmuch as the practice of using incense ash for the purpose was characteristically Chinese.

Correspondence with the Reverend Gallagher requesting clarification yielded more definitive results. He proposed that the difficulty with the passage derived from the words *tormentorum* (*tormentum*) and *fomites* (*fomes-fomites*). The basic meaning of the Latin word *tormentum* is any machine or device for twisting or turning an object. Inasmuch as the common means of turning or filtering ashes was a grate, he applied this word in his published translation.

Consultation with the Reverend Joseph Sebes, S. J., Professor of Chinese History in the graduate school of Georgetown University, resulted in a translation even more precise than the foregoing. Sebes interpreted the word *fomites* in Trigault's text to mean "tinder, touchwood used for the purpose of igniting gunpowder." He proposed that the words *tormentorum nostrorum* in the same passage were intended to mean "our guns," and rendered the passage as follows:

In those [instruments] which are operated by fire, time is measured by an odoriferous ash, like unto the tinder-sticks or touchwood of our guns.

As his authority for this translation, the Reverend Sebes mentioned that Fr. Johann Adam Schall von Bell, S. J. (1591–1666), a Jesuit missionary in China in

---

[49] Needham *et al.*, *Heavenly Clockwork*, p. 155, n. 6.

[50] *Ibid.*, Needham, *SCC*, vol. 3, p. 330; vol. 4, part 2, pp. 509, 57; vol. 4, part 3, pp. 509–10.

1665, used the term *tormenta bellica* to mean "machines of war" or "guns."[51]

Although the passage in Ricci's memoirs may have referred to the use of graduated incense sticks marked with time segments, it seems likely that the missionary was in fact referring to the use of the incense seal. Trigault's version leaves no doubt that an incense seal was intended and that the reference to "reversible grates" identified the template used to form the incense trail.

Not to be overlooked is a passing reference to the several primitive forms of Chinese time measurement in the report of John Nieuhoff (1618–1672), of the embassy of Peter de Goyer and Jacob de Keyser. The embassy was sent in 1665 by the East India Company of the United Provinces of the Netherlands to the Grand Tartar Cham, Emperor of China.

The Chinese, although most ingenious and subtle, have not near [accurate] instruments to indicate time, and those which they have are as imperfect as can be made. Those which indicate the hours by means of water resemble in some ways our pounce boxes, and those which are by means of fire resemble our wicks (or fuses). There are some who attempt to make sundials, but with so little success that it is pitiful.[52]

Another traveler who observed the use of graduated incense sticks for time measurement was the Reverend Gabriel de Magalhaens, S. J. (1609–1677).[53] He described it several years later, in 1668, in his written account of his experiences in China. One of the most outstanding of the Jesuit missionaries, he arrived in China in 1640 and spent the last thirty-seven years of his life working among the Chinese. He constructed several clocks for the Chinese emperor, and he naturally expressed interest in the indigenous forms of timetelling which he observed among the people. He wrote:

The Chinese have also found out, for the regulating and measuring of the parts [Watches] of the Night, an Invention becoming the marvelous Industry of this Nation. From a certain Wood, which they grated and pounded to reduce it into a Powder and make from it a Paste, they form Cords and Sticks of various shapes. Some are made from more precious woods, such as Sandalwood, "Eagle wood" [*bois d'Aigle*] and other odorous Woods, with a length of about one finger or thereabouts, which the rich and literate personages caused to be burned in their chambers. There are other, more ordinary forms, of one, two or three cubits in length, and also of one, two or three ells, and a little more or less larger than a goose quill, which they burn in front of their Pagodas or Idols.

They use them also as a Match [wick or fuse] to bring fire from one place to another. They make cords of powdered wood of equal circumference, of a uniform thickness, by passing them through a draw-plate or a trough specially made for the purpose. Then they

[51] The translation by the Reverend Sebes was verified by the Reverend Neil Twombley, S. J., Latinist in the School of Languages and Linguistics of Georgetown University.

[52] Nieuhoff, *Embassy*, p. 159.

[53] *Nouvelle Biographie Générale*, vol. 32, cols. 662–63.

coil them into a round form by beginning at the bottom and forming a spiral or conical figure which enlarges with each turn up to one, two or three palms in diameter, and even more, and which will have a duration of one, two or three days or even more, in proportion to the size, because one sees them in the temples which last ten, twenty or thirty days. These machines or matches resemble a fisher's bow net, or a cord wound around a cone. They are suspended from the top, and they are lighted at the bottom end, from which the smoke issues slowly and faintly, following all the turns which have been given to this coil of powdered wood, on which there are ordinarily five marks to distinguish the five parts of the evening or night. This method of measuring time is so accurate and certain that no one has ever noted a considerable error. The literate, the travelers, and all those who wish to arise at a precise hour for some affair, suspend at the mark at which they wish to arise, a small weight, which, when the fire has arrived at this point, invariably falls into a basin of brass which has been placed below it, and which awakens the sleeper by the noise which it makes in falling. This invention takes the place of our alarm clocks, with the difference that they are very simple and extremely inexpensive that one of these devices [machines] that can last for a period of twenty-four hours, costs no more than three *deniers* [pennies or farthings], and that the clocks which are composed of a quantity of wheels and other pieces and are consequently very expensive that they cannot be employed except for the rich personages.[54]

In this passage Magalhaens provided one of the earliest Western descriptions not only of the incense stick, but also of the incense coil, as forms of Chinese incense timetelling devices. Both of these were in common use in China and French Indochina [Viet Nam] through the centuries and survive in modern times in great profusion. They have been noted again and again in the accounts of travelers.

In his account of the Chinese capital, the Roman Catholic Bishop of Peking, Monsignor Alphonse Favier (1837–1905), included a sketch of a Chinese interior in which were shown not only a European weight-driven wall clock, but also an incense coil placed under the table. (In this instance however, it is the more likely that it was used to dispel insects than to tell time.)[55]

The incense coil is customarily hung from a ceiling rafter, and the interior of Chinese and French Indochinese temples even at the present time frequently display great numbers of them (Fig. 9).[56]

The incense spiral was used also on the table, suspended from a special bracket over a bronze platter supported on feet. The bottom end of the spiral was lighted and presumably the platter served as the alarm when an attachment to the incense coil dropped at the designated hour.[57]

---

[54] Magalhaens, *Nouvelle Relation*, vol. 1, pp. 124, 153–54; Needham, *SCC*, vol. 3, p. 330.
[55] Favier, *Peking*, p. 364.
[56] Lewis, *Dragon*, p. 177.   [57] Planchon, *L'Horloge*, pp. 254–56.

Incense was mentioned also in a report on contemporary Chinese timekeeping methods submitted to the U.S. Patent Office in 1851 by Dr. D. J. Magowan, a medical missionary in China:

Time is often kept with tolerable accuracy in shops and temples by burning incense sticks made of sawdust, carefully, but slightly, mixed with glue, and evenly rolled into cylinders two feet long [61 cm], and divided off into hours. They consumed without flame, and burned up in half a day.[58]

Alice Morse Earle reported that as late as 1899 the guardian of the "copper-jar-dropper," the famous water clock which stood in the watch-tower at Canton, sold "time-sticks" or incense sticks marked for timetelling. She described them in virtually the same words as had Magowan half a century earlier.[59]

[58] Magowan, "On Chinese Horology," p. 340; Magowan, "Modes," p. 431.
[59] Earle, *Sun-Dials*, p. 54. The similarity between her text and Magowan's is to be noted, suggesting that his work was one of her sources.

III

THE INCENSE SEAL
IN JAPAN

IO

Japanese *kōbandokei*

The Japanese adopted the simple incense-burning methods which were utilized for time measurement in China, the graduated incense stick and time cord, and then proceeded to develop three devices utilizing incense which were peculiarly their own. Categorized by the generic term *kōbandokei* (incense-clocks), these consisted of two types of the incense seal and a specialized device incorporating the graduated incense stick.

The incense seal appears to have been introduced into Japan from China by Buddhist missionaries not long after the Chinese adapted the seal of Avalokitésvara for time measurement. This may have occurred during the latter half of the eighth century, early in the Nara period. In its primitive forms, the Japanese incense seals were crudely made in simplified versions of the more-elaborate Chinese incense seal. Those presently preserved in the Shōsōin, the Imperial Treasury at Nara, although of the same period and brought from China, were by contrast elaborately and richly made as gifts of state.[1]

The earliest account of the use of the incense seal in Japan by a foreign traveler was provided by the Reverend João Rodrigues Tçuzu, S. J. (1562–1633) in a manuscript work, *Historia da Igreja do Japão*, which was not published until modern times. In 1578, at the age of fourteen, Rodrigues left his native Portugal for Japan as a Jesuit missionary. He remained in Japan for forty-five years, and during this period he became so proficient in the Japanese language that he earned the title of Tçuzu meaning "interpreter," to distinguish him from another missionary of the later period also named João Rodrigues. His skill as a linguist brought him to the favorable attention of the Japanese military leaders and statesmen, Toyotomi Hideyoshi (1536–1598) and his successor, Tokugawa Ieyasu (1542–1616). He died at Macao in 1633.[2]

In his *Historia* he wrote:

The Japanese do not possess ordinary clocks with which to tell the time, but the [Buddhist] bonzes have very ingenious fire-clocks to measure the time both on long and

---

[1] Descriptions of the examples in the Shōsōin are provided in chapter 6.
[2] *Catholic Encyclopedia*, vol. 8, p. 109 d.

short days in order to know the hour of prayer and when to ring or sound the bells in their temples. And for this they have fixed measurements that depend on the length of the day, which, whether it be long or short, is always divided up into six hours.

They make the clock in this way. They take a square wooden box and fill it with a kind of fine sifted ash; this ash is very dry and they make the surface of it very flat. On this surface they draw a continuous line of furrows of a determined length, breadth and depth in the form of a square, and they fill these furrows with a dry scented powder, or flour, obtained from the bark of a certain tree. They set light to the end of one of the furrows and the fire continues to burn very slowly so that one of the squares is burned up every hour. They can thus measure the time very accurately for they know from experience how to regulate it so that the fire continues to burn in the same way and at the same rate. The furrow is made proportionately longer or shorter according to the length or shortness of the day and night.[3]

The noted nineteenth-century Russian vice-admiral and navigator, Captain Vasili Mikhailovich Golovnin (1776–1831), also commented on the use of incense seals in Japan. Born in the province of Ryazan, he was educated at the Cronstadt naval school. He served as a volunteer in the British Navy under Admirals Horatio Nelson and Sir William Cornwallis, and upon returning to Russia in 1806, he was given the command of the sloop *Diana*. During the next four years, he surveyed the coasts of the Russian empire and circumnavigated the globe. In 1810, while attempting to survey the coast of Kunashiri, he was captured by the Japanese and imprisoned until late 1813. He made a second circumnavigation in the corvette *Kamchatka* in 1817–1819, and published an account of his adventures several years later.[4] In his account of captivity in Japan between 1811 and 1813, Golovnin observed and described a primitive form of the incense seal:

To measure time they employ a small beam [block] of wood, the upper part of which is covered with glue, and whitewashed; a narrow groove is made in the glue and filled with a vegetable powder, which burns very slowly; on each side of this groove at certain distances, there are holes formed for the purpose of nails being put into them. By these holes, the length of the day and night hours is determined for the space of six months, from the spring to the winter equinox. During the other six months, the rule is inverted, the day becoming night hours, and the night day hours. The Japanese ascertain the length of a daylight hour, and mark it off with nails; they then fill the groove with powder, set light to it at noon, and thus measure their time. The beam is kept in a box, which is laid in a dry place; but the changes of weather have, notwithstanding, a great influence on this kind of timekeeper.[5]

---

[3] Rodrigues Tçuzu, *Historia*; reprinted in *Daily Life and Customs*, pp. 231–32.

[4] *Nouvelle Biographie Générale*, vol. 21, cols. 129–32.

[5] Eyries, *Galownin*, p. 68. The account of Golovnin's voyage and captivity was first published in 1824 and translated into French, English and German.

The "nails" mentioned by Golovnin were time markers made of wood, bamboo or metal inscribed with the characters for the hours and supported on metal pins inserted along the incense trail. Golovnin's account did not mention a bed of ash, and he indicated that the incense trail was formed not with a template but by means of a groove incised on the surface of the wooden block, suggesting that the version he had observed was a primitive form, or perhaps that he may not have been able to examine the device in detail.

Later in the nineteenth century the use of *kōbandokei* in Japan was noted also by a French traveler, Edouard Fraissinet. In a chapter on the subject of contemporary Japan, he noted:

> Formerly, that is to say in very ancient times, use was made in this country of ignescent clocks [*horloges ignées*]. These were maintained by means of a fulminant powder extracted from the bark of *Illicium religiosum*. The guardians [keepers] of the clock sprinkled this substance into a series of furrows laid out on a bed of ashes; and it was the progressive burning of the ignited powder that indicated to the guardians the passage of the hours by the ringing of the bell. In order to regulate the movement as much as possible, the entire device was enclosed within a box ventilated only by a limited number of holes. But towards the middle of the seventh century, this ingenious procedure was replaced by clepsydrae (water clocks) which were recognized to be more exact.[6]

Known in the Western world as "star anise" belonging to the magnolia family, and as the *shikimi* tree in Japan, *Illicium religiosum* has special significance in Buddhist religious rites. The fragrant evergreen branches are placed in bamboo cups upon one of the shelves of the household Buddhist altar (*butsudan*) as a necessary component of the offerings, or else ground into powder as incense.[7]

Fraissinet's statement that clepsydrae replaced the incense seal in the seventh century is confusing, for in fact incense seals continued in use, as evidenced by their presence in temples at the time of his visit.

The accounts of the three Western travelers provide a useful description of the Japanese incense seal in its early form. Several decades ago an American traveler visiting Japan reported having observed several such devices in a private collection in Tokyo.[8] The several early examples he reported are of two types. One consists of a solid cube of hard wood, possibly teak or other closely grained wood, measuring 30 to 36 cm, the upper surface of which has a narrow groove carved into it in an intricate continuous pattern. The groove is formed as a square channel approximately 0.5 cm in width. It was filled with powdered incense,

---

[6] Fraissinet, *Japon contemporain*, p. 170.

[7] Jisaburo Ohwi, *Flora*, pp. 468–69; Everett, *Encyclopedia*, vol. 6, p. 1787.

[8] Letter from Edwin Pugsley to Brooks Palmer, October 18, 1952, *NAWCC Bulletin*, pp. 323–25.

which burned at a uniform rate. Markers inserted at intervals along the incense trail indicated time intervals. The lengths of these were adjusted according to the winter or summer season.

What may be a somewhat later version was provided with a removable cover in the form of a "roof" having several tiny chimneys along the path. When the smoke was emitted from a particular chimney, it presumably indicated the conclusion of a particular time segment. These incense timekeepers were said to be used only in Buddhist temples.[9]

Over the years the Japanese improved the incense seal into a more sophisticated device, formed upon a bed of finely sifted wood ash to provide greater consistency in the burning rate of the incense. It evolved in two forms which were uniquely Japanese. These were the "permanent (or constant) incense board" (*jōkōban*) and the "time measuring incense board" or "time incense tray" (*jikōban*). The two versions were basically the same in structure and general appearance although not in function. The *jikōban* was used exclusively in Buddhist temples in religious rites, and the *jōkōban* was used in community life as well as in the temple.[10]

Three basic differences distinguish Japanese incense seals from the Chinese versions. While most of the later Chinese incense timepieces were made of metal, in Japan they were made exclusively of wood without metal parts. Nor did the Japanese utilize the seal-characters of augury which were used by the Chinese to form the incense trail; instead, two patterns were commonly used, and several others were used less frequently. Finally, although the incense trail of Chinese incense seals was formed upon the surface of tamped wood ash, that of the Japanese incense seals was laid into a furrow fashioned in the bed of wood ash by means of a special utensil. Japanese incense seals most resemble the earlier forms used by the Chinese derived from the manner of forming the incense seal of Avalokitésvara.

The *jikōban* was originally devised not as a timepiece for the measurement of time, but as a means of continuing the burning of the sacred fire in Buddhist temples for limited time periods, such as through the night hours or for a portion of the day. It continued to be used for that purpose, but gradually it came into use also for the measurement of segments of time for prayer. By the Tokugawa period it served chiefly for timing the ringing of the temple bell.[11]

---

[9] Communications from Edwin Pugsley, August 25 and November 3, 1959.

[10] It is to be noted that the nomenclature is not consistent in the historical accounts of these devices by the late Asahina Teiichi and by Takigawa Seijirō. Other sources established that the *jikōban* was designed for measuring limited periods of time and used in Buddhist temples, while the *jōkōban* was used in the temples and the community for measuring extended periods of time.

[11] Asahina, "Jikōban ni tsuite," pp. 19–34; Kiyoshi Yabuuchi, "Kōban," pp. 8–10; communications from Professor Ryuji Yamaguchi, November 2, 1958 and from Professor Yabuuchi Kiyoshi, June 11, 1960.

Both forms of the *kōban* were made with two parts. An upper section consisted of the incense tray in a square wooden box. The lower section was usually a cabinet having one or two drawers for the storage of utensils and an incense supply (Fig. 111).

Several utensils were required for preparing the incense trail of both the *jikōban* and the *jōkōban*. All were made of wood. A tamper (*hainarashi*) was used to flatten the bed of ashes. A template (variously known as *kōuke*, *kōgata* or *kōin*) was used for forming the furrows of the incense trail through its openings. A utensil used with the *jikōban* is a wooden spatula (*osae*) used to impress a groove in the ash bed through the opening in the *kōin* (Fig. 112).

The pattern of the incense trail of the *jikōban* consisted of six segments, based on the Buddhist division of the day into six parts. It was used in one of two forms, each of which featured six straight furrows, connected at alternate ends. In one version known as *kōnoji*, the furrows were parallel, and in the other they were spread to form a zigzag design resembling an inverted letter "W". The time required for the incense to burn the length of one furrow was equivalent to the time elapsed in one-sixth of a day. A pattern found in rare examples utilized a continuous trail formed as a spiral with square instead of round corners. Yet another pattern infrequently encountered formed a trail by means of the outline of three interlinked diamond shaped forms, providing a total of twelve straight segments (Fig. 113).[12]

The bottom of the *jikōban*'s incense tray was covered with a bed of finely sifted dry wood ash, slightly dampened. After it had been compressed with the tamper, the template was placed upon it. The reverse side of the *jikōban*'s template was marked with graduations of one *sun* (approximately 2.5 cm) which were impressed on the surface of the bed of ashes to render more easily the determination of incense consumption. Approximately fifteen minutes are required for the consumption of one *sun* of incense.[13]

The *kōin* of the *jikōban* had a continuous groove cut through it in the shape known as "Genji incense" (*Genjikō*). It has been suggested that a relationship existed between the template of the *jikōban* and the form used to record the ancient incense-sniffing game of "Genji incense" (*Genjikō*) named after the chapters of the *Genji monogatari*, in which the players were to identify each of five varieties of incense being burned.[14]

In a later period the *jikōban* was called *museirō* (noiseless clepsydra) because, when compared with the clepsydra, it made no sound.[15]

Designed for measuring extended segments of time, the *jōkōban* was similar in

[12] Hashimoto, *Nihon* pp. 213–15.
[13] Yabuuchi, "Kōban," pp. 8–11.
[14] Yabuuchi, "Kōban," pp. 8–10; Yabuuchi, "*Chūgoku no tokei*," pp. 19–25.
[15] Takabayashi, *Tokei hattatsu-shi*, p. 126

construction and principle to the *jikōban*, but it differed in that the length of the incense trail was greater and formed in a different manner. The incense tray was made to swivel on a pivot projecting from the lower storage section. A round hole on the underside of the incense tray segment fitted into a wooden peg projecting from the center of the upper surface of the base, so that the incense tray could be swiveled while forming the incense pattern.

The incense tray of the *jōkōban* was generally supported upon a small chest having two drawers with brass pulls; the only metal used. The drawers served for storage, the upper one for the incense supply, and the lower one for the utensils. A characteristic of *jōkōban* used in rural areas was that the bottom drawer often also contained a chessboard and chessmen.

The trail of the *jōkōban* was formed in the shape of the swastika with double crampons, requiring a somewhat different procedure. The *kōin* used to form the groove of the incense trail of the *jōkōban* somewhat resembled a wooden comb, basically consisting of five straight lines connected at one end by shorter straight lines at right angles. It was made with a projecting frame to fit over the edge of the incense tray at each corner. To form the trail the *kōin* was fitted over the edge of each corner of the incense tray in succession. A tracing spatula (*ōsae*) was inserted through the *kōin*'s openings to create a furrow in the ash bed. Each time, it formed a quarter segment of the whole incense pattern. The incense tray was swiveled on its pivot to bring the next corner forward, until the pattern was completed. The spatula was firmly held in place while the powdered incense was being added. It was then tamped firmly with the tamper and the *kōin* was removed. The complete trail marked in this manner was 80 *sun* or 240 cm in length, because each groove of the *kōin* was exactly 20 *sun* or 60.6 cm long. Inasmuch as each *sun* of the *kōban* required 14.4 minutes to burn, the complete trail burned for approximately twenty hours (Fig. 114).[16]

The late Asahina Teiichi noted that the *jōkōban* which has survived in the Inoue clan house and the *jikōban* in the Kaisendō of Tōdai-ji temple both measure twenty hours.[17]

The incense trail was ignited at a designated point – in the *jikōban* at the terminal of the far left segment, and in the *jōkōban* at the elbow of the lower left segment of the figure of the swastika. It was said that *nerikō* (paste incense) was used for ignition.

[16] The swastika (*svastika* in Sanskrit, *manji* in Japanese) is a mystical diagram symbolizing good luck, and it is also a symbol of Buddhist esoterics. Its origin is traced to the Greek *gammadion* in Troas, anterior to the thirteenth century B.C. Some theories claim that it was transmitted westward through Iceland, and eastward to Tibet and Japan. It serves as the basis for the key-pattern and other decorative elements. Chamberlain and Mason, *Handbook*, p. 95.

[17] Takigawa, "Kawachi," pp. 8–10.

Small time plates, made of bamboo or metal, supported on wooden or metal rods, were inserted at regular intervals along the trail, each bearing the character for one of the twelve hour segments.

As already noted, except for the design of the incense pattern, the general appearance and component parts of the *jikōban* and the *jōkōban* are similar. Often they were made of the wood of cryptomeria or of *Paulownia tomentosa steud*, known to the Japanese as *kiri*, a tree introduced from China which is found throughout Japan. The light and soft wood is yellowish gray to light brown in color, and does not warp or crack, and is incombustible. It is widely used for making musical instruments, toys, gift boxes, furniture and linings for steel safes. Another wood frequently used was *keyaki* (*Zelkova serrata*), a tree native to Japan, China, Korea and Manchuria. The wood is used in Japan also for boat-building and the construction of temples. Because of its high oil content, the wood resists moisture and is greatly favored for cabinetmaking and inlay work.[18]

The exterior of many *jikōban* was left in polished natural wood or lacquered in black, while that of the *jōkōban* was generally lacquered in black or red, or a combination of both. Occasionally, some were decorated also in gilt. The primary component of the Japanese incense seal is an incense tray in the form of a square box with a wooden latticework cover, with openings for the emission of heat and smoke, which is fitted over the box frame. In the *jikōban* the incense tray is supported either upon a wooden pedestal with a simple base or a pedestal terminating in a small chest of one or two drawers.

Inasmuch as the rate of burning of incense depends to some degree on the humidity and heat of the environment, the measurement of time by means of either the *jikōban* or the *jōkōban* is not entirely precise (Figs. 115–17).

The use of the swastika for the incense trail of the *jōkōban* has considerable significance. The swastika is a variant of the form of the cross. It is a symbol of considerable antiquity, which was common to many civilizations but whose origins are unknown. It is a mystic figure used by several sects of India, and was equally well known to the Brahmins and the Buddhists. The symbol existed long before it was named. The term "Swastika" is derived from the Sanskrit for well-being or good fortune, *su* meaning "well" and *as* meaning "to be;" or, "to be well," symbolizing resignation of spirit, and at the same time denoting happiness, pleasure and good luck. The Japanese name for swastika is *manji*. In India it was the monogram of Vishnu and Siva, and in Scandinavia it was known as *fylfot*, a word derived from old Norse and was erroneously said to represent the hammer or battle axe of Thor.

The swastika is often found at the beginning of Buddhist inscriptions on manu-

---

[18] Bailey, *Cyclopedia*, pp. 3540–41; Hora, *Encyclopedia*, p. 153; Leathart, *Trees*, p. 131; Everett, *Encyclopedia*, pp. 3586–87; Jisaburo, *Flora* pp. 791–92; *Forestry of Japan*, p. 217.

scripts as well as on coins. It signifies *wan* meaning "many," "10,000," "longevity" or "infinity," and is supposed to be in itself an accumulation of 10,000 felicities. It is ordinarily accepted as the accumulation of lucky signs possessing 10,000 virtues.[19]

The symbol was to be found also in Peru in ancient times and was used as well by North American Indians. It appears to have been transmitted with Buddhism from India to China and Japan, possibly as a variant of the meander used in diaper patterns. Ordinarily the crampons of the symbol are directed to the right; in this form it is the first of the sixty-five auspicious signs on the footprint of Buddha. The fourth sign is the "Suavastika," with the crampons directed to the left. The use of double crampons appears to be uniquely limited to use in the incense pattern, it has not appeared elsewhere. It may have been devised as a variant of the swastika to provide a longer incense trail than was possible with the standard representation. A double swastika with double crampons is included among the designs of Chinese incense seals of Ting Yün (see chapter 8).

The use of the incense seal was infrequently noted in Japanese literature, but it is mentioned in several Japanese classics, including Lady Sei Shōnagon's *Pillow Book* as well as in Lady Murasaki no Shikibu's *Genji monogatari*. It was also featured and illustrated in the seventeenth-century novel *Kōshoku gonin onna* (Five Women who Loved Love) by Ihara Saikaku (1642–1693). The novel was first published in Osaka in 1686 and illustrated with wood block prints by Yoshida Hanbei, the leading artist of the Kyoto-Osaka region (Fig. 118).

In this work the *jōkōban* was featured in the tale of Oshichi, the daughter of a greengrocer. After a fire one late December night which devastated part of the city and came near her home, Oshichi and her mother ran with others for refuge to the Buddhist temple, the Kichijō-ji, in Komagome. There the displaced refugees were forced to remain for a few days. During this interlude the young girl was attracted to a handsome young man named Kichisaburo who was also staying in the temple, and cast longing eyes on him.

One night Oshichi sought out the young man, and left the temple's guest room. In the kitchen she was informed by the cook that Kichisaburo was sharing the bedchamber in which a young Buddhist novice of the temple slept. Hoping to find him in the priest's quarters, she made her way there. She did not find Kichisaburo, but only the novice preparing a *jōkōban*, forming the incense trail in the shape of the double-armed swastika. The story continued:

> It must have been about the eighth hour that the bell which served as a reminder of the perpetual incense burner fell with a ring that sounded throughout the temple for some time. Apparently the novice was supposed to be on duty. He got up, put the bell back on a

[19] Wilson, "Swastika," pp. 763–901; Loewenstein, *Swastika*, pp. 5–28.

string, and threw fresh incense on the fire. Then he sat at the altar for what seemed an eternity to Oshichi, who was impatient to enter the bedchamber. Acting on a sudden inspiration, as women are wont to do, she pulled down her hair, made a dreadful face, and approached the novice menacingly. But imbued with the calm courage of Buddha, the novice was not the least bit frightened by her.[20]

Another translation of the same text noted that,

It was about two in the morning. The small bell of the jokoban falling made a sonorous sound. It must have been the duty of the child priest to refill the incense pan, for he added more of the incense and then to O-Shichi's annoyance remained in the room. The waiting became intolerable and as it seemed as if he was never going to leave, O-Shichi acting impulsively rumpled her hair and tried to frighten him by making a terrible face in the darkness. Like a priest, however, he was not at all frightened or surprised but with glaring eyes fiercely scolded her.[21]

There is no evidence, however, of a bell as part of an incense seal that drops and make a sonorous sound; it is possible that Ihara may have known of a special device not elsewhere recorded and which was not illustrated by Yoshida Hanbei. In a footnote, the translators of the second version attempted to explain the incense seal alarm, which may in fact have been yet another form of the incense timekeeper, although no example of it has come to notice:

The jōkōban is a scale in one pan of which incense is burned and in the other a small bell. As the incense is consumed in a half or one hour's time, the incense pan rises, allowing the bell to fall. This serves as a timepiece.[22]

A Japanese refinement in the use of incense seals, not found in China, was the development of an extensive variety of specialized incense recipes and odors for indicating the individual twelve hour intervals. This practice undoubtedly was inspired by the popular "sniffing parties" and the value which the Japanese placed on the skill of detecting the identity of incenses by their scents.

The markers inserted along the incense trail to denote the time intervals by the Chinese and the Japanese customarily consisted of small wooden, bamboo or metal labels on wire pegs identifying each time interval. Eventually the Japanese refined this aspect of the incense seal and devised markers in the form of hard paste tablets or pastilles, each made of different components so that each had an individual odor. In this manner, as the incense trail burned progressively along its way, and a marker was reached by the fire, it too burned and the subtle emanation of a different scent alerted the priest that another time interval had elapsed. He

[20] Ihara, *Five Women*, pp. 170–73.
[21] Fuji and Perkins, transl., *Gonin Onna*, p. 7.
[22] *Ibid.*, p. 7n.

was able to identify the particular time interval if the order of odors was prescribed.

The twelve time intervals utilized the signs of the Terrestrial Branches, namely, rat (*ne*), cow (*ushi*), tiger (*tora*), rabbit (*u*), dragon (*tatsu*), snake (*mi*), horse (*uma*), sheep (*hitsugi*), monkey (*daru*), cock (*tori*), dog (*inu*), and wild boar (*i*). The twelve markers, called "cards or plates to tell time" (*jikoku fuda*), were inserted at appropriate intervals beside the incense path.[23]

In the collections of the Kamiguchi Domestic Clock Conservation Organization (Kamiguchi Wadokei Hozon Kyōkai) in Japan is preserved a *jōkōban*, maintained in a box which is said to have been purchased by Kinokuniya Izaemon in "the fourth Kaei year," or 1851.[24]

In November of the fifth year of the Meiji (1872) the Japanese government issued a proclamation which contained an article specifying that the lunar calendar would be replaced by the solar calendar in January of the following year. It specified also that henceforth the day was to be divided into twenty-four hours. Consequently the time plates of the incense seal became useless and were eventually discarded, except in ceremonial use.[25]

Nonetheless, the *jōkōban* continues to be used in temples in Japan. Catalogues of several modern establishments specializing in accessories for Buddhist temples offer for sale altars, candlesticks, incense burners, patens, and other items, including *kirikuji kōro* and the *jōkōban*, in several sizes. The *jōkōban* is made in three sizes, 36 cm, 45 cm, and 54 cm square, and in two qualities, "better made" and "lacquered." They are available decorated in "Chinese red or gold leaf attached to black lacquered surface."[26]

[23] Communication from Professor Ryuji Yamaguchi, November 7, 1958.
[24] Asahina, "Tokei," pp. 576 ff.
[25] Asahina, "Jikōban," pp. 19–29.
[26] Catalogues of the Yokohama Buddhist Shrine Accessories Store, Yokohama, 1988, p. 72, item 227.

In Japanese Buddhist rites

Buddhism, which was introduced into Japan in the 6th century, received imperial recognition and became the state religion towards the end of the reign of Empress Suiko, who ruled from A.D. 593 to 628.[1]

As in China, in Japan incense was burned in Buddhist household shrines as well as in the temples. The *kōro* (incense burner or censer) was usually placed upon a polished hardwood or lacquered stand in a recess in the main wall of the room (*tokonoma*) at the side of a hanging scroll (*kakemono*) with a vase of flowers. Made of bronze or lacquer, the *kōro* was usually global in shape, three-footed, and lined with silver or copper; it had two ear-like handles, and a cover terminating in a knob. The sides were plain or worked in relief with an auspicious poetical theme. *Kōro* made of pottery were generally cylindrical or hexagonal in form. The *kōro* often was made also of carved hardstones for the wealthy.[2]

As in China, before the entrance of every Japanese Buddhist temple was to be found a huge bronze or iron urn decorated on its outer surface with dragons, lotus and other appropriate symbols. It was kept half-filled with wood ash into which worshippers inserted burning incense sticks which they purchased from the temple keepers and which were ignited from a pure flame provided nearby. Smaller censers, usually without decoration, were used on the temple altar. The earliest incense censers in use in Japanese temples during the Nara period were lotiform, made of gilt metal attached to a long handle with a supporting crook at the opposite end.[3]

The inexpensive, common incense burned by the poor and pilgrims before Buddhist icons in Japanese temples was known as *ansokukō*. It was of very poor quality and contrasted sharply with more expensive varieties burned in temples of the wealthy. In Buddhist rites the three types used by the priests were an odorous ointment called *zukō*, many varieties of ordinary incense known as *kō*, and a fragrant powder called *makkō*, which may be identical to powdered sandalwood.[4]

Soon after the introduction of Buddhism into Japan, the incense seal was used

---

[1] Fraissinet, *Japon contemporain*, pp. 170–71.
[2] Casal, "Incense," pp. 58–59; Armstrong, *Buddhism*, pp. 19–20.
[3] Casal, "Incense," pp. 52–53.   [4] Hearn, *Ghostly Japan*, pp. 26–27.

for religious rites as well as for the community's measurement of time. The *jikōban* became an important feature of Buddhist temples in Japan during lengthy events such as the Buddhist mass, the "parting of the seasons," and the "hungry-ghosts-feeding" rites. It measured the time for striking the great bell, and informed those living near the temple of the current hour.

Tōjō-ji, the temple of the Shingon sect at Niihari-gun, or Niihari-mura in the Ibaragi Prefecture, contains a *jōkōban* of large size whose drawers are filled with Chinese anise leaves and incense powder. It was used long ago to indicate the time for the bell to ring, which informed the people of the time in the morning and evening. Smaller *jōkōban* are to be found in several other Shingon temples in the southern region of the Ibaragi Prefecture, including the Tōkō-ji at Tsukuba-shi, Mabuchi; the Anpuku-ji at Tsukuba-shi, Nishi-Ohashi; and the Fugen-in at Tsukuba-shi, Hanari; as well as another in the City Museum at Tsuchiura-shi, Chuō.

Several contemporary published works include references to *jikōban* and *jōkōban*. The *Umezo no nikki* contains a chapter on "Burning incense, knowing time" which notes that the *jōkōban* was used as a time measurer from ancient times, and that this method of timekeeping is referred to as "incense time" (*kōkoku*). According to another work, the *Shinpen Kamakura shi*, various words and phrases of rules relating to maintenance of the *jōkōban* and a formula using incense (*kō o motte jōshiki to nasu*) are carved on boards in the Jōraku-ji temple, such as "time of the first watch of the night and midnight" (*shogoya no toki*).

After the suppression of Buddhism during the Meiji Reformation, knowledge of the original function of these time measurers appears to have been lost, although *kōbandokei* were made as late as the Tokugawa period, and continue today to be reproduced for Buddhist temples.

The preservation of *jikōban* and *jōkōban* in a number of the major Buddhist temples in Japan at the present time suggests that others survive in places of worship, in addition to the handful which belong to public museums and private collectors. There is evidence that *kōbandokei* may be found in a number of country temples but that the priests in attendance are not aware of their original function. According to a survey conducted several decades ago, there were more than eleven Buddhist sects in Japan with approximately 73,000 Buddhist temples. Incense seals presently known to exist in Japan are found in temples of at least six sects, including Tendai, Kegon-shu, Shingon, and Zen, suggesting that the use of incense in Japanese Buddhist ritual was common to many sects. Japanese writings mentioning incense seals generally state that they were used in "esoteric" Buddhist temples, indicating that they may not have been used in all places of Buddhist worship.[5]

---

[5] Descriptions of individual temples are based on entries in *Japan: Official Guide*, pp. 141–50; *Baedeker's Japan*, and other guide books.

The organization of periods of prayer in Buddhist temples, each interval signaled by the striking of the temple's great bell, was described by Golovnin:

The prayers are repeated three times in a day; at daybreak; two hours before noon; and before sun-set; as the matin, noon and vesper masses are performed with us. The people are informed of the hours of prayer by the ringing of a bell. Their method of ringing is as follows: after the first stroke of the bell, half a minute elapses; then comes the second stroke; the third succeeds rather quicker, the fourth quicker still; then come some strokes in quick succession; after a lapse of two minutes, all is repeated in the same order; in two minutes more for the third time, and then it ends. Before the temples there stand basins of water, made of stone or metal, in which the Japanese wash their hands before they enter. Before the images of the saints lights are kept burning, made of train oil, and the bituminous juice of a tree, which grows in the southern and middle parts of Niphon [Japan].[6]

The *jikōban* and the *jōkōban* both were used in Japanese Buddhist temples. The *jikōban* was the traditional form used for measurement of short periods of time, designed for and understood by initiates of the priesthood, and for the annual sacred water-drawing ceremony (*omizutori*).

The *Shinpen Kamakura shi* notes under the caption "Jōraku temple" that:

... the first and second watches of the night [services] will be fixed by means of incense ...

confirming that an incense seal was used at Kenchō-ji Rinzai-shū the foremost of the five great Zen temples of Kamakura, where the famous monk Doryū Rankei lived.[7]

The rule of the Buddhist temples of the Nara and Heian periods was extremely strict. Since services were required to be performed at six specific times during the twenty-four hour day – in the early morning, at midday, sunset, early evening, midnight and past-midnight – a means of dividing the time was essential to ensure that the schedule of worship would be announced accurately. It is likely that each of the temples was provided with a *jikōban* or a *jōkōban* (or both) for this purpose.[8]

The known examples of *jikōban* surviving in Japanese Buddhist temples at the present time are no longer used for their original purpose. One carefully preserved in a temple in Kobe is known as a "permanent incense furnace" (*jōkōro*). Another, at Shingon-shū Buzanha Shain, Nagahama-shi, Miyamae-chō, is more

---

[6] Golovnin, *Memoirs*, vol. 3, pp. 57–59.
[7] The temple at Kenchō-ji Rinzai-shū was founded in 1253 by Hōjō Tokiyori (1226–1263), regent of the fifth Kamakura Shōgun, for the Chinese priest known in Japan as Daigaku Zenji. Daigaku Zenji was one of the many Chinese priests who took refuge in Japan after the fall of the Sung Dynasty. The original temple building was destroyed by fire in 1415 and the reconstruction was ravaged repeatedly in succeeding years by civil wars. Protected by the Tokugawa, the priest Takuan (1573–1645) did much to restore it once more. The temple contains many important cultural treasures.
[8] Takigawa, "Kawachi," pp. 8–29; Armstrong, *Buddhism*, pp. 4–96.

than a century old, and yet another is to be found in the Shimizu Kannondō in Tokyo.[9]

A *jikōban* in the Japanese Shintō shrine, Mikumari-jinja, located in the Prefecture of Nara, is described as a *kōinban*, and bears the date 1490 and an inscription.[10]

A *jikōban* maintained in the Buddhist Taima-dera temple, also in the Nara Prefecture, bears the date 1556 as well as an inscription.[11]

Another example of the Edo period, of the sixteenth or seventeenth century, is in the Myōshin-ji temple in the Prefecture of Kyoto.[12]

In the Saikōin-ji temple of the Tendai sect in Nara City is found a *jikōban* having an incense trail of unusual form. Made of a very dark lacquered wood and supported on a cabinet with two drawers, it is decorated with Buddhist symbols inlaid on each of its four sides, including the wheel of the law and the thunderbolt. It measures 31.5 cm square and 42.5 cm in height. The incense trail is in the form of a squared spiral instead of one of the customary forms (Fig. 119).[13]

Until recent times, an early *jikōban* existed in the Shōgaku-ji temple at Ichinomiya, near the former national shrine at Keta on the Noto peninsula, which was founded in the early Heian period. In former times, the shrine had many pagodas, but with the suppression of Buddhism during the Meiji Revolution and Reformation, these were removed, and now only the Shōgaku-ji temple remains. Among its treasures is a *jikōban* similar to the one of the Nigatsudō of the Tōdai-ji. Visitors have reported that the priest in attendance was unaware of the *jikōban*'s function, although the purpose is described in the temple's guidebook:

The incense clock
It is said that about three hundred years ago, when Lord Maeda Kagatsune donated the hanging bell of the temple, he also donated this incense clock so that the temple would be

---

[9] Communication from Professor Yabuuchi Kiyoshi, November 11, 1960; Hirashi, "Kara-kuri," pp. 70–72; Takeo, "Incense Clock."

[10] The Mikumari-jinja Shrine in the Yoshino hills, built in 1604, is a typical example of the shrine architecture of the Momoyama period.

[11] The Taima-dera Temple, situated near Shimoda, was founded in 612 and moved to the present site in 684. Its chief treasure in the Mandala Hall, rebuilt in about 1243, is the famous Mandala painting of the Buddhist paradise executed in 763. Measuring fifteen feet square, legend states that it was the work of Chūjō-hime, daughter of Fujiwara Toyonari, the Minister of the Right. She became a nun in this temple.

[12] The temple of Myōshin-ji was founded between 1336 and 1338 on the site of an imperial villa of Emperor Hanazono, but the existing buildings are of later date. It contains a bell said to have been cast in 690 or 698. The Myōshin-ji and its subordinate temples are extremely rich in art objects.

Information concerning the three foregoing incense seals was provided by Dr. Shunichi Sekine in a communication of August 15, 1987.

[13] Communication from Professor Akio Gotoh, October 10, 1987.

able to announce the time accurately to the people of the area. It is considered a most rare object.[14]

The *jikōban*, although always consisting of the same components, was made in varying degrees of elaboration (Fig. 120). Most surviving examples are simply made, with minimum if any decoration, such as the seventeenth-century example in the Nigatsudō. Simplicity of exterior seems to have been the rule, while elaborate examples were the exception. Consequently, one *jikōban* decorated with extensive carving is deserving of special attention. Dating from the seventeenth century, it is made of a relatively dark wood, and assembled from approximately 140 separate cubical units which interlock with plain lateral strips of graduated length. The exterior is in natural finish without lacquer. Wooden pins are used at intervals to keep the blocks correctly spaced. A square wooden bolt in the bottom holds the entire assembly together. If the bolt were to be removed, the entire structure would disassemble. The sides confining the incense tray consist of four solid wooden panels elaborately carved; each panel representing a different design of waves, clouds, lotus flowers and artemisia.

The section containing the incense tray, which is somewhat charred, is 6 cm deep. The compacted bed of ash is from 3 to 4 cm in depth and the incense trail is 0.5 cm wide, 8 cm deep, and 172 cm long.

The template for the incense trail, which forms a variation of the parallel linear design, is assembled from six parallel units, the ends of which are dovetailed all the way through the sides, with the ends made flush with the outer sides (Fig. 121). Each is held in place, somewhat loosely, by means of a single wooden pin. The looseness is deliberate, to ensure that the stencil lies closely on the bed of wood ash. The openings, consisting of seven courses, provide a continuous channel for depositing the incense to form the trail. The compacted ash bed could be as deep as approximately 4 cm. The entire trail is approximately 18 cm long, 0.5 cm wide and 0.8 cm deep. The overall measurements of this *jikōban* are 366 cm square and 30.5 cm high. The inside measurements are 28 cm square.[15]

Another elaborate example, made of lacquered reddish wood with an equally elaborate matching base, is preserved in the Nara-machi Museum in Nara. It is 39 cm square and 51 cm in height overall (Fig. 122).[16]

Although the *jikōban* was almost invariably made of wood, at least one ceramic example has come to notice, and undoubtedly there are others. It is preserved in

---

[14] Takigawa, "Kawachi," pp. 8–29. The Maeda clan was at one time the most powerful in Japan, with extensive property holdings particularly in the Noto Peninsula, and owned castles in Takaoka and Toyama. Recent inquiries addressed to the abbot at Shōgaku-ji were not acknowledged.

[15] In the collection of the writer. Formerly owned by the late Mr. Edward S. Jones.

[16] Information courtesy of Professor Akio Gotoh.

the Sanzenin-ji, an old temple of the Tendai sect in the village of Ohara, outside of Kyoto. The temple owes its origin to the noted priest Dengyō Daishi (767–822).[17]

The ceramic incense seal probably was made in the second half of the seventeenth century or the early eighteenth century, of Arita porcelain, which was traditionally produced in the region. It is decorated in colors, and measures slightly under 31 cm in diameter. The bottom of the vessel has a deep groove for the incense trail, in the form of the seven parallel lines. The channels may have sufficient depth to accommodate a bed of ash upon which the incense is placed, but it appears unlikely. The ceramic incense seal is displayed on a low table just inside the entrance of the temple[18] (Fig. 123).

The porcelain incense seal appears to be an example of the ceramic wares of Kyoto and Takamatsu. A pottery was established at Omuro by Ninsei, a celebrated seventeenth-century potter, where in addition to other ceramic wares, he produced incense burners for temples. A notable surviving example made of white clay and decorated in red, blue and green, features sacred designs representing the Buddhist thunderbolt (*vajra*) and the wheel of the law (*dharam-chakra*). It is inscribed, "Ninsei, with the priestly rank of Harima Nyūdō, respectfully presents [this burner] made by him for the holy Buddha this fourth day of the fourth month 1657." Ninsei made burners of this type for the Ninna-ji and other temples. It is conceivable that the *jikōban* at Ohara is the work of Ninsei.[19]

Although the *jikōban* appears no longer to be used in most Buddhist temples, the device is still used in the sacred water-drawing ceremony each springtime at Nara. It was for this event, known as the *omizutori*, that the earliest records relating to the use of an incense seal in religious rites in Japan survive. This ceremony has been held in the Nigatsudō of the imperial temple at Nara, the Tōdai-ji, from the time that it was founded.

The *omizutori* (water-drawing ceremony) is also known as the *taimatsu shiki* (pine torch ceremony). Another name for it is *Shunie*. It was initiated in 752 upon

---

[17] The temple at Sanzenin was rebuilt in 860 by the priest Jōun at the command of Emperor Seiwa. The second prince of Emperor Horikawa became its first imperial abbot with the title of Saiun. The main hall, called "The hall of paradise in rebirth," was built in 985 by the priest Eshin (942–1017), and the other buildings were reconstructed from the materials of the Ceremonial Hall in the Imperial Palace during the Keichō era.

[18] Information from Kirby R. Vining. Recent inquiries made to the abbot have not been answered. A Korean potter named Yi Sampei, is believed to have first discovered clay deposits suitable for making porcelain in 1616, at Izumiyama near the village of Arita in the province of Hizen. He was one of numerous Korean potters captured during the invasion of Korea in 1592 and brought back to Japan. From about 1750 to 1830 a fine Arita porcelain ware was produced in the region as well as other types of porcelain including Imari, Kakiemon and Nabeshima. Cox, *Pottery*, vol. 1, pp. 256–63.

[19] Sato, *Kyoto Ceramics*, pp. 77–83. The large incense burner is in the collection of the Fujita Art Museum in Osaka.

the completion of construction of the Nigatsu-dō (second month hall or February hall), in which the ceremony takes place and for which it was named.

The Nigatsudō, which forms part of the imperially founded and patronized "Great East Temple" (*Tōdai-ji*) is situated on a hillside overlooking the Yamato Plain and is reached by a paved street lined with the houses of the temple priests. The interior of the Hall is built in concentric zones around the main shrine.[20]

The Tōdai-ji was the principal temple of the Kegon-shū sect of Buddhism and had been built for the Emperor Shōmu, who reigned from A.D. 724 to 749, by a Buddhist priest named Jitchū, a disciple of Rōben, the first abbot of the Tōdai-ji. The temple was twice destroyed by fire, in the late twelfth century and again in 1567, and was rebuilt each time. The last reconstruction was completed in 1669, in the Edo era, but greatly reduced the temple in size.[21]

The Nigatsudō is dedicated to the deity Kannon. A legend relates that a tiny copper image of the deity was found on the site and discovered to possess the miraculous quality of warmth like that of living flesh. It was enshrined in the Nigatsudō and a special ceremony relating to it was held on the eighteenth day of each month. Ever since then a special series of services called the *dattan no myōbō* have been held in the deity's honor during the first half of the second month of the year. It is from this practice that it is said that the temple derived its name Nigatsudō or "Hall of the Second Moon or Month."[22]

The official name for the water-drawing ritual at the Nigatsudō is "The memorial service of Nigatsu Hall for the eleven-headed Kannon." It has taken place annually for more than a millennium. It continued to be held even during World War II, when the use of light at night was prohibited throughout Japan because of the danger of bombing. This annual event has always been given considerable importance by the Japanese.

The ceremony originally took place during the second month of the Chinese lunar calendar, but with the adoption of the solar calendar the festival has continued to be celebrated every year during two weeks beginning with March 1. Each year, preparations for the *omizutori* are begun on February 20 in anticipation of the public ceremonies which begin on March 1 and are held continuously until the morning of March 15. Although historical records do not exist, the procedure has remained unchanged from the Kamakura period, according to the chief priest of the Tōdai-ji temple, Tsutsui Eishun, and to Zenka Kenshū.

On February 20 all the priests who are to participate in the ritual take up residence in a special section of the monastery. There they remain secluded under

[20] Aoyama, *Nara*, pp. 4–14.
[21] Piggot, "Hierarchy," pp. 45–77; Aoyama, *Nara*, pp. 8–29.
[22] Chamberlain and Mason, *Handbook*, p. 366.

strict regulations and kept apart from the secular public while they are trained and prepare for the events.

It is during the first two weeks of March of the current calendar that water begins to flow from a sacred well in the Nigatsudō, an event upon which the ceremony is based. It is believed that an intermittent natural spring lying far back in the hills above the temple overflows at this particular time due to the added burden of melting snows in the mountains, activating the temple well.

According to legend, when the temple was dedicated, the god of Ō nyu in the province of Wakasa asked for the privilege of providing the holy water. Thereupon a huge black and white cormorant flew out of a nearby rock and disappeared, and water gushed from the spot it had occupied. From that time the stream that had flowed past the shrine of Ō nyu dried up and its water was transferred to the Nigatsudō.[23]

The sacred well, called Wakasa-i, is sheltered under a small building within the Nigatsudō premises. As the well begins to flow, the water is collected and dispensed as miraculous waters among the pilgrims who have come from great distances to partake in the event. It continues to be dispensed during the two-week period. One year it was estimated that seven and one-half wagon-loads of water had been taken away by the pilgrims in two and four-ounce bottles.

On March 1, the participants move into Nigatsudō Hall, where they walk about ceaselessly chanting *sutra*, striking bells, blowing horns of conch shell, rattling their rosaries and clanking iron staves in an ever increasing crescendo of rhythmic sound as an accompaniment to the fire and water ceremonies taking place outside the Hall until March 14. The ceremony's climax comes on the night of March 12, when gigantic torches are lit and water is first drawn from the Wakasa-i beneath the floor of the hall in a secret ceremony which brings the spring festival to a close.

Because of the complicated procedures of the ceremony, it is necessary for participants to have awareness of time lapsed, and these time intervals are measured in the Nigatsudō by means of a *jikōban*, which is placed in the southeastern side of the *shumidan* in the inner room of the temple during the ceremonies. According to all available evidence, such a *jikōban* was used during the spring rites from the time that the ceremony was originally initiated. The incense seal presently in use dates from the seventeenth century, when the Hall was reconstructed (Fig. 124).[24]

Made entirely of wood, the *jikōban* is of large size, in the form of a square box consisting of two sections, one placed upon the other. The sections are measured

---

[23] *Ibid.*, pp. 366–67.
[24] Communication from Rev. Exsei Kawaguchi, Yokohama.

in *shaku* and *sun*; one-tenth *shaku* is equivalent to 1 *sun* and one-tenth *sun* is equivalent to 1 *bu*. The upper section measures 1 *shaku* and 2 *sun* on each side, and it measures 3 *sun* and 4 *bu* high. The bottom board is 5 *bu* in thickness and the bed of ashes which is laid upon it is to be 1 *sun* and 5 *bu* in depth. The lower container is 3 *sun* and 4 *bu* in height and is used to contain incense supplies and such items as the ash leveling tool and the pattern tray. One *sun* is equivalent to 3 cm.

The pattern of the incense is formed with a template (*kōgata*) having openings in a zigzag shape approximately resembling an inverted letter "w". A channel in this form is perforated in the template with which the path is formed. The underside of the incense tray is carved with eleven parallel gutters which are 3 *bu* in width and occur at intervals of 7 *bu*. When the pattern grid is pressed into the leveled bed of damp wood ash, eleven ridges are formed by means of the gutters. These ridges remain on the surface of the bed of ashes and interrupt the incense trail at intervals of 1 *sun*. Powdered incense made from a specified formula is poured upon the template and sifted into its openings. After the *kōgata* is removed, the incense trail remains in this pattern. Small metal markers attached horizontally to metal peg-like stands are inserted along the trail to indicate time intervals (Figs. 125–28).

The time system used with the *jikōban* for scheduling the individual parts of the event is known as *rokuji*, or "six hours," by means of which the day is divided into six segments. The daylight hours and the night hours each consist of three periods: the daylight intervals are *shinchō* (morning), *nitchū* (noon), and *nichibotsu* (evening), and the periods of the hours of darkness are *shoya* (early night), *chūya* (midnight), and *goya* (after night). During the *ōmizutori* the midpoint in the *rokuji* is called hanya (half night) instead of *chūya* (midnight).

This time system is described in several early works of reference, and appears to have been used elsewhere in Japan for other purposes, Oda Tokunō noted that the *rokuji* time system was also known as the requirement to "admire Amida Buddha at the six hours of the day" (*rokuji no raisan*) in Buddhist worship, specifying prayer to Amitabha six times daily. He referred to the use of the phrase, "Although I am kept extremely busy, continually praying to Amitabha at *rokuji*, I dare to take this pleasure. . ."[25]

Participants in the water-drawing ceremonies include Buddhist priests, civil officials, and pilgrims. Eleven *bōzu* or priests known as *rengyōshū* assume assigned positions identified by the traditional titles *wajō*, *daijōshi*, *shushi*, *dōtsukasa*, *kitazashū no ichi*, *nanzashū no ichi*, *kitazashu no ni*, *nanzashū*, *no ni*,

---

[25] Asahina, "On the Jikōban," pp. 19–29. This series of six periods are described by T. Oda, the *Amidakyō*, and *Saiikki*, and Masutani Fumio, all cited by Asahina Teiichi.

chūdō no ichi, gonshosekai, and *shosekai.* The priests are prohibited from speaking during the course of the ceremony except for the traditional ritual phrases which they are assigned. Participating civil officials are designated the hereditary offices entitled *shokō, dōdōji, kushi, komori, ōi,* and *inji.* The titles given to participating pilgrims are *kakubugyō, tsubonebugyō, nakama,* and *dōshi.*

At the end of each evening, the inner room of the *Nigatsudō* is swept clean. The incense in the *jikōban,* which has been burning since the previous evening, is cleared away and replaced with new incense.

At 6:00 P.M. the *daishōya* announces the time to the *shokō* by striking the great bell of the temple. The *shokō* reports the time to the *dōtsukasa* in the *sanrō shukusho* and then proceeds to the *Nigatsu-dō* where he obtains kindling coals from the *shosekai.* The *shokō* makes a ceremony of lighting the sacred lamps in the Hall. The fire which is burning in the sacred lamp of the inner room (*naijin*) is transferred to the slender pieces of the oil-rich parts of the pine root (*matsu no jin*). The same fire is used to ignite the sacred lights in the Hall. As part of the same rite, the incense trail of the *jikōban,* which has been remade with new incense, is lighted from the *matsu no jin* by the *shōsekai.*

The *shokō* then proceeds from the temple down to the room in which awaits the *dōtsukasa,* or *dōtsukasa shukusho,* to inform that dignitary that the temple has been lighted. The *shokō* thereupon returns to his house. The *dōtsukasa* dispatches *kakubugyō* to the *Nigatsudō,* and the latter runs up the north steps to the temple entrance with a flaming pine torch in his hand. As he places the torch at the temple entrance, he recites the words "Time incense is × *sun* × *bu* in length before *shoya*" (*jikō shoya made nan sun nan bu*).

An important part of the ritual is the "announcement of the incense time" (*jikō no hōku*) in a question and answer exchange. The question asked is "How long to *shoya*" (*shoya made nan sun nan bu*)? The *kakubugyō* then returns to report the time to the *shokō,* who informs the *daishōya* with the same words, "*jikō shoya made nan sun nan bu*"; thereupon the *daishōya* strikes the temple bell. Approximately one hour has elapsed and the time is now 7:00 P.M.

Upon hearing the bell, the *dōtsukasa* sends the *kakubugyō* to the temple hall to announce "*yōji no annai!*" to the *shosekai.* Following orders, the *kakubugyō* completes the preparation of the *naijin* in the inner room for the impending arrival of the *rengyōshū* in the hall.

The *dōtsukasa* again commands the *kakubugyō* to announce to the *shosekai* that "*shusshi no annai.*" When the bell rings once more, the members of the *rengyōshū* enter the temple in order, with the *dōshi* carrying a torch of green bamboo. The *dōshi* bears the torch to the front floor of the *Nigatsudō,* brandishing it as traditionally prescribed in the manner of a rolling wheel. The torch is shaped like a basket and is somewhat more than 4 *ken* in length (1 *ken* is equivalent to 1.82 m) and weighs 20 *kan* (1 *kan* is equivalent to 3.75 kilograms).

This type of torch is used during the particular three of the days of the ceremony, namely, March 12, 13 and 14. The handling of the torch during the ceremony requires considerable skill. The spectacle that is created as the lighted torch is kept in motion is dazzling. The sparks falling from it resemble the petals of cherry blossoms, according to Japanese observers, who find it extremely exciting. Soon after all the members of the *rengyōshū* have entered the temple, the *shoya no gyōhō* begins.

The *omizutori no gyōji*, the ceremony of taking water from the sacred well, begins in the middle of *goya* on March 12, or more precisely, at approximately 2:00 A.M. on March 13. The five dignitaries who actively participate at this time include *kitazashū no ichi*, *nanzashū no ichi*, *chūdo no ichi*, *gonshosekai*, and *shosekai*. They follow *shushi* down the stone steps to the southern part of the temple hall to scoop water from the well. Thereupon the *dōshi* steps forward with a *hasu*, a large torch about 1 *jō* in length, in his arms. Scooping water from the sacred well is the hereditary function of the *dōshi* families, and they have kept the details of this well a closely guarded secret.

Accordingly, during the three days of March 12, 13 and 14, the special ceremony of the *dattan no myōbō* follows the event of the "after-night" previously described. The *dōshi* participate in scattering water scooped from the well, and at the same time a large pine torch weighing approximately 20 *kan* and measuring 1 *jō* in length is revolved rapidly, showering sparks.

In the ceremony to ignite the *jikōban*, a wooden spill (*tsuketake*) placed at the left side of the incense trail is used to transfer the flame to the *jikōban* from the sacred fire kept constantly burning in the inner room (*naijin*) of the Nigatsudō.

The spill is made from the wood of the *shidebō* tree which grows in the precincts of the Tōdai-ji. The *shidebō* tree, so called by the participants in the ceremonies, is probably a corrupted form of the original name "Fire containing tree" (*hideboku*), for it appears as such in the ancient records of the *dōshi* families.[26]

Examination of the wood has proven it to be *inushide*, and consequently the tree or wood (*boku*) of the *shidebokes* may also have been corrupted to the word *shidebo*. Two varieties of the tree grow in Nara, the *inushide* and the *aoshide*. Among the *shide* groves of Nara are a number of large, withered trees, the trunks of which have become fragile and porous. The spills used in the ceremonies are cut from these old trunks and dried by the *doshi* by dehydrating them in an earthen pan.[27]

---

[26] Asahina, pp. 576 ff. The *shide* trees are large deciduous shrubs or trees of the Betulaceae, the Carpinus family (*kumashide zoku*). *Inushide* is the species *Carpinus tschonoskii Maxim* (*Carpinus yedoensis Maxim*), and is one of a number of species of this genus which grow in Japan, China and Korea. Ohwi, *Flora*, p. 371.

[27] Asahina, "Tokei," pp. 575–76.

The torches are employed also at the beginning of the *omizutori*, during the *ittokuka*, which is held during the early morning of March 1. During the ceremony, the Ittoku, heir of the Inagaki family which has ministered to the Sho Kannon of the Nigatsudō since the Tenpyō era, ignites a spill by means of flint and steel. The fire from the spill is transferred to the perpetual sacred light (*jōtōmyō*) in the inner room of the Nigatsudō.

Although the schedule of the Shunie varies somewhat from day to day, it is always arranged and timed by means of the *jikōban*, which therefore becomes an important factor in the water-drawing ceremony. It undoubtedly played an even more important role in ancient times.

Ironically, although the *jikōban* continues to be an important feature of this age-old ceremony, the actual timing of the events today is in fact achieved by means of a modern pocket watch or wrist watch![28]

An interesting study made by Iwaki Masao over a six-year period concerning methods of firemaking in ancient Japan revealed that in several of the older shrines there still exist ancient fire-making equipment. At Izumo-taisha, a shrine with one of the oldest histories, a traditional fire-drill continues to be used in a ritualistic fire ceremony held in the autumn. The Ise-jingū shrine, also among Japan's oldest, continues to use an ancient pump-drill on a daily basis, as do the shrines of Ōkunitama-jinja in Fuchu City, Tokyo and at Toro, Prefecture of Shizuoka.[29]

[28] Communication from Dr. Akio Gotoh, Nara.
[29] Iwaki, "A Report," vol. 16, 1977, pp. 91–93.

12

In Japanese civil life

In Japan's civil life the incense seal played an important role at all levels of society. It was used at the Imperial Court, in the palaces of the nobility, and in government offices. It is known to have been in use during the Nara and Heian dynasties as well as during the period of northern rule of the Kamakura Shogunate (1192–1333). The form used for such purposes was the "time incense tray" (*jōkōban*). Although closely resembling the *jikōban*, it served for time measurement over long spans of time rather than for shorter periods, such as were required for ceremonial use.[1]

In Book 3 of the *Shinpen Kamakura Shi* (The New Edition of the Kamakura Record), under the heading of "The Regulator of the Jōraku-ji [Temple]," confirmed that "the first and second watches of the night will be fixed by means of incense." The *Konchi ji* of Baiennikki included a similar statement, suggesting that the *jōkōban* was used from an earlier period.[2]

Early records of Japan are not always specific concerning the use of the incense seal in community life, but as in China, it was used in association with the public clepsydra as well as for private use. Excerpts from early official documents combine to provide some concept of public time measurement in Japan and the use made of the incense seal.

As documented in the "Institutes of the Bureau of Divination," official time was established by the Bureau of Divination, and the year was divided into 40 periods. It was noted in the *Enki onyōryō shiki* that during the Enki period (901–902)

From the thirteenth day of the "Great Snow" the sun rises at X hour and sets at Y plus six. From the first to the twelfth of "Small Snow" the sun rises at X4, sets at Y, etc.[3]

The hours were fixed for the sounding of the first drum to announce the opening of all gates, for the sounding of the second drum to signal the opening of

---

[1] Takigawa, "Kawachi," pp. 8–10.

[2] *Ibid.*, pp. 8–10; Asahina, "Tokei," pp. 576–79, notes that according to Baiennikki's *Konchiji*, "*Imamo Jōkō o takumono no tok io hakaru kōto ari.*"

[3] *Enki onyōryō shiki*, cited by Takigawa, "Kawachi," pp. 8–29.

the Great Gate, for the drum announcing that officials were to leave court, as well as for the drum to announce the closing of the gates.[4]

In the major Japanese cities, such as Nara, the capital of the Heijō period, and Kyoto, the capital during the Heian period, the official time was announced publicly by the striking of a gong and a drum in the tower. The community's bell tower served several functions. It not only sounded the passing hours, but also gave notice to the populace in times of danger, calling for assistance and advising the people in the event of fire, or warning of an approaching enemy.[5]

It is probable that the same facilities existed for announcing the official time and warnings of danger in or to the community, in most of the outposts of the Japanese empire; undoubtedly a community water clock also was provided.

An entry in the last of the Three National Histories, the *Sandai jitsuroku*, noted that on "the twenty-third day of the eighth month of the thirteenth year of Jōgan during the reign of Emperor Seiwa" (A.D. 871), a water clock was erected in Deha province, for the same reason that one was required at Chinju-fu. There is ample evidence that in the mid-Heian period, water clocks were installed in all the seats of the national government, and it may be assumed that *jōkōban* were widely used in cooperation with them, not only in Kyoto but in the provinces as well:

It seems probable that *jōkōban* we use. The sound of the thrusting of the time posts (*toki no kui*) is also charming.[6]

The "time posts", also known as "time labels" (*toki no fuda*), were tablets which the court pages displayed in the small courtyard of the Seiryō palace to announce the time. The Imperial Archives Keepers and Chamberlains organized the day's events by means of these time labels. The Imperial Guards of the Sovereign's Private Office as well as the Inner Palace Guards also divided their watches by this means. An entry for the *Ch'eng men lang* (city gate keeper) in the *T'ang liu tien* from the Chinese T'ang dynasty stated:

When the water clock's notch appeared the bell began striking and continued until the label for the notch appeared.[7]

It is likely that the *toki no fuda* were similar to the labels of the Chinese Water Clock Bureau, and that *jōkōban* were maintained in numerous government offices from the Great Council of State on down through the various levels of the bureaucracy.

In all probability *jōkōban* were used also for scheduling the imperial express riders. The Imperial Rescripts from the Great Council of State were dispatched to

---

[4] *Ibid.*
[5] *Chūyūki*, entry for 1127, cited by Takigawa, "Kawachi," pp. 8–10.
[6] Sei Shōnagon. *Pillow Book* (c. A.D. 991), quoted in Takigawa, "Kawachi," pp. 8–11.

Dazai-fu, the government headquarters at Kyūshū, and to other provinces by the hand of trusted express riders (*ekishi*).[7]

The express riders were racing against time and were ordered to avoid any possible delay, and it was customary to note on the Rescripts not only the date but also the hour of departure. The *Yōrō no kōshiki rei* (Public Institutes of the Yōrō Period) mentioned the necessity to record "... the day, month, year and hour of the Rescripts' acceptance."[8]

The *kōshiki reigikai* (annotated code of public ceremony), containing regulations for express riders, also specified that it was necessary to "Record the year, month, day [hour not mentioned] and name of the guard by the clock."[9] For doing so, however, there was not sufficient time for the Secretary of the Great Council of State to obtain the exact time from the Bureau of Divination, so undoubtedly he maintained a *jōkōban* in his office.

The *jōkōban* appears also to have been used in the Eastern and Western Markets Offices outside the Imperial Palace, as ordered in the *Yōrō no kanshirei* (Yōrō Code Concerning Markets), which required that government employees "Gather at noon in the Market ... and at the third *tsu* disperse,"[10] while the *Yōrō no gokurei* (Yōrō code concerning prisons) reported:

All grave crimes were dealt with in the marketplace after ... and execution of royalty was to be carried out the Hour of the Sheep.[11]

In the Ōchō era (1311–1312), according to these texts, the death penalty was enforced in the marketplace where public crowds gathered, after the Hour of the Sheep, or two o'clock in the afternoon. It is implicit that the work of the Eastern and Western Markets offices could not have been conducted at specified parts of the day without a *jōkōban* to measure time.

After the Taika Reforms, which reorganized the bureaucratic system, the time for beginning and ending work by government officials was established by law. The earliest known mention in a historical record of the time of arrival at work and departure of government officials is an entry for the Taika third year (A.D. 647) in the *Kōtoku-ki* (Chronicle of the Emperor Kōtoku), which specified,

... all with rank ... will gather outside the South Gate and await the hour of the sun's rise.[12]

This protocol was obviously based on the Chinese practice of the T'ang period,

---

[7] It was through Kyūshū, the third largest of the main islands, that the cultures of China and Korea reached Japan.

[8] *Yōrō no Kōshiki rei*, quoted in Takigawa "Kawachi," pp. 8–29.

[9] *Kōshiki reigikai*, quoted in *ibid.*, pp. 8–29.

[10] *Yōrō no kanshirei*, quoted in *ibid.*, pp. 8–29.

[11] *Yōrō no gokurei*, quoted in *ibid.*, pp. 8–29.   [12] *Kōtoku-ki*, quoted in *ibid.*, pp. 8–29.

where, in the capital city of Ch'ang-an, an area was set aside in which officials waited for the gates to be opened.

With the establishment of the Japanese penal code and legal system during the reign of Emperor Monmu, who reigned from A.D. 697 to 707, the beginning and ending hours of work of government officials were made into law with punishments provided for latecomers, as indicated in the *Yōrō no kyūerei* (Yōrō Code Relating to Palace Guards):

It was specified that government officials and artisans were to assemble before the Main Gate [Chōdō] before the sounding of the second "gate-opening drum," which varied depending on the length of daylight, although it was struck generally at the Hour of the Hare or the Hour of the Dragon – 6:00 or 8:00 A.M.

According to the same source,

In opening the gates the first sounding will open all [small gates], the second sounding will open the Main Gate. At the sounding at the end of the audience, when the water-clock is run out the gate drum will be sounded and all gates [will be] closed.[13]

The *Yōrō no shokusei-ritsu* (Yōrō Era Service Regulations Law) provided that

All who are late or not where they should be, those who were not in time for the roll call held immediately after the opening of the bureau offices, are sentenced to ten lashes as punishment. If some were absent an entire night, punishment is thirty lashes, or fines equal to them.[14]

Although tardiness was as much a crime for officials as well as for office employees of lesser rank, they were not subjected to physical punishment. Instead they were fined one *kin* (approximately 100 grams) of copper coins, and the crime was not to be recorded in their personnel file.

Officials holding government positions would be required to leave their homes upon hearing the first "gate-opening drum" in order to arrive in time for the second drum signaling the beginning of work. In order to do so, the officials of higher rank would have to have *jōkōban* in their homes, although it was unlikely that officials of lesser than the sixth rank would also own them. Since the households of the imperial princes above the fourth rank, and officials of the third rank, included numerous stewards, assistant stewards, official retainers, scribes, manservants, maidservants, and other servants, it was necessary to have a *jōkōban* in such establishments.

Just as the dialogue, "How many *sun* of time incense until Early Night?" was held between the *shōkō* and the *daishōya* during the ceremony in the Nigatsudō, a

---

[13] *Yōrō no kyūerei*, quoted in *ibid.*, pp. 8–29.
[14] *Yōrō shokusei-ritsu*, quoted in *ibid.*, pp. 8–29.

similar dialogue probably occurred in the households of the nobility while preparing to appear at court or at work, between the steward keeping watch over the *jōkōban* and the servants preparing the ox cart, or with a maidservant readying the court dress: *Koji ruien* "So many *sun* until the second gate drum."[15]

In the *Koji ruien*, an encyclopedic history of Japan, is contained the *Ryūmei shō* ("Ryūmei Anthology"). Included in the section on divination is a somewhat cryptic passage which appears to relate to the use of the *jōkōban*:

> Neither watching the burning of the incense,
> Nor listening for the call of the watch,
> Neither inquiring after the position of the stars,
> Nor considering the advancing of the moon;
> Even if he says, "It is time to go now,"
> It may be either the Hour of the Rat or Hour of the Ox.[16]

An important seat of government in Kyūshū during the Nara and Heian eras was at Dazai-fu. Just as Nara and Kyoto were divided into a left and a right capital, a central avenue divided Dazai-fu into eastern and western districts. Its government was a miniature version of the Emperor's "Eight ministries and one-hundred offices." In Dazai-fu as well as in the Kanzeon-ji, time was measured by a clepsydra, in cooperation with a *jōkōban* and announced by the customary gong and drum. Even at the present time the people of Dazai-fu refer to a raised wooden area at the right side of the ruins of the ancient bell tower as "The plateau of time" and state that it was the site of the gong and drum from which the time was announced.[17]

The inland post of Chinju-fu in the eastern provinces corresponded to Dazai-fu, and in "the fifth year of Emperor Kōnin's Hōki era," (A.D. 774), a water clock was constructed in that city.

Dazai-fu is situated a short distance from Fukuoka, Japan's tenth largest city and the administrative, economic and cultural center of the island of Kyūshū. From the early seventh to the fourteenth centuries, Dazai-fu served as the seat of the Governor-general of Kyūshū, and was closely associated with Sugawara Michizane (845–903), claimed to be the greatest scholar of Chinese literature of his time. After having achieved the highest post at the Kyoto court, he was demoted, through a rival's intrigues, to the post of Vice-Governor-general of Kyūshū. On his way into exile he stopped on Mount Tenpaizan to pray for Emperor Daigo. Arriving at Kyūshū, he never again left his residence and devoted his life to study. At the Dazai-fu Shrine, called the Dazai-fu Tenmangū, is an

---

[15] *Yōrō shokusei-ritsu*, quoted in *ibid.*, pp. 8–29.
[16] *Koji ruien*, quoted in *ibid.*, pp. 8–29.
[17] *Sandai jitsuroku*, quoted in *ibid.*, pp. 8–29. Translated by Dr. Anthony J. Cannon.

ancient apricot tree said to have followed him into exile from his residence in Kyoto because he loved apricots. Michizane was posthumously deified and worshipped throughout Japan as the God of Literature.[18]

It seems probable that *jōkōban* were also maintained in the Kanzeon-ji as well as in Dazai-fu. The former, a temple of the Tendai sect, was built in A.D. 746 and was repeatedly destroyed and rebuilt. The present edifice was constructed in 1690.

Just as the formal codes required that the time of departure of the express riders be recorded on the Imperial rescripts, the time of arrival of express riders bringing urgent reports to the capital also were recorded.[19]

Due to Edo uprisings which occurred occasionally, dispatch riders were sent from inland Chinju-fu to the capital, necessitating the presence of a water clock in the city. The *Zōkki* (Continued Chronicles) contain the following entry for "the eleventh month of the third year of Hōki" (A.D. 774):

The hour of dispatch should be noted and so there is a water clock . . . and in the province . . . there should be one also.[20]

One of the most important uses to which the *jokōban* was applied in Japan was in agriculture, for measuring the distribution of water from ponds and other local bodies of water to the fields for irrigation. Water was a highly prized life-giving necessity in agricultural communities, and it had to be strictly controlled to ensure that every family received its proper share, particularly in times of drought. A secure method of water distribution was established in Japanese communities from ancient times by means of the *jokōban*.

Much of what is known about this application in agriculture is due to the efforts of Takigawa Seijirō, who supervised the preparation of a history of the city of Hiraoka in Kawachi province. At the time that the project was initiated in about 1960 by the chamber of commerce of that city, a local search was made for documents and artifacts surviving in the ancient households. One of these documents of the Edo period relating to water rights made repeated use of the term *kōsun* (incense inch).

The term *kōsun* is not included in either the *Nōminshi goi* (Glossary of the History of Farming) by Professor Ono Takeo nor in the *Ike no bunka* (Pond Civilization) by Professor Suenaga Tadao, and apparently its use was limited only to documents of the late Edo period relating to the history of Hiraoka.

Further investigation by querying some of the elders of the town revealed that modern Hiraoka originally comprised a number of small villages, each of which had its own small water reservoir. The right of each household in a community to

---

[18] *Japan, The Official Guide*, p. 823.
[19] *Kōshiki reigikai*, quoted in Takigawa, "Kawachi", pp. 8–29.
[20] *Zōkki*, quoted in *ibid.*, pp. 8–29.

draw water for its own fields from the reservoir was limited by a specified amount of time, including the process of opening the pond gate and allowing the water to flow into the fields. This period was measured by the length of time required by powdered incense to burn along one groove of the "incense tray" (*kōban*) specified to be one "incense inch" (*kōsun*). For this reason the right to draw water from the reservoir in this region came to be known as *kōsun*, and accordingly the status of each household was established by the size of its fields and the number of *kōsun* allotted.

Rice paddy cultivation in Kawachi province was of ancient origin. It was the province in which naturalized immigrants settled in the greatest number in the most ancient times. It was in that region that the advanced agricultural methods of irrigating the paddies by opening trenches or ditches from reservoirs were first developed. Techniques were presumably brought from the homeland of the immigrants and applied for these endeavors. Confirmation is to be found in the *Sūjin-ki* (Chronicle of Emperor Sujin), a ruler who according to traditional Japanese history reigned from 97 to 30 B.C. Under the entry for his sixty-second year is noted:

Agriculture is the basis of the world and the brotherhood of the people. Now Kawachi is plentiful in chasms and mountains but arable fields and water are scarce. Therefore the people of this area are troubled in their farming. Let there be many ponds and irrigation ditches opened up.[21]

The use of naturalized immigrants for excavating the ditches is noted also in the *Ōjin-ki* (Chronicle of Emperor Ōjin), who reigned from A.D. 270 to 310. The entry for "the seventh year, the autumn, the ninth month," noted

Koguryŏ, Paekine and Silla people ... were set to work building a great tank which was then known as the Korean tank.[22]

In the *Nintoku-ki* (Chronicle of Emperor Nintoku), who reigned from A.D. 313 to 399, the subject of irrigation is again mentioned. The entry for "the eleventh month, in the winter of the fourteenth year" of his reign (327), reported that an order was made to:

Further, dig a canal for Kanku and lead the Ishikawa [river] in to enrich the upper and lower Suzuka and upper and lower Toyoura and all the places therein ... 40,000 and more of land on which the people may settle and will know no bad years.[23]

It was inevitable that after the reservoir had been completed and was ready for

---

[21] *Sujin-ki* (Chronicle of Emperor Sujin), quoted in *ibid.*, pp. 8–29.
[22] *Ōjin-ki* (Chronicle of Emperor Ōjin) quoted in *ibid.*, pp. 8–10.
[23] *Nintoku-ki* (Chronicle of Emperor Nintoku), quoted in *ibid.*, pp. 8–10.

use, conflicts would arise concerning the distribution of water between the farmers whom it would supply. The problem of controlling water rights existed as early as the late fourth century during the reigns of Ōjin and Nintoku, and conflict continued long thereafter. It is likely that as a consequence the settlement of water rights was first achieved by means of the *jōkōban* and that it was first used for that purpose in Kawachi province before the procedure was established in other parts of Japan. It appears to have been the earliest civil use made of the *jōkōban* in Japan.

The *Nihon shoki* (Chronicle of Japan) of A.D. 720 provided an account of the construction of the reservoir at Sayama in Kawachi and of the Kanku canal in Ishikawa prefecture in Kawachi.[24]

It is no longer readily clear, however, which areas in the present day correspond with "upper and lower Suzuka" and "upper and lower Toyoura." However, Namegawa Naga in his *Kawachi-shi* (Kawachi Record) tentatively identifies Toyoura and Izumoi of Hiraoka City with upper and lower Toyoura.[25]

As questions concerning the distribution of water among the farmers arose not long after the completion of the irrigation ditches, it was logical to resolve the problem by means of measuring time. At first, the measured burning of incense in the form of the graduated stick may have been used. Even before the Heian period a better solution was found in the use of the *jōkōban* in Kawachi.

However, a problem continued to exist during the Kamakura period in those remote communities, where neither incense sticks nor the *kōban* were known or available.

That it became impossible to obtain powdered incense and incense sticks in remote regions of Japan was reported in local documents, including the *Bokumin kinkan* (Golden Mirror of the Herdspeople), *Chihō ochiboshū* (Collection of Local Gleanings), *Zōho den'en ruisetsu* (Supplementary Notes of Fields and Gardens) and *Chihō bonrei roku* (Local Record of Common Events).[26]

As a consequence, a crude form of homemade water clock was used for timing water rights during those periods. It consisted of a tub of specified capacity, through the side or bottom of which a hole had been drilled. The tub was filled with water and the time required for the vessel to empty was designated as one unit. A specified number of units were assigned to each household for the flow of water from the pond or reservoir. This form of water distribution was known as *tarumizu* (barrel water). It is believed that the *tarumizu* method was based on an

---

[24] *Nihon shoki* (Chronicle of Japan), A.D. 720, quoted in *ibid.*, pp. 8–10.

[25] Namegawa Naga, *Kawachi-shi* (Kawachi Record) quoted in *ibid.*, pp. 8–10.

[26] *Bokumin kinkan, Chihō ochiboshū, Zōho den'en ruisetsu, Chihō bonrei roku*, quoted in Takigawa, pp. 8–29. For information about water control for irrigation in a modern Japanese village, see Beardsley, Hall and Ward, *Village Japan*, pp. 126–38.

example used in more agriculturally advanced home provinces, where the volume of water was measured by the time required for draining one bucket. *Tarumizu* is described in the "Golden Mirror of the Herdspeople," the "Collection of Local Gleanings," the "Supplementary Notes of Fields and Gardens," and in the "Local Record of Common Events" previously noted.[27]

At the time that a history of Hiraoka was being contemplated, a search was made for a *jōkōban* of the early period that had been used in the community for regulating water rights. This brought to light a *jōkōban* in the home of the Inoue family. Although described to have been similar in structure to the one featured in the Nigatsudō, its incense trail was probably in the form of the swastika with double crampons. According to an Inoue descendant, these devices were used for controlling water distribution in Kasuka village as late as the Taishō period and possibly until the beginning of World War II.[28]

Each year when the drought came to rural areas and it became necessary to draw water from the community reservoir, the same age-old ceremony took place. Several of the villagers were designated to carry the *kōban*, as the *jōkōban* was known to the village, to the Hiraoka shrine. It was borne upon their shoulders, suspended from one end of a pole while the matting in which it was stored was suspended from the other. Villagers and farmers joined the procession. Upon reaching the precincts of the shrine, those in the procession first washed their hands and then engaged in prayer to the shrine's deity, promising to observe the *kōban* faithfully. Worshipping of the deity symbolized a sacred oath that each of the elders would watch over the *kōban* with an honest heart and do nothing that was in the least dishonest, such as incorrectly calculating the time.

The fire used to light the *jokōban* was taken either from the holy fire at the shrine or from a lantern lighted from the holy fire and brought out from the shrine. Upon conclusion of worship, the *kōban* was placed in an appropriate location within the confines of the premises of the shrine and sheltered by means of the matting, which was disposed in such a manner as to divert any movement of air and keep the incense trail alight.

When all was ready, one of the elders lighted the end of the incense trail in the *kōban*, and another simultaneously struck a bell to notify the keeper of the reservoir, or pond-keeper, which was some distance from the shrine. The keeper then promptly lifted the gate of the conduit, allowing the water to flow. The elders kept a watch over the *kōban*, and when a *kōsun* of incense had been completely burned, they again struck a bell and the pond-keeper closed the water gate. Most often the farmer would have brought several witnesses with him to be present at

[27] Takigawa, "Kawachi," pp. 8–10.
[28] *Ibid.*

the water gate, in order to ensure that the pond-keeper performed his duties promptly and in accordance with the regulations.

Meanwhile, since the vigil of the *kōban* was extremely boring, the elders engaged in a game of checkers or chess. A chessboard and chessmen or checkers were stored in the bottom drawer of the *jōkōban* for that purpose. This the elders removed at the beginning of their vigil, and played until their duties had been fulfilled. Needless to say, such a chessboard was not part of the *jikōban* or *jōkōban* used for religious or other purposes, such as ceremonies at the Nigatsudō.[29]

The practice of employing time measuring devices for allocating water to farmers in periods of drought or in regions where the water supply was constantly limited, was used also in the Western world from at least the Middle Ages. In eastern Spain two of the methods of time measurement for regulating water rights had been derived from the Muslims. One consisted of dividing the water source, such as a river, proportionately between canals and individuals, based upon the amount of land requiring irrigation. In this manner an equitable distribution could be assured without the need of measuring the flow of water. In smaller areas of eastern Spain as well as in North Africa, fixed time units were imposed for water flow. The unit was described as lasting from sunrise "until a man's shadow attains a length of eight feet [2.4 m] . . . which will be more or less two hours after sunrise," and the measurer paced out the distance with his own feet.

In medieval Spain three timing devices were used to control water distribution for irrigation: the sinking bowl clepsydra; the *gadus*, which was a crude form of water clock, and the sand clock. The sinking bowl clepsydra had been in use in Egypt and Babylonia for regulating water rights since before 1500 B.C. and continues to be used in the Near East and North Africa in modern times. The *gadus* was the bucket of a noria which was punctured in such a manner that it could serve as a makeshift outflow clepsydra. A later device was the sand clock, which first appeared in the fourteenth century and was much used by astrologers and mariners.

By the late fifteenth century, the mechanical clock was adapted for regulating irrigation rights. One was installed by the local lord, Père Bou, in his castle in the village of Collosa, near Alicante. It was made by the Valencia clockmaker, Micer Rubi, and proved useful to the villagers because for the first time "the water for irrigation is divided into hours." In the 1490s a mechanical clock installed at Granada was used to govern water flow for irrigation by the striking of the "Irrigation Bell." Despite the advent of the mechanical clock, however, the practice of timing of shadows persisted.[30]

---

[29] *Ibid.*, pp. 8–29.
[30] Glick, "Medieval Irrigation Clocks," pp. 424–28; Hilton-Simpson, "Further Notes," p. 430;

The use of incense for time measurement in Japan is not a practice limited to the past, but has modern applications as well. In a relatively recent study, Dr. Suzuki Seitarō reported his year-long experiments on the effect of weather on combustibility in Japan. In his experiments he used incense sticks in preference to other means because they burned always at the same rate. These sticks were nearly 1.5 mm in diameter, and made of pulverized fragrant wood bark and pine resin formed into a paste which was then hardened. Setting several ignited incense sticks out of doors on a screen from January 14 to December 3, 1924, he noted that the rate of combustibility varied. He also experimented with artificially induced weather conditions within a closed room.

Dr. Seitarō determined that ignited incense sticks changed weight constantly when exposed to the open air, a change assumed to be due to the amount of moisture retained by the stick and the variation of relative humidity in the surrounding air. The stick's weight was reduced to a minimum when kept in a dry state for five days in a desiccator filled with concentrated sulphuric acid, and it recovered its old value when exposed to the outdoors. For example, five sticks were reduced to 1.159 g in the desiccator, and after exposure to outside air for three hours and forty minutes, reached a maximum weight of 1.259 g, showing an increase of 0.1 g due to the increase of humidity in the air. Using 15 degrees Centigrade as the mean room temperature during the drying process, the relative humidity of the desiccator is far less than one percent.[31]

One of the most specialized uses of incense for time measurement in Japan from the Edo period until recent times was the computation of the cost of entertainment in the "Flower and willow world," the name given to the licensed quarter in which geisha houses (*okiya*) were to be found. The name was derived from that of the licensed quarters for courtesans in China, and was adopted in Japan to specifically identify the geisha quarters.

The word *geisha* is a Sino-Japanese word meaning "a skilled person," and was applied to the professional singing and dancing girls of Japan. The geisha came into existence in the mid-nineteenth century, first in the purlieus of the shrine of the war god Hachiman in Fukagawa, on the eastern side of the Sumida River. Small boats from this quarter sailing upriver were accustomed to mooring near a tributary of the Sumida. In time the boat houses developed into tea houses and were frequented by the Fukagawa geisha. It was not long before the outlet at Yanagibashi developed its own geisha community, which subsequently became

Garrido-Atienza, "Los Alquezares," p. 65; Hilton-Simpson, "The Influence of Its Geography," pp. 27–29; Archivo del Reino de Valencia, *Gobernaçion*, 2377, 46th hand of 1485, fol. 13ᵛ, June 1, 1486; Scott Moncrieff, *Irrigation*, p. 118.
[31] Suzuki Seitarō, "The Fires," pp. 2–22.

centered at Edo (Tokyo). In time geisha flocked wherever people gathered, and generally were to be found near public resorts.

The true geisha underwent a period of strenuous training, with lessons in dancing and singing, which sometimes began as early as the age of seven. A geisha was required to be proficient in music and to possess some knowledge of the *samisen*, to be able to sing, dance and discuss the most trivial topics, play games such as *ken*; all designed to while away the time pleasurably for her guests.

There were three types of geisha. The first was the *jimae* who was her own mistress, living in her own home; another type was the *kakae*, who entered service in a geisha house for a specified term under contract and was supervised by a house mistress, the third type, the *tateki-age*, was financially independent but associated herself with a geisha house because she may have lacked introduction to the tea houses of the locality in which she was a stranger. Almost as free as a *jimae*, she was required to earn enough to bring the house a fair profit. In addition there were the geishas who were too young, called *hangyoku*, and had to be trained; they earned half as much as the older ones. The capitation per month on a geisha was 2 yen, and 1 yen for apprentices.[32]

A geisha's earnings were of two types. The first was the regular fee for entertainment (*gyoku*), which was added to the tea-house bill. The fee was based on time, and in the better establishments in one period it was twenty-five *sen* per hour. The time was calculated on the basis of the burning duration of an incense stick, variously reported to be from twenty-five to thirty minutes. The time spent by a geisha with a guest during an entire evening was generally calculated to be equivalent to four incense sticks. The geisha houses compiled their accounts monthly and paid the geishas on this basis, but in fact most of a geisha's income was derived from gratuities received from her guests.

Several systems of compensation for the geisha existed. In the early period a girl was apprenticed to a geisha house, the owner of which paid the parents or guardian a fixed sum. For a stated period of apprenticeship the novice was provided training, clothing, and room and board at the cost of the mistress. After her training had been completed, her earnings were divided with the geisha mistress. This division often consisted of seven parts for the employer and three parts for the geisha; by another system the earnings were divided half and half. In yet another arrangement, a trained geisha attached herself to a prominent geisha house, and there she would live, paying all her own expenses and arranging an independent system of sharing her earnings with the house mistress.

In more recent times, geishas were no longer apprenticed. If a girl decided to become a geisha, she would at the age of eighteen make application to a central

---

[32] Inouye, *Sketches*, pp. 52–62.

registry office after having satisfactorily completed secondary school training. She was required to appear for an interview before a selection board consisting of several of the senior geisha mistresses, to demonstrate proficiency in one of the several arts noted. If accepted, and after being registered, she would be required to pay a monthly fee and was expected to attend a school conducted by the central registry office. The geisha lived at home and traveled to the geisha house in which she was registered.[33]

Formerly, the period of engagement of a geisha with a guest was calculated not by the clock, but by the time required for an incense stick to be consumed. The geisha's working day consisted of ten hours. The charges for her services were always listed on the bill as the money amount for the number of "sticks" consumed.

The record of a geisha's entertainment was maintained by means of an "incense stick clock" (senkōdokei) a container generally made of cryptomeria wood (sugi) in a flat or tented rectangular shape. The use of the word senkōdokei is a misnomer, for it is neither a clock nor a time measurer in the true sense, but merely a device for recording the passage of time. The upper surface is perforated with twenty round openings from which project metal or bamboo holders for incense sticks. Each stick holder or opening in the senkōdokei was assigned to one of the oiran in the establishment and marked with a tag bearing her name, which was attached to an iron prong beside the opening. A drawer encompassing the full interior of the container was used for storage of complete incense sticks and remnants of those which had been consumed (Figs. 129–31).[34]

As each geisha became engaged during an evening to provide entertainment for a guest, an incense stick was dropped into the bamboo holder in the opening labeled with her name, and ignited.

When, during the course of the evening, the burning incense stick reached its end, it was removed and replaced with a fresh stick from the supply in the drawer, and the burned remnant set aside with the geisha's name. An accounting of the sticks burned during each geisha's service was maintained by the mistress of the establishment by means of an abacus (soroban). As late as 1924 geisha entertainment was still being computed and charged on the basis of incense sticks consumed (Fig. 132).[35]

The use of the senkōdokei was described in a recent work by Ernst Jünger from details he obtained from Giko Takabayashi:

a wooden box in which nine holes are drilled: one for each of the girls living in the house. Each time one of the girls retired with a guest, the host (or senior in charge of the geisha

[33] Scott, *Flower and Willow*, pp. 54–58, 111, 143–44, 168–70: Inouye, *Sketches*, pp. 48–63.
[34] Asahina, "Tokei," pp. 576 ff.
[35] Giko Takahashi, quoted by Jünger, *Das Sanduhr*, pp. 49–50.

house) would light an incense stick and set it up for her ... The duration of burning corresponded to a definite amount. A flower girl could say, "Yesterday I earned six sticks!" Today such boxes are no longer in use. But the Japanese word for the sticks, "*senkō*" still retains its meaning, so that even today one asks what is the price in this or that district for a "flower girl incense stick." To such an elegance of the blossom language, we Western barbarians would never penetrate.[36]

[36] Jünger, pp. 47–52.

13

The fire in the smoke

Time measurement with incense, and particularly with the incense seal, was at no time a frivolous diversion in East Asia. In addition to the ordinary incense stick, which was commonly used for time measurement in all Far Eastern countries from an early period, the incense seal was used continuously in China and Japan until the end of the nineteenth century, a period of more than twelve hundred years. Time measurers of incense served utilitarian purposes in religious and civil life, and the incense seal in particular captured the imagination of Chinese scholars and writers through the centuries.

In due course Chinese and Japanese incense seals were occasionally purchased by foreign travelers and dealers, and found their way to other countries as curiosities. New England shipmasters and sailors acquired examples, usually in the belief that they were incense burners or opium stoves, and brought them home from their voyages during the years of the China trade in the nineteenth century.

In time a few of these found their way into museums, such as the collection of the East India Society in the Peabody Museum in Salem, Massachusetts. Others were relegated to New England attics where descendants of sailing families have from time to time discovered them and disposed of them in the shops of local dealers in antiquities. Some have been brought to the American market by importers and dealers on the West Coast. Inasmuch as incense seals were classified under the applied arts and were never considered *objets d'art*, they proved to be of little interest to art collectors and museums in general.

Their time measurement function was not generally known even to museums prior to the mid-twentieth century, and therefore they failed to arouse the interest of clock collectors. It was not until the publication of an account in English of the use of incense for time measurement that Chinese and Japanese "incense clocks" attracted the attention of collectors and others, and became desirable acquisitions.[1]

It is unlikely that many have survived in Buddhist temples in China, however. With the formation of the People's Republic of China in 1949, Buddhism and

---

[1] Bedini, *Scent*.

Taoism, which had already suffered serious setbacks from the Nationalist government, were stripped of their lands and temples by land reforms. Priests were forced into so-called "productive labor," and temples were converted into public meeting places, warehouses, and granaries. Anti-religious campaigns were launched to demonstrate to the peasants the antisocial and superstitious nature of ancestor worship and associated religious ceremonies.

Under the auspices of the Communist regime, an association was formed to control Buddhism in Tibet and Inner Mongolia and to provide to countries of southeast Asia the impression that observance of religion was permitted in mainland China. Christian churches and missionaries suffered a similar fate, and it was not until 1978 that a number of churches were permitted to reopen. As a consequence of the usurpation of Buddhist lands and the closing of the temples, a great number of the Buddhist ritual items not only in the temples but also in the ownership of private families were confiscated, and then destroyed or stored in government deposits.

When mainland China was opened up to foreign commerce and traders were allowed into the country to make purchases, often they were required to buy up entire miscellaneous lots including numerous items of the same nature; materials which they may not have considered to be particularly marketable and in which basically they had no interest. These lots included silver jewelry, decorative boxes, needle cases, ink stones, *I-hsing* pottery, incense seals, and similar items. It is by this means that they were brought out of the country in substantial numbers. From about 1975, discreet quantities of these items, including incense seals, found their way into shops of dealers in antiquities in Hong Kong, and to a lesser extent in Macao and other localities outside of mainland China. Dealers first purchased these lots in Canton and later in Peking until by about 1979 supplies began to diminish. By 1986 the Chinese closed their wholesale warehouses due to depletion of stock and the fact that there was no more confiscation from the people. The government wholesale sources have transferred the remainder of their inventories to government retail outlets in the larger cities such as Peking and Shanghai, where antiques are now sold in "friendship-stores," which also purchase antiques from private family collections for resale. Private families in China have retained family heirlooms and are now collecting antiques. In some instances items of value are being smuggled out of China for sale elsewhere to realize higher prices. Incense seals are occasionally acquired by dealers outside the mainland from government-run antiquities shops. As a consequence of the diminished supply, prices for these items have escalated substantially.

At first, dealers generally were unfamiliar with the purpose and value of the incense seals, and they were sold for modest prices. Most of the examples that made their way into the shops were coated with a stubborn heavy tarnish resulting from dampness and other deleterious storage conditions to which they had been exposed over a period of years. Dealers made them marketable by offering to have the surfaces buffed to make them bright, at the same time removing all patina.

Identification of incense seals as a form of timekeeper soon followed, rendering them immediately desirable and sought after by clock collectors. During a visit to Hong Kong, one American collector was able to purchase as many as twenty-six Chinese incense seals at one time, in various states of disrepair, all of them nineteenth-century examples in a variety of metals and forms of the type favored by Chinese intellectuals. Similar experiences were reported by other visitors to Hong Kong during the same period. The major supply from mainland China was eventually exhausted, and only a few examples have since become available from time to time.

Although they are no longer used for the purpose originally intended, incense seals still survive in some Japanese Buddhist temples, preserved with other Buddhist memorabilia of the past. During the past several decades Japanese incense seals, categorized as *kōbandokei*, have appeared on the antiquities market in Japan only infrequently. Inasmuch as they were primarily used in temples and government offices, and were not used by Japanese scholars and other intellectuals as they had in China, the Japanese incense timekeepers were originally produced in limited numbers and consequently are much scarcer than Chinese examples.

The burning of incense always has been a vital part of life in Far Eastern countries, and it has been the subject of philosophical interpretation as well as of practical use. In his *Hyaku-tsū-kiri-kami* (Hundred Writings), the Shinshū priest Myōden described the Incense Paradise, stating that in "the thirty-second Vow for the Attainment of Paradise of Wondrous Incense" it was written

That Paradise is formed of hundreds of thousands of different kinds of incense, and of substances incalculably precious; – the beauty of it incomparably exceeds anything in the heavens or in the sphere of man; – the fragrance of it perfumes all the worlds of the Ten Directions of Space; and all who perceive that odor practise Buddha-deeds.

Myōden went on to note that in ancient times certain individuals of considerable wisdom who as a consequence of having taken their vow, achieved perception of the odor. This perception was not possible for men of later times born with less wisdom and virtue. He added, "Nevertheless it will be well for us, when we smell the incense kindled before the image of Amida, to imagine that its odor is the wondrous fragrance of Paradise, and to repeat the *Nenbutsu* in gratitude for the mercy of the Buddha."

Myōden also compared the life of man to the smoke of incense, and quoted from the Buddhist work *Kujikkajō* (Ninety Articles):

> In the burning of incense we see that so long as any incense remains, so long does the burning continue, and the smoke mounts upward. Now the breath of this body of ours, – this impermanent combination of Earth, Water, Air and Fire – is like the smoke. And the changing of the incense into cold ashes when the flame expires is an emblem of the changing of our bodies into ashes when our funeral pyres have burned themselves out.[2]

[2] This and the two preceding quotations are from the *Kujikkajo* "Ninety Articles," quoted in Myōden, "The Hundred Writings;" Hearn, *Ghostly Japan*, pp. 28–29.

APPENDICES

APPENDICES

Appendix A

Chronology

DYNASTIES OF THE EMPIRE OF CHINA

Prior to 2205 B.C.	Legendary age	535–556	Western Wei
2205–1766 B.C.	Hsia	550–577	Northern Ch'i
1766–1123 B.C.	Shang or Yin	557–581	Northern Chou
1122–249 B.C.	Chou	590–618	Sui
722–480 B.C.	Ch'un Ch'iu	618–906	T'ang
483–221 B.C.	Warring Kingdoms	907–960	Five Dynasties
221–207 B.C.	Ch'in	907–923	Later Liang
206 B.C.–A.D. 7	Earlier Han (Western)	923–936	Later T'ang
A.D. 25–220	Later Han (Eastern)	936–947	Later Chin
220–265	Three Kingdoms (Shu, Wei, Wu)	947–950	Later Han
		951–960	Later Chou
221–264	Shu	907–1125	Liao
220–265	Wei	960–1126	Northern Sung
222–280	Wu	990–1127	Hsi-hsia
265–317	Western Chin	923–936	Southern T'ang
	Southern Dynasties	960–1126	Pei Sung (Northern Sung)
317–420	Eastern Chin	1127–1279	Nan Sung (Southern Sung)
420–479	Former Sung [Liu]		
479–502	Southern Ch'i	1279–1368	Yüan
502–557	Southern Liang	1368–1644	Ming
557–589	Southern Ch'en	1644–1912	Ch'ing
	Northern Dynasties	1912–1949	Republican Period
386–535	Northern Wei	1949–	People's Republic of China
534–550	Eastern Wei		

RULERS OF JAPAN

660 B.C.–A.D. 708	Semi-historical Period	781	Kanmu
710–784	Nara Period (Capital at Nara)	806	Heijō
		794–1185	Heian Period (Capital at Kyoto)
715	Genshō (Empress)		
724	Shōmu	1192–1333	Kamakura Shogunate: period of northern rule (capital at Kamakura)
749	Koken (Empress)		
759	Junnin		
765	Shōtoku (Empress)		
770	Kōnin		

1338–1568	Ashikaga Shogunate (capital at Kyoto)	1603–1867	Tokugawa Period
		1615–1868	Edo Period
1568–1603	Muromachi and Azuchi-Momoyama Periods	1867–1912	Meiji Period
		1912–1926	Taishō and Showa Eras

DYNASTIES OF KOREA

57 B.C	Kingdom of Silla	1494–1506	Yonsan-gun
c. 1392–1776 A.D.	Yi Dynasty (Kingdom of Chosun)	1506–1544	Chungjong
		1544–1545	Injong
Reigns:		1545–1567	Myongjong
1392–1398	T'aejo	1567–1608	Sonjo
1398–1400	Chongjong	1608–1623	Kwanghae-gun
1400–1418	T'aejong	1623–1649	Injo
1418–1450	Sejong	1649–1659	Hyojong
1450–1452	Munjong	1659–1674	Hyonjong
1452–1455	Tanjong	1674–1720	Sukchong
1456–1468	Sejo	1720–1724	Kyongjong
1468–1469	Yejong	1724–1776	Yongjo
1469–1494	Songjong		

Appendix B

Documents

1. Pasquale D'Elia, S. J., ed., *Fonte Ricciane*, vol. 1, p. 33:

Gli loro horiuoli sino adesso furno di acqua e di fuoco con certe pipite odorifere, fatte tutte della stessa grandezza; fanno anco altri con ruote mosso di arena; cose tutte che di se tengono molta imperfettione. De' solari, solo hanno l'equinotiale, ma non lo sanno ben collocare conforme ai luoghi dove il pongono.

2. Nicolas Trigault, S. J., *De Christiana Expeditione Apud Sinas*, p. 22:

Horis metiendis vix habent instrumenta, quae habent, vel acqua vel igne mensurantur. Aquea sunt velut intes clepsydrae; Ignea ex odorifero cinere confecta, tormentorum nostrorum fomites imitantur. Pauca etiam alia conficiunt rotulis ab arena velut acqua circumactis, sed omnia as nostratia artificia, sunt umbra, et fere multum peccant in ipsa metiendi temporis symmetria. E sciotericis solum norunt illud, quod ab aequatore nomen accepit, sed neque illud pro ratione locorum didicerant collocare.

3. João Rodrigues Tçuzu, *Historia da Igreja do Japão*, edited by João de Amaral Abranches Pinto. Collecção Noticias de Macau, vol. 2, ch. 15, pp. 129–30:

Os Japoens não tem relogios ordinarios para medir as horas, porem os Bonzos em seus templos para saberem as horas de rezar, e tangrem, ou darem sinal a hora, tem huns *relogios de fogo* muy artificiozos com que medem as hoas assim dos dias grandes como pequenos, tendo para isso suas medidas determinadas conforme a quantidade do dia com que repartem sempre em seis horas quer seja dias comprido quer pequeno. Concertão o relogio deste modo: Tem huma caixa de madeira quadrada cheya de certa laya de cinza muy fina peneirada, e muy seca, cuja superficie estã muy plana, e nella com certa medida que tem fazem huns regos de certo comprimento, largura, e profundidade, continuados a modo de quadrado, e os enchem de hum certo pö, ou farinha, feitos de cascade huma certa arvore muito secos, que juntamente tem algum cheiro, e per huma das cabeṣas lhe poem o fogo com que vão ardendo muy de vagar, queimando em cada hora hum delles com que mede a hora muy arrezoadamente por terem ja experiencia do mod que se regem para o fogo continuar, e sempre do mesmo modo, e o mesmo compasso sendo o rego mais, ou menos comprido conforme ao dia, e noite, mayor, ou menor com sua proporção.

Os Chinas tem relogio do Sol ... Tem tambem relogios de agoa com que medem as horas ...

4.   P. Gabriel de Magalhaens, S. J., *Nouvelle Relation de la Chine*, pp. 153–54:

Les Chinois ont trouvé, pour regler & mesurer les parties de la nuit, une invention digne de la merveilleuse industrie de cette Nation. Ils mettent en poudre un certain bois en le rapant & le pilant; ils en font une espèce de pâte, dont ils forment des cordes & des batone de diverses figures. On en fait quelques-uns d'une matière plus pretieuse, comme de sandal, de bois d'Aigle, & d'autres bois odorans, & de la longeur d'un doigt ou environ, que les personnes riches & les Lettrez font brüler dans leur chambres. Il y en à d'autres plus ordinaires, d'une, de deux & de trois coudées de longeur, & même d'une, de deux & de trois aunes, & un peu plus ou un peu moins gros qu'une plume d'Oye, qu'ils brulent devant leurs Pagodes ou Idoles. Ils s'en servent aussi comme d'une méche pour porter du feu d'un lieu a un autre. Ils font ces cordes, de poudre de bois d'une grosseur égale, en les passant par une siliere, ou un trou fait exprés. Ensuite ils les entortillent en rond, en commençant par le centre, & forment une figure spirale & conique, qui s'élargit à chaque tour, jusqu'à une, deux, & trois palmes de diamétre, & même davantage, & dure un, deux & trois jours, & plus encore, proportion de la grandeur qu'on luy a donnée: car on en voit dans des Temples, qui durent dix, vingt & trente jours. Ces machines ou méches ressemblent a une nasse de pescheur, ou à une corde entortillée autour d'un cone. On les suspend par le centre, & on les allume par le boit d'enbas, d'ou le feu tourne lentement & insensiblement, suivant tous les tours qu'on à fait faire à cette corde de poudre de bois, sur laquelle il y à ordinairement cinq marques pour distinguer les cinque parties de la veille ou de la nuit. Cette maniere de mesurer le temps est si juste & si certaine, que jamais on n'y remarque aucune erreur considerable. Les Lettrez, les Voyageurs, & tous ceux que veulent se lever à une heure précise pour quelque affaire, suspendent à la marque qui indique l'heure a laquelle ils veulent s'éveiller, un petit poids, qui, quand le feu est arrivé a cet endroit, ne manque pas de tomber dans un Bassin de cuivre, qu'ils ont mis au dessous, & de les réveiller par le bruit qu'il fait en tombant. Cette invention supplée à nos Horloges à réveil, avec cette difference, qu'elle est très-simple & a si bon marché, qu'une de ces machines qui peut durer Vingt-quatre heures, ne Coûte qu'environ trois deniers, & que les Horloges sont composées de quantité de roues & d'autres pieces, & sont si chères, qu'elles ne peuvent estre employées que par des personnes riches.

Appendix C

Writings about Ting Yün
(Ting Yüeh-hu)

Following are translations by Kirby R. Vining of the dedicatory prefaces written by the friends and associates of Ting Yün (Ting Yüeh-hu), during the reign of Ch'ing emperor Te Tsung (1875–1908). These were collected and published in the memorial volume *Hsiang lu t'u p'u*, which was perhaps also known as *Yin hsiang t'u p'u*.

1. By Shih Yün-sheng (Courtesy name Lan-pin):[1]

Where the Yangtze River meets the sea, at a place called Mai Yü Wan [Fish-mongers Bay] some twenty or thirty miles north of Lang Mountain [Wolf Mountain], lived a reclusive scholar by the name of Ting, called Yüeh-hu [Moon Lake] who has been my friend for longer than I can remember. Mr. Ting has a passion for things ancient, writes very well, and in addition writes in the *chang ts'ao* script[2] and the *mo yin* script,[3] is an accomplished painter, versed in verse and prose styles – all to a high degree of refinement. He is also skilled in the [drinking] game of *t'ou hu*.[4] It comes as no surprise that Mr. Ting has many friends and contacts among the gentry, nobility, and the learned who often make gifts to him of books and money when he visits government posts south of the Yangtze River. Mr. Ting, however, is humble, having no ambitions for fame or official career, acclaim or criticism, and has been content to wield brush and ink for more than fifty years of his life. I am Mr. Ting's junior by twenty years, still rather uninformed and of only superficial opinions, still unworthy to be equal to the glory of my elder, Mr. Ting. However, there is between us a mutual respect and friendship, of which I am unworthy, that has developed in the "short time" we have known each other.[5] In the spring of 1876 we happened to be speaking of something called an "incense seal." I indicated that I knew of it as a rather crude device, possessing no refinement and unworthy of a gentleman's appreciation. Mr. Ting, upon hearing this, promised that if I should live yet a while longer, that he would

---

[1] This preface was written in a very clear *k'ai shu*, or standard formal calligraphy, although phrased in a stilted classical vocabulary.

[2] A form of cursive writing that originated in the Han dynasty.

[3] A seal script patterned on one used in the Ch'in Dynasty.

[4] An ancient drinking game in which arrows are tossed into a bottle placed at a great distance.

[5] Shih Yün-sheng cannot recall how long she has known Ting Yün, suggesting a long period of time, but as Ting's junior, it would be inappropriate to indicate a long time from Ting's point of view.

bring a pattern[6] from a new, elaborate incense seal for me to see. This pleased me greatly. Mr. Ting set to work with great energy and concentration, designing patterns. [Each] one was stranger than the previous one, and he created them in worked metal, earthenware, forged metal, hammered metal, cut metal, in the end producing items of such high refinement that people came from far and near to have a look at them – rarely seen treasures. I asked Mr. Ting why he did not put together a book of these patterns that all might see them. He agreed, and assembled a book of one hundred of the patterns, arranged for comparison, also allowing all to see his accomplished writing hand and get a glimpse of the refinement which permeates all of his work. He seeks no official praise for this work, but merely presents this collection for the benefit of those, like himself, who find joy in books and music which can dissipate sorrow. I believe there are many who will appreciate this, and like myself, remember seeing this book as an auspicious occasion.

> 1878, autumn, eighth month of the lunar calendar. Shih Yün-sheng, courtesy name Lan-pin, respectfully presents this preface.[7]

2. By Ch'en Lang-wen:

If a man is negligent and frivolous, and does not make proper use of his virtue, he will end up spending his life aimlessly indulging in passions, becoming more and more depraved. The ancients who acted with understanding did so by acting in accord with virtue, letting it guide them. Without occupation to organize and clarify one's thoughts, actions of true understanding and virtue are not possible. Confucius relates this, saying it would be better to do nothing at all than to act not in accord with virtue and understanding. He says that when men of understanding and learning rule the world, even the simple villagers will be attracted to them. The great man uses his understanding and virtue to benefit the people, while tradesmen and merchants grow lazy [as to virtue] thinking only of personal gain. Those whose ambitions are aimed at obtaining a career as a bureaucrat use their understanding in pointlessly shallow, stupid arguments, heedless of the importance of spiritual matters. Those people are doomed to remain in their narrow niche.[8]

Only [this] reclusive scholar has neither ambitions for an official career nor the meanness of a trader's or merchant's life, as he remains simple, tranquil and balanced. He spends his days playing the lute, composing poetry and sitting in meditation before burning incense. Although it would seem to some that this is not a proper application of virtue and understanding, sitting in meditation before one of Moon Lake's burning incense seals, you will come to see the correct application to which he has put his understanding.

The reclusive scholar Moon Lake is possessed of peerless talent, intelligence and learn-

---

[6] Rubbing or other image.

[7] Two seals: "Lan-pin" (author's courtesy name) and "Shih Yün-sheng *yin*" (Seal of the author).

[8] The first paragraph consists of a mixture of quotations from Confucius to form a somewhat typical note of praise for Ting Yün.

ing, such that if his skills were divided among ten other men, each would still be haughty and proud of his skills.

Not seeking a bureaucrat's career, he retired to the mountains where he leisurely imitates the ancients in fashioning his book on incense seals. This work is profound in thought and execution, containing not the smallest flaw. I would say he has used his virtue and understanding towards beautiful ends. It would be sad if this, his life's work, were not circulated. He has exerted himself not only to record all he knows on the subject, but to carefully see that it was all correctly presented. I am sure this work of his will be well received, but probably there are those who will say he has wasted his virtue and understanding, time and blood, making little use of his great talent. But those who say this do not truly know the joy that can come of the correct application of virtue and understanding.

Second month, 1878, Hsiao Shan (Mountain),
Ch'en Lang-wen.[9]

3. By Kan Po:

Moon Lake of Yü Wan is talented – remarkably so. He has had a disposition towards scholarly activities for his entire life, excelling in calligraphy, poetry and painting. His prose and music sing of South China. He is an engraver of ancient seal scripts and adept at various instruments. As broad as is his learning, he remains humble. He showed me his book of incense seal diagrams and asked me to add a preface to it. I was quite surprised at [the book]. The Han Dynasty scholar Yang Hsiung said that in his era there were only unskilled scribblers who were not emulating the brilliant works of the past, and that a gentleman of taste could not be expected to tolerate the products of such hands. Such is not the case with Moon Lake's works, which move those with discerning taste to remark at their delicate beauty. There are not fifteen scholars living who could surpass Moon Lake's skill. The incense seal diagrams are of pure and elevated design, not the crude rustic sort that are so common. The delicate designs, swirling inward and outward through literary phrases he has selected for execution are uniquely satisfying – a delight to the mind and the eye, a joy especially when feeling listless or bored.

When Moon Lake was young he enjoyed traveling all over the country to famous mountains and great rivers. After [retirement], he looked back upon all his past fifty years, encumbered as they were with duties and obligations, as so much trivial bother, and retired to the mountains to spend his time remaining in contemplative silence. He developed a deep interest in the *Book of Changes* and Taoist learning. May those unskilled scribblers of our own era look upon this achievement of Moon Lake's as a monument, a record for emulation.

Fraternally submitted by Kan Po,
Last week of the seventh month, 1879.

---

[9] Neither the author, Ch'en Lang-wen, nor his home, Hsiao Shan Mountain, have been identified.

4. By Kan Po:

In the seventh month of 1879, Moon Lake visited me at my home to show me his book of incense seal diagrams and seek a preface for it from me. Not thirty days after I had sent the preface to him, he passed away. I shed bitter tears to think of this man's work cut short by death.

> I breathe in the cool incense smoke from the metal brazier,
> While thinking about a poem (I am writing)
> On the death of my dear friend Lu Wa.[10]
> My sandalwood-hearted companion spits out plum blossoms (of smoke),
> Looking like the cloudy fog of the other world.
> Perhaps it's the soul of my friend the old mountain man
> In the smoke's dense patterns?[11]

> Kan Po, in memoriam. [Undated]

5. By Ma Wen-hsi:

Ours is not to reproach the mundane world literati who still cling to the "flowers of emptiness" and have not penetrated it to gather what wisdom may be found therein. [On the contrary, in] sympathy (with the above mundane literati), one hundred meandering inscriptions composed of elevated thoughts are (therein) unraveled to provide accompaniment for the lute![12] These in your [Ting's] spare time you have written (to the author of this preface) that such knick knacks [incense seals] were made idle amusements.

The *"T'ien kung fu"*[13] and this multi-talented art bring to mind the basket lanterns of West Garden,[14] listening to two new songs, leaning on a table while a wave of incense smoke floated ten thousand miles down the Yangtze River [from the] Cloud-Viewing Pavilion.

When I think back to that time, the incense seal is turned to ash, the seal script is cold, I

---

[10] Su Tung-po wrote a poem entitled "Burning Incense" for Lu Wa.

[11] "The cloudy fog of the other world" is commonly used in prose and poetry to refer to the haze descending over consciousness at death, thus a poetic reference to death itself. The short poem appears to be a derivative of a poem by Su Tung-po concerning the death of a friend, also a recluse.

[12] "Accompaniment for the lute [*ch'in*]." Elsewhere it was mentioned that Ting Yün played the lute, and this phrase implies that he enjoyed burning incense while playing the instrument, or that he timed his playing by means of incense seals. Among designs of incense seals in his compilation are some in the form of a lute.

[13] "T'ien kung fu" – fu is an ancient Chinese verse form, indicating that this is the name of a poem.

[14] "Basket lanterns of West Garden" – basket lanterns may refer to the basket censer for scenting clothing. A zoological park called "West Garden" existed in Lo-yang, the capital of the Eastern Han Dynasty (25–220 A.D.).

close the book and cry for the fate of ancient tradition, saying today is the Calyx Pavilion Generation responding to the Golden Thread Song.[15]

> Written for sixth Elder Brother, Moon
> Lake's Incense Seal Catalogue Memorial
> Manuscript.
> 1880, ninth month.
> Fondly, Ma Wen-hsi, Shanghai.[16]

6. By Hsü Lü-t'ung:

I remember the existing occasion when we met on the Ch'ing Huai Yin Tao.[17] Years ago we [Ting and the author] took a boat trip on the Yüan River,[18] little suspecting that this would be the last time that we would meet. On that trip I asked Ting to stop and paint the fabulous mythical seabird called the I[19] and the Li,[20] a fantastic black horse, the harbinger of spring.

When I again visited the South-of-the-Flowers Society,[21] I learned that Ting had ridden off to heaven. At the time of Ting's funeral, entering the hall and gazing on the funerary portrait [of Ting], smoke from the incense seal was wafted by the funerary banner.

When I finally arrived, two months too late, he was already amongst the immortals.[22] Ting passed away in the fall and I arrived in the middle of winter. I am saddened to think that had I not been so frugal and delayed this journey, I might have seen him alive again.

The *tang chi* tea had not yet been brewed, nothing remains at Chao [indecipherable word] Tower.[23] For ten years I was concerned and without news of Ting. During our ten-year separation I cherished several of Ting's poems which I feel are truly of rare quality. I tried in vain to correspond with him.

Surrounded by autumn's sadness, one of Ting's incense braziers helps me cut through the rain's cruelty. When I open Ting's poems, my tears pour fourth. As the deadline

---

[15] "Calyx Pavilion Generation" and "Golden Thread Song" appear to be the names of poems or songs. These phrases may relate to poems previously exchanged between Ting and the author of the preface. It was common for friends to write poems to each other using the verse pattern and rhyme of a poem written earlier by the other.

[16] Three seals – the seal of Ma Wen-hsi; another unknown; and one of "The Peaceful, Persevering Retired Scholar."

[17] The first section, ending with the parenthetical remark about Fish Bay Literati, is written in pentameter rhymed verse, thereafter in essay form and style. The phrase "the covered road in/to Ch'ing Huai" has not been located.

[18] Yüan River, in Kiangsu Province.

[19] I, a mythical fabulous sea bird, the symbolism of which is unknown, was often mentioned by poets.

[20] Li, a black horse.

[21] "South-of-the-Flowers Society," probably the name of a local literary society of which Ting Yün was a member.

[22] This appears to imply that the funerary display lasted for a long period after Ting's burial.

[23] These may be two references to the death of a friend, but their source and full meaning are uncertain.

quickly approaches, that fateful day when the Fish Bay Literati[24] will present their prefaces, I write this having just returned from a ten-year stay in Anhwei Province.

Sixth elder brother, Moon Lake,[25] had brilliant talent. All his life he upheld righteous, moral principles. Given to song, he was carefree and had no bureaucratic ambitions. He lived out his quiet life in the Kiangsu area [a few words indecipherable],[26] and the ten years during which I knew him passed as if they were one day. I remember like a dream one autumn when we went to [an unknown city in] Anhwei Province, threw away our cares and spent time in the country of wine, indulging in the flowers, reveling in abundant spring.[27]

I recall him as a reincarnation of Tu Mu,[28] riding east with his disciples on his cart drawn by a white horse [meaning unknown].

Cousin Lan-pin[29] looked over the new edition of the "Catalogue of incense seals,"[30] and showed me this ingenious and excellently wrought work. What a tragedy that [word indecipherable] the grandson of heaven[31] did not have an opportunity to converse with [this man who could] point out a Ch'an precept with a simple knock on the floor.

Cousin Lan-pin [which was her courtesy name] Kan Chu,[32] was so successful in her editing work, it was as if she relied on the talents of Ou-yang Hsiu[33] and Tu Tao-i[34] in preparing this work for dissemination.

So now I have returned [home] to compose my feelings into this prose-poem to honor and pay tribute to the memory of [Ting].

[Written at] Yang Lake[35] [possibly], Shang Stream.
[Signed] Hsü Lü-t'ung presents this preface.
[Undated].[36]

[24] The meaning of "Fish Bay Literati," is not known, but presumed to refer to the group of individuals whose dedicatory prefaces are included in this work.

[25] "Sixth Elder Brother, Moon Lake" is probably a term of familiar respect rather than having a sibling meaning. More than one writer of prefaces speaks of Moon Lake as the "Sixth Elder Brother." The literary group of which they were all members may have assigned such "ranks" or formal seniority titles to its members.

[26] Kiangsu Province and environs.

[27] Indulging in the flowers in wine country is almost certainly a euphemism for philandering and drinking.

[28] Tu Mu was a famous poet of the T'ang Dynasty.

[29] Cousin Lan-pin, literally "eldest daughter of [Ting's] older brother [named] Lan-pin." This is the name of the author of the first dedicatory preface and may lack a surname, which may be Ting.

[30] *Hsiang lu t'u p'u* "Catalogue of incense seals" is the collection of pattern designs by Ting Yün, his own preface and the dedicatory prefaces by others.

[31] The author's meaning is not clear.   [32] Courtesy name of Shih Yün-sheng.

[33] Ou-yang Hsiu (A.D. 1007–1072) was a famous official and literary figure of the Sung Dynasty, apparently known for his editorial expertise.

[34] Tu Tao-i has not been identified.

[35] Yang Lake is situated in Wujin District, approximately fifty miles southwest of the Nan T'ung City area, where it seems that Ting Yün lived.

[36] There are two seals, not identified, one of which may read "Shang Stream," possibly the name of this writer's home.

7. By Ch'i Hsüeh-chung:

Moon Lake, the scholar from Ch'ung Chuan[37] in the foothills of the Wen Mountains, lived at [the estate] called Yü Wan [Fish Bay]. In his retreat called Incense Ocean and in his bamboo farmhouse he played the lute to the moon, kept bananas in his garden and pursued his studies. He entertained admirers of high social rank who came to visit him. Their joyful meetings were as pleasant as the song of birds, the sound of the flute. There are flowers there for enjoying,[38] and there was no wine which he would not buy for his guests. They were as Wang Wei and his wine-reveling band of poets.[39] I first made the acquaintance of Mr. Ting more than ten years ago when he [words missing] [condescended to allow me] to visit him at Lang Mountain.[40] His great work, "Incense seal diagrams," is the product of his profound wisdom and lifelong studies. Having a thorough understanding of the symbolism of the *I ching*,[41] he has cleverly inscribed these symbols in the designs of the [incense seal] patterns, and they resemble the grain of wood in their sinuous flow. The patterns coil and wind about like images of swirling, curling dragons. This he has accomplished through studying ancient bronze inscriptions to a high degree of mastery. The expression "Auspicious Blessings," "Good Fortune,"[42] burns for six [Chinese] hours, moving through the rectangular pattern of its design like birds flying wing-to-wing, like the intertwining pattern of crossed tree branches. During the day the brazier serves [this writer] as a brazier in sacrifices/prayers to heaven. Its thick, luxuriant smoke [pouring out] like beads of longevity [in a rosary]. I will keep [this incense clock] in my tower and never forget it. If I cease to use it, I will keep it and treasure it as I would another fine art work. Such a refined object should be preserved for posterity, and [now] there is this book [about it], should the [incense seals themselves] vanish [as testimony of the creator's skill, etc.].

> 1880, tenth month, fourth day, written as my guest brought me this profound manuscript by my friend Moon Lake, from whom I have been separated for "100 days." May this book reach the eyes of those who can appreciate it. Written at age seventy-six by Ch'i Hsüeh-chung at Hsing-chiang [Star River].

8. By Wang Ting-hsiang:

This brilliant one-hundred page booklet is no hack imitation of the ancients. It is, on the contrary, one of the few new voices on the subject of incense burning devices since the

---

[37] This place name has not been identified. It may be a town on the large island at the mouth of the Yangtze River called Ch'ung Ming Island, near Shanghai.

[38] This may refer to the provision of sing-song girls by Moon Lake as entertainment for his guests.

[39] Wang Wei was a poet of the T'ang Dynasty.

[40] Lang Mountain is the site of a major Buddhist temple.

[41] "The book of changes" of Confucius.

[42] Possibly as inscribed on a specific incense seal.

work of Ting Huan.[43] Moon Lake is a recluse scholar, a man of leisure living in the "Mai Yü Wan" [Fish-monger's Bay], preserving the classical scholarly traditions and shunning public life. It is as though he were a reincarnated "Commissioner of Altars and Rituals" returned to revitalize these arts.[44] These various, unique, eccentric metal incense braziers are the greatest that have been cast in some time. Moon Lake participated in this creation [passage not clear] and asked that I, an old man, keep this manuscript for a while, and examine it. I was sent a book with a fresh scent, like the petals of a flower, but I fear I am returning it with the foul stench of my own writing, destroying its beauty.

Postscript

Mr. Ch'iu Chi-ch'un[45] gave me this manuscript [Ting Yün's designs] when I met him at Hai-ch'ang[46] and asked that I add a preface to it. I hope that Mr. Ch'iu will mail me more prints [copies] of the illustrations containing tetrameter verse in the days to come, as I enjoy both the poems and the calligraphy.

Wang Ting-hsiang, also known as "The Old
Man of Mount Miao Wan."

9. By Hsü Ch'i:

Ambergris,[47] a harmonious distillate of the numerous excellent mists of blue seas, [in which] magpie's tails[48] interlock around the seven precious green bamboo tubes of Mei-shan,[49] burning in winter windows during school days, official documents beside hundred-word chants[50] in the Persian sandalwood city[51] in the country of [word indecipherable]. Fragrant brocade sleeves washed in fern dew find it hard to disperse the vastness of spring. Again, the reeds whistle, pine trees sway, still spring does not lose its strength, and there is a touch of peace remaining in my heart. Skillfully written palindromes beneath the

---

[43] Ting Huan is mentioned in a "History of the Han dynasty" as the inventor of two types of incense burners. The first called the "mattress scenter" or "quilt scenter" was made in a flat form for deodorizing beds. It is noted that the method of fabrication is now lost, but that the incense burned in a spiral which traveled four orbits. The second was the *Po-shan hsiang lu* made in the form of a covered footed bowl, the cover of which was decorated with images of animals and described as having "automatic movements."

[44] This may have been an ancient official title of religious importance.

[45] Not identified.

[46] Hai-ch'ang, a location not identified.

[47] Ambergris is a waxy substance found floating near tropical shores and is believed to originate in the intestines of the sperm whale. It is used as a fixative in perfumery.

[48] "Magpie's tails" (incense burners) may refer to incense seals made in the form of the *ju-i* scepter.

[49] The phrase "Seven precious green bamboo tubes of Mei-shan" has not been identified. Mei-shan is a mountain and district in Szechuan Province. All of the above appear to describe incense burning in some of Ting Yün's incense seals.

[50] "Hundred word chants" may be a reference to Taoist magic formulae.

[51] "Persian sandalwood city" is not identified. Certain types of incense ingredients were believed to have been imported from Persia or the regions along the Chinese Silk Road.

clothes,[52] the baskets need not be paired and stacked to display their uniqueness. As [the incense] burns down slowly, so may you live a long life and sing until you have no more ideas for incense censers.

Mr. Ting [with the personal name] Moon Lake, this truly skillful craftsman, has cultivated a skill in transforming the mundane [old style] incense censers with [clumsy] animal designs[53] into new designs, forming seal script patterns. [He transformed the old] precious duck design[54] old tiles and ancient scripts, creating a lasting impression [by] spreading incense throughout [the negative part of the design], like casting money at night for fortune-telling purposes,[55] [visually patterned] like leaves and bamboo scattered, fallen on the forest floor, calming the heart and pulse. In the same way, [Ting fashions designs of] gourds.

The autumn when we last parted, Ting wrote down a profound Ch'an saying, penetrating the seven profound fluttering butterflies,[56] going about drunk looking at the flowers.[57] Passing through the flowers, the married couple are overturned and plunged as if in a dream. From [Ting's] resulting feelings, he auspiciously drew forth beautiful verse praising fortune and long life. (To this point I have composed this preface from [and about] phrases from tiles Ting has used as censer [incense seal] inscriptions in his catalogue.)

All the creations of this multi-talented craftsman become treasured rare objects in the five capitals,[58] making their way into the not-so-secret heavenly libraries[59] like the thinnest of needles. How could he keep such treasures to himself? A gentle chanting sound engraved on the silence, as when opening Yü Hsin's *Lung ku* scroll,[60] ancient colors and rare incense fill the senses.

Throughout the catalogue, every name and every image bear testimony to this honest, erudite, retired scholar who lives at Fish Bay and whose talents are as numerous as the scales on a gold carp.[61] The geese that brought me this [manuscript] left footprints of the

---

52 Palindromes may refer to the "meandering" inscriptions in Ting Yün's designs for incense seals. The phrase "beneath the clothes" apparently relates to the practice of placing braziers of burning incense underneath the baskets of clothing to be deodorized or perfumed.

53 "Clumsy animal designs" refers to the zoomorphic forms of early *Po-shan hsiang lu*.

54 "Precious ducks" were the early duck-shaped censers.

55 "Casting money (coins) at night for fortune-telling purposes" may refer to a popular form of fortune telling by tossing coins, reading the results in the *I ching*; however, it is unclear in this context.

56 The reference to "seven profound fluttering butterflies" is unclear.

57 The phrase "becoming drunk on the flowers's fragrance" generally means to be captivated by women, but in this context is unclear.

58 The "five capitals" may mean "widely collected and appreciated." If the phrase refers to a group of five capital cities from periods when China was divided into many independent kingdoms, it is unclear in this context.

59 Not identified.

60 The painter and poet Yü Hsin (A.D. 513–581) was a native of Hsin-yeh in Honan whose work was much admired by Tu Fu. He was a commander of cavalry, a high military appointment. The scroll mentioned has not been identified. Giles, *Dictionary*, p. 956.

61 Literally "correspondence – increase – gold carp," probably intended to bring praise for Ting's talents and production.

hung.[62] (I had a friend hand carry this catalogue south to me that I might view it.) I found the catalogue boundlessly interesting. The excellent forms [designs] give new life to the ancient words which have been thus coaxed out of their hidden niche in the history of literature, each and every one superb.

Some of the [texts of the inscriptions] are of the *p'ien li* type,[63] which are given new interest. The author of this catalogue would surely have no fear to attach his own name to it.[64] Sage Ting's "Incense"[65] itself serves as the real preface to this work as is appropriate. (In Yeh Tao-lu's incense catalogue, Yeh notes that "Chicken's Tongue Incense" is also called *ting tzu hsiang*,[66] as is appropriate.)

> Eighth day, middle third [second ten-day period]
> of the eighth month, 1879, written by Hsü
> Ch'i,[67] also known as "The Farmer of
> Hangchow,"[68] in the Shou hu t'ang,[69]
> Hangchow.[70]
> "Great seal of Hsü Ch'i."[71]
> [Another unidentified seal].

10. By P'an Feng-t'ai:

The *Pu shih kuan*[72] states, "The impulse among the learned and skilled to pass on their arts and thereby acquire recognition and fame must be equated to greed and opportunities. Works created by such motives cannot be said to be intelligent. Learned men such as Kuo Chien and Li Chui[73] who gloated over their accomplishments and stank of wealth did

---

[62] The *hung* is a mythical bird, a harbinger of auspicious occasions. The reference implies that this manuscript was delivered by another unnamed person, and was favorably received. The reference to geese relates to travel or correspondence.

[63] *P'ien li* (inscriptions) are euphemistic sentences or clauses, each usually containing four or six words, which may be contained in this text but have not been detected. This was a popular literary form among writers of the sixth and seventh centuries.

[64] This phrase implies that both "Ting Yün" and "Moon Lake" are pseudonyms, which is suggested also by the fact that he is not mentioned in any of the major works of Chinese biography of Ch'ing dynasty artists and writers.

[65] *Ting tzu [chih] hsiang* or "Sage Ting's incense" appears to be a pun. In a list of incense recipes, "Chicken's tongue incense" is given the alternate name of *ting tzu hsiang* or "Clove incense," which is the second half of the pun.

[66] Yeh Tao-lu is unidentified but is probably synonymous with Yeh T'ing-kuei, also unidentified but noted in the recipe for "Clove Incense" in the *Ch'en shih hsiang-p'u.*

[67] Hsü Ch'i, the author of this preface, has not been identified.

[68] "The farmer of Hangchow" is this author's pen name or nickname.

[69] *Shou hu t'ang*, meaning "Hall of blessings" is otherwise unidentified.

[70] Hangchow, a large city in China popular among poets for centuries, is one hundred miles southwest of Shanghai.

[71] The author's seal.

[72] Possibly the title of an ancient work entitled "Admonition to officials."

[73] Scholars and/or officials not identified.

not know the damage done by their 'cleverness.' " [Passage following is not clear, possibly written at Shih Kang (Stone Harbor)].

These [illustrations in the book] are rubbings of incense seal patterns done and collected in book form by Ting Yüeh-hu. [He studied] the excellent nine-tiered examples of *Po-shan* braziers made during the Hsüan Te period of the Ming Dynasty which were brilliant, lustrous and fabulous beyond compare. [He also studied] Han Dynasty *pi hsieh* braziers[74] in which incense smoke escaped through the eyes, ears, and the nose of the evil-averting ornamental animal exterior. [Ting Yün] used the tracing method[75] to transfer details of *li* and *chou* script,[76] cast the impressed designs in sand in unusual shapes such as the square *kuei* and round *to*.[77] The script in these true incense seals is as interesting as the skills of the traditional arts of wood carving, jade polishing, ivory and stone carving and metal-working – a veritable flower garden sprung from his imagination.

Rather than cataloguing his knowledge of ancient sacrificial vessels and aromatics in a manner to praise precious ancient works, [Ting Yün] has used this knowledge and understanding to write seal script inscriptions [on these incense seals]. One could say that [Ting Yün] was a master of the literary classics and histories, poetry and ancient literature, and especially talented in calligraphy and painting. Still he will not allow fame and fortune to compel him to race about in the mud for a day's glory, and he will not descend into the newfangled Western artistic modes so in vogue in this Emperor's reign, but keeps his own counsel on refined standards of beauty and taste. He has practiced the arts of ink, brush, and stone inscription, reading and writing since his early years. [Here follows a passage only partially intelligible, probably describing travel and living in the Shanghai area and possibly some foreign travel].

I cannot smile at the so-called foolish use to which [Ting Yün] has put his talents – one look at his inscriptions makes it clear that he works in the path of truth. Thus it gives me pleasure to write this preface.

Second lunar month, first ten-day period, 1879,
at Wu Shan,[78] by P'an Feng-t'ai.[79]

11. By Chu Yü-chin:

It is said that arranging jade and composing pearl [i.e., editing] [requires] finer skill than carving dragons [i.e., original literary compositions], and that parting the clouds to discern the moon is more valuable than creating a phoenix.[80]

---

[74] Incense burners in the shape of mythical beasts, a popular motif in the decorative arts of the Han Dynasty and said to ward off evil.

[75] A tracing technique for copying calligraphy.

[76] Two ancient scripts, precursors of the seal script.

[77] The *kuei* is a jade shape with pointed end and square base, and the *to* is a bell-shaped design, a symbol of the teaching profession.   [78] Name of a town or estate, not located.

[79] Author of this preface has not been identified. The signature is followed by two stamps which appear to state "(From the) collection of Feng-t'ai."

[80] That is, discrimination and understanding of the thoughts of the ancient is more valuable than any new contributions.

Thus, no ingenious scholar of temperament would fail to understand that skill in editing is analogous to the care which must be taken in making a fox coat so that no gaps remain to allow the elements to enter. Such is the skill required of one who would penetrate the secrets of past masters. Without perseverance, one's efforts are in vain.

Such is my friend, the recluse Moon Lake. He follows in the scholarly tradition of the "Two Ting Brothers,"[81] the learning of the "Two Yü Mountains,"[82] and the philosophy of Yang Hsiung,[83] the masterful calligraphy of Hou Pa[84] and Wei Tan,[85] the embroidered phrases of Ts'ao Jui,[86] carving these words in intricate, ornate scripts.

Whether painting in the "boneless drawing style"[87] or singing chants to heaven, in each and every elegant instance showing his skill in Ch'ing T'an,[88] he further attested to his brilliance. On quiet occasions he would sometimes burn incense to keep company with the moon and write brilliant calligraphy in the red clouds (of moonlight pouring through the incense smoke) from an unusual nine level Po-shan incense burner[89] in which the reversed swastika[90] appears on [a bed of] ashes.

[81] The "two Ting Brothers" were Ting Ping (1832–18??) and Ting Shen (dates not known). Ting Ping was an outstanding scholar who was recommended for the position of County Magistrate of Kiangsu Province, which he declined. He was an author, editor, and collector of rare books. His older brother, Ting Shen, also was a collector of rare books, whose collection was said to contain more than ten thousand volumes. He edited a number of local histories and philosophical commentaries in collaboration with his younger brother. They donated their collections to the city of Shanghai to form one of the first collections of printed books in South China.

[82] These are two mountains in Hunan Province, one of which contains a cave called Ta Yu Tung, in which is supposedly preserved a library of some ten thousand volumes. It was said that they had been collected and stored there in the Ch'in Dynasty by a scholarly recluse. The cave is one of the Taoist Sacred Caves.

[83] Yang Hsiung (53 B.C.–A.D. 18) is mentioned in other prefaces as well. He was the author of "Model sayings," an imitation of the "Analects" of Confucius, and "Classic of the supremely profound principle," an imitation of the *I ching*.

[84] Hou Pa (d. A.D. 37) was a student of Yang Hsiung and tradition states that he mourned his teacher's death for three years. He served as governor of Liu-huai in the Han period. Giles, *Dictionary*, p. 269.

[85] Wei Tan, a writer and calligrapher active in the Wei Dynasty who excelled in literature and calligraphy. He also made brushes and ink sticks that were highly regarded.

[86] Ts'ao Jui (A.D. 205–240), possibly Ts'ao Mo, succeeded his father, Ts'ao P'ei, in 227 as second emperor of the Wei Dynasty. Another, also named Ts'ao Jui or possibly Ts'ao Mo, was a warrior of the Spring and Autumn Period (722–481 B.C.).

[87] "Boneless drawing style" was the name given to a school of painting of old China in which paintings were created with areas of color without line drawing delineation, somewhat similar to modern water color.

[88] *Ch'ing t'an* or "pure talk" is Taoist-based high-minded philosophical discussion which originated in Wei and Chin Dynasties.

[89] *Po-shan* (incense burner), which is frequently mentioned in the dedicatory prefaces of Ting Yün's work, is either a term of praise for the quality of a particular incense burner, or it is a specific type of censer the cover of which is formed like a mountain-topped goblet embellished with precious stones and with shapes of zoomorphic birds and beasts.

[90] The reversed swastika is a symbol of good luck.

Moon Lake continuously reworked the old [scripts] picking among them to create an outstanding new work, through which lines [of script] incense burned like a coiling spring earthworm, like a twisting autumn snake, spreading and expanding, the smoke drifting softly, knotting and gathering like emerging banana leaves, mercilessly dispersing any poverty of spirit like the shattered fibers of the lotus root, leaving the cloud seal's scar[91] by which to appreciate and praise this creation precious enough to grace the table in Pomegranate Hall[92] or to adorn any lady's boudoir.

Thus, although there are those of little refinement who dismiss these things as mere curious toys, the refined will seek out and collect these pages as if they were a rare Sung Dynasty manuscript.

> Written midsummer 1879.
> The Panpipe Scholar,[93] Chu Yü-chin,[94] wrote
> this preface at Fish Bay Village [possibly in or
> at] Ch'ueh Shu Hsieh.[95]

12. By Chu En-hsi:

> Moon Lake delighted in rare metalwork,
> His fame increased with each brazier he fashioned.
> Square and round forms,
> Scripts from Chou Dynasty *ting*[96] and Han Dynasty tiles.[97]
>
> He assembled the ancient thought of Ou Lao[98]
> And modeled the pure scripts of Fu Shih-chün.[99]
> This poem is my memorial to the deceased Mr. Ting
> And I am grief-stricken to recall his virtues.

[91] That is, the image remaining visible in the censer.

[92] "Pomegranate Hall" is the name given to the library and study, situated in Kiangsu Province, of Wang An-shih (A.D. 1021–1086), the controversial prime minister of the late Sung Dynasty.

[93] Sheng Shih, possibly the "Panpipe Scholar," may be a nickname chosen after an ancient Chinese wood instrument. It also may mean the maternal surname Sheng as opposed to the usually written paternal surname.

[94] Not identified.

[95] Two seals, the first of which has not been identified and may be the author's name. The second is *Chen Nan*, meaning "Real south," which possibly is Buddhist terminology which could be a transliteration from Pali or Sanskrit.

[96] Chou Dynasty (1122–249 B.C.). A *ting* is an ancient form of bronze vessel shaped as a tripod having two ear-like handles, variously rectangular or cylindrical in shape, often inscribed with the ancient writings mentioned, and generally used for sacrificial purposes.

[97] Han Dynasty (206 B.C.–A.D. 220). One of the sources of Ting Yün's inspiration were the scripts forced into the shape of the end of tiles, which often were round and used to decorate and complete the bottom eaves of elaborate tile roofs.

[98] The first two lines of this verse are extremely obscure. Ou Lao is probably Ou-yang Hsiu (1007–1072), a poet and statesman of the Sung Dynasty.

[99] The first and third of these words would form the name of a Ch'ing Dynasty official title. The middle word means "stone" or "rock."

So preoccupied was Mr. Ting in compiling this book of Incense Seal Script that he worried that it would not be done before flowers grew on his grave. Lan-pin and others assembled the work and sent it to the printers (to fulfill Moon Lake's wish) [but] feigning it were only in order to prepare this memorial and complement it with eulogies.

Fourth month of 1881
[Signed] Chu En-hsi at Yüan Hu.[100]

13. By Shen Yu-jen:

In a remorseful evening's writing, I attempt to [pay tribute] to ten years of work in metal worthy of the emperor by this scholar who died before his time. He leaves behind only some profound calligraphy which fuses the calligraphic styles of the Ch'in Dynasty,[101] steles of the Han Dynasty and tiles.[102]

[The patterns on the braziers] move continuously in a circle, each one equally auspicious. They are spreading incense throughout Kiangsu[103] and are greatly appreciated by the people.

The lute, wineskin and wooden musical blocks[104] accompanied him and brought him peace and refined delight. Among his intimates was his friend and neighbor Ai Wu-lu.[105]

He is gone now, become a Buddha, reborn in boundless heaven. When will he return incarnate? While marking time in the heavenly realm, it is as if he were still here, having bequeathed to us his gift of a burner of incense.

Respectfully presented, Spring of 1881, to the memory of my fraternal friend Moon Lake [for] the "Incense seal memorial volume."

Preliminary draft written by Shen Yu-jen of
[illegible] *shui*. [Seal possibly of Chou li].[106]

14. By Ts'ai Wei-hsing:

I fondly remember [Moon Lake] from a meeting with him ten years ago while detained in Yüan P'u.

(I met Moon Lake in 1868 while stopping at Hsü Lang-hsüan).[107]

---

[100] "Yüan Hü" is a place name, not located, literally meaning "Mandarin Duck Lake." Two seals, the first, *En-hsi*, is the personal name of the author. The bottom seal consists of two characters, the left of which is not known and the other meaning "bereaved" or "in mourning."   [101] This preface is written in partly rhymed heptameter.

[102] These steles and tiles are considered to be among the ancient calligraphic masterpieces.

[103] This might refer to the city of Soochow, fifty miles west of Shanghai, traditionally a gathering place for poets and scholars.

[104] The lute and wineskin were long considered to be morally uplifting scholarly pastimes. The purpose of the musical woodblocks is not known.

[105] *Ai Wu-lu* is the name which appears on the printer's block at the end of Tin Yün's volume in my own collection.

[106] Shen Yü-jen, name of an individual not identified. "... *shui* ..." Place name, the first word of which is undecipherable. The second word means "water," commonly found in place names. Possibly Chou li (seal), which may be a pseudonym of the author of this preface.

[107] Name of a studio not identified.

Pure as water, he was, with an attitude favoring the simple and rustic,
He followed an eccentric way of life – as free as smoke.
Snail-like seal scripts wiggle and crawl,
Like a glossy pearl passing through an ants's nest [the flame burns].
Unexpectedly he left a book [of this work] increasing [my] excitement.
(Moon Lake died in the winter of 1879).
The petal-shaped incense used in Buddhist worship leaves a lasting scent as a
    reminder.
Unexpectedly summoned by the Crown to Japan.

Coming and going from Yü Wan,[108] his reputation gained admirers.
Who would have expected his current fame?
As strange as T'ai-po riding off on a dolphin.[109]
Magically transformed into vigorous, fine calligraphy frozen in metal,
The quality of craftsmanship is as superb as the design.
His descendants will have something of which to be proud – they should be
    poets,
That they might offer prayers and offerings fitting to his memory.[110]

> Respectfully presented Summer 1880 [as a
> preface for] Mr. Moon Lake's *Yin hsiang t'u*
> *p'u t'ung hsiang*.[111]
> [Signed] Ts'ai Wei-hsing.[112]
> [Two seals].[113]

15. By Chang Yüan-k'ai:

How perfectly serene, incense burning quietly in its censer,
And how hard to come by is a censer of Po-shan quality.[114]
I turn the pages of this new book of patterns,
I feel it is brilliant and without rival.
Many examples of cast metal, in a hundred different designs,
Edited and sent to the printers – doubly enticing;
With Han and Ch'in ornate seal-scripts, Hsüan-te painting images,[115]
Most of which were gleaned from original scrolls.

---

[108] *Yü Wan* is "Fish Bay," where Ting Yün lived.

[109] T'ai-Po, also known as Li Po (A.D. 701–762), a poet of the T'ang Dynasty, considered to be one of China's greatest writers. There is a legend describing how he rode a dolphin out to sea, like Arion in Greek mythology.

[110] Written in rhymed heptameter.

[111] Tin Yün's memorial volume does not have a formal title or title-page. *Yin hsiang t'u p'u* is another version of the title that is used.

[112] Ts'ai Wei-hsing, the author of this preface, has not been otherwise identified.

[113] The second seal appears to be "Chao Mu-lan," probably a personal name of the author or of a collector.

[114] The Po-shan region in Shantung Province is known for its early production of fine incense burners.

[115] The Hsüan Ho era (A.D. 1119–1126) was the last reign of the Northern Sung Dynasty.

[Meditating briefly?] to purify the heart, I know unbounded joy,
And this house has the reputation of being the dwelling of a benevolent man.
With double-hung door-curtains – just the thing for detaining an honored
     guest,
It's a peaceful place, suitable for studying unusual writings,
Sweeping the grounds, entertaining peaceful thoughts just after dawn,
Calling out, praying earnestly to heaven at the advent of night;
Oh, for a glimpse of my native town,
Not even the Imperial Libraries could be better![116]

> Respectfully presented, early summer, 1879, to
> my respected "elder brother," Moon Lake,[117]
> [written at] Ying-cheng, Wu-lin,[118] by Chang
> Yüan-k'ai.[119]

16. By Liu Jui-fen:[120]

Writing about the carving of elaborate seal scripts, Yang Hsiung[121] said that a true scholar does not create such works simply for his own benefit, but creates for an elite audience. A true scholar can create such things, but one who is not a true scholar can not only not create such refined things, but cannot even appreciate the Tao revealed through such things. For such eyes, good and bad become all mingled together as one and craftsmanship is in vain.[122]

Confucians of the Sung Dynasty established the neoclassical essay as an art form, devoting themselves to it and other trivial amusements. Their art was a combination of flights of vision and moral purpose.

As for those of the K'ang-hsi period,[123] they were also unsuccessful at equalling the [literary/artistic] heritage they inherited, and left only artless works for posterity. Men of discrimination thought [their efforts] mediocre and unworthy of the appreciation of a true scholar.[124]

---

[116] The entire preface is written in an A-A-B-A rhyme pattern in heptameter verse.

[117] The term meaning a close friend, an older friend, not a blood relative.

[118] Ying-cheng and Wu-lin are place names not identified.

[119] Chang Yüan-k'ai, not identified.

[120] Liu Jui-fen (1827–1892) was a native of Kuei-ch'ih, Anhwei. He was Intendant of the Su-sung-t'ai Circuit, Kiangsu Province and later served as acting governor of the same province. In 1885 he was appointed Chinese minister to England and Russia, and in 1887 he served in the same capacity to Belgium, France and Italy. Upon his return to China in 1889 he was appointed governor of Kwangtung. His collected works were published under the title *Yang-yün shan-chuang ch'üan chi*. Hummel, *Eminent Chinese*, pp. 522–23.

[121] Yang Hsiung was a poet and scholar of the Han Dynasty, born in the Szechwan Province. It was said that he led a simple and unconventional life. His scholarship and thought were widely known and highly regarded.

[122] The source of this quotation is not mentioned in the text.

[123] K'ang Hsi period (1662–1723), second reign in the Ch'ing Dynasty (1644–1911). Emperor K'ang Hsi was responsible for the creation of a citation dictionary in forty-two volumes, still highly regarded as a reference tool on the origin of Chinese words.

[124] True scholar, loosely translated as "aesthetic heavyweight."

Mr. Ting, [also known as] "Moon Lake," however, has made eager and thorough studies of the hundred classics and histories.[125] He is accomplished at literature and poetry, calligraphy and painting, the composition and playing of music – none of which are mastered without considerable practice.

His family has means but without rank of government position.[126] He has never been given to bragging, and has an unrestrained, generous spirit. Scholars of this nature are not often met with (on earth). Finally, today, [a son or attendant] has brought me his *Catalogue of Incense Seal Designs*.

He [Ting or the deliverer, not clear] asked me to write a preface to the book, to which I gladly consented, observing that this book is not merely another text, though it far from captures the vitality of Ting's life's work and ambitions.

When I look at this [short section illegible] not since the end of the Sung Dynasty [have there been] carvers of the [seal script] stone and bronze. There were many examples of clothing-scenting braziers[127] which [had as their design] parallel-couplet proverbs.[128]

These were called Po-shan incense braziers[129] and were for the most part the property of the Crown Prince.[130] Popularized copies of these, basically of the clothing-scenting type, were rarely mentioned in writings of the period, and incense in these was not patterned into any configuration before burning. [Only] Li Hou-chu had among his treasures several dozen kinds [of incense seals?].[131]

Among these seals, Li had several different Lotus-handled Phoenix-of-the-clouds-subdues-the-Blessed-Lion-of-the-Heavens Incense Seals, which were made of stone.[132] Mr. Ting's cast incense seals surpass even these in the novelty and skill with which he re-uses the ancient forms.

Mr. Ting's designs move inward and outward, convoluted like the movements of a flying dragon, like the twisting of clouds and lightning, like the twisted grain of wood. [One phrase illegible].

In these designs, Mr. Ting has transformed the ancient calligraphic scripts into a scholar's chamber amusement. In the author's preface, Mr. Ting further [illegible passage

---

[125] Not a specific collection, as well as can be determined; to be translated as "myriad."

[126] Undoubtedly the reason that no published biographical account of him can be found.

[127] Clothing-scenting braziers or censers were tent-shaped wire cages enclosing burning incense over which articles of clothing were draped for deodorizing or perfuming.

[128] The reference is not clear. If "parallel-couplet proverbs" is intended, it would be a standard form of Chinese literary expression, often four or five words per line, frequently containing a dramatic contrast between the two lines.

[129] As described elsewhere, the Po-shan incense brazier was cone-shaped with a small goblet-like pedestal, the cover decorated with figures of mythical animals. The name derives from Po Shan in northeast China, the region famous for producing this form.

[130] No specific crown prince was mentioned.

[131] Li Hou-chu (A.D. 937–978) was the last emperor of the Later T'ang Dynasty (A.D. 923–936). He was acclaimed for his literary genius, but blamed for encouraging the practice of footbinding among Chinese women.

[132] The meaning is not clear. These may resemble large stone seals, such as those preserved in the Shōsōin (see chapter 6). The phrase may be a colorful descriptive name for the brazier, or a name based upon the words inscribed in the seal itself.

probably concerning lofty understanding and enlightenment] making the incense take the shape and appearance of calligraphy on a scroll, [worthy of] sitting and studying in silent contemplation, cleansing the mind, as though free among the clouds, just as the negative and positive areas [of the seal pattern] integrate yet remain distinct. Such is the art of this carving.

Alas, his talent is gone, and those who remain and those who will follow us [short phrase illegible].

And if this essence is destroyed [the book, his skill or the seals themselves?], this art will be mistaken for the Tao itself, and his work will be scorned as being without artistic merit [?]. Alas, though Mr. Ting produced poetry, music, and writing of various kinds which could be passed down, why are such things not here [in this collection], giving the world reason to believe he passed his days in silent emptiness?

> [Closing] Sixth Year of the Kuang Hsü period
> [1880], middle of the [first?] month.
> [Signed] Liu Jui-fen, *Pu cheng shih*,[133] Su sung-
> t'ai Circuit[134] respectfully submits his preface.[135]

AUTHORS OF THE DEDICATORY PREFACES

Name	Date	Place
1. Shih Yün-sheng	8th month 1878	Hsing-chiang
2. Ch'en Lang-wen	2nd month 1878	Hsiao Shan
3. Kan Po	7th month 1879	N.P.
4. Kan Po	N.D.	N.P.
5. Ma Wen-hsi	9th month 1880	Shanghai
6. Hsü Lü-t'ung	N.D.	Yang Lake
7. Ch'i Hsüeh-chung	8th month 1880	Hsing-chiang
8. Wang Ting-hsiang	N.D.	Mt. Miao Wan[?]
9. Hsü Ch'i	8th month 1879	Hangchow
10. P'an Feng-t'ai	2nd month 1879	Wu Shan
11. Chu Yü-chin	Summer 1879	Ch'ueh Shu Hsieh
12. Chu En-hsi	4th month 1881	Yüan Hu
13. Shen Yu-jen	Spring 1881	. . . shui
14. Ts'ai Wei-hsing	Summer 1880	N.P.
15. Chang Yüan-k'ai	Summer 1879	Ying-cheng, Wu-lin
16. Liu Jui-fen	1st month 1880	Shanghai[?]

[133] A provincial official responsible for civil and financial administration.

[134] Su-sung-t'ai Circuit is one of the major administrative districts of China, a division above the Provincial level. It included Kiangsu Province and the region surrounding it to the south of the Yangtze River. It was a very powerful post due to the extensive commerce in the port of Shanghai.

[135] Two seals, the first of Liu Jui-fen, official and diplomat born in Kuei-ch'ih, Anhwei. He served as Intendant of the Su-sung-t'ai Circuit, Kiangsu Province. Later he became governor of Kiangsu Province. His collected works were published in 1893 with the title *Yang-yüin-shan-chuang ch'uan chi*. The second, in ancient script, has not been identified.

THE *HSIANG LU T'U SHIH* OF TING YÜN

Of the three copies found of Ting Yün's *Hsiang lu t'u shih* [*Yin hsiang t'u p'u*] in the United States, Copy "A" is in the collection of the Asian Division of The Library of Congress (H788/T49), Copy "B" is in the author's collection, and Copy "C" is in the collection of the C. V. Starr East Asian Library of Columbia University.

A page by page comparison of the three copies reveals some slight discrepancies, indicating that they are not of the same identical edition. There is also variation in the woodcut illustrations. Twelve of the designs in Copy "B" are not included in copy "A", for example, and four of the designs in copy "A" are not contained in copy "B". These variations add to the uniqueness of the known extant copies.

Copy "A" is classified as a rare book.[136]

---

[136] Communication of June 15, 1987 from Dr. Chi Wang, Head of the Chinese and Korean Section, Asian Division, The Library of Congress.

Appendix D

Notes for museums and collectors

CHINESE INCENSE SEALS

It is unlikely that examples of primitive wooden Chinese incense seals will appear on the market. If so, however, they should be treated in the same manner as Japanese *kōbandokei*.

To build a fine representative collection, museums and private collectors are advised to acquire only examples which are complete, in original condition, and if possible, which include the required utensils. The existence of old repairs, if well made, are not detrimental to the value of the item.

Although Chinese incense seals were most commonly made of paktong, copper, bronze and pewter; they were made also of gold and silver. Examples of gold and silver seals have not been available for study, although it has been reported that several seals of silver were sold in recent times.

The basic components of an incense seal are an incense tray, a perforated cover, one or two tiers for storage of incense and utensils, and a template. In order to make a sale, merchants may have assembled incomplete incense seals from disparate parts which to the unwary customer appear to be matching. A purchaser should ascertain that the template for forming the incense trail is original. It was originally made to fit snug within the incense tray, and should not be smaller.

A rule of thumb for determining whether an incense seal is complete and consists of the original number of sections is that generally, square or round examples are approximately as high as they are square or as the diameter of a round example, and often somewhat higher.

The utensils generally consist of a small shovel, a tamper, and occasionally a pointed piece of metal for tracing the incense trail; often the latter is incorporated in the handle of the shovel, which in such instances is made with a triangular point.

Although incense seals were made in an extremely wide range of shapes and decorative motifs, identical duplicates exist, in each case presumably duplicated by the same maker. For example, a number of identical seals of the *ju-i* scepter form which were produced by or for Ting Yün are to be found in private and public collections.

Examples bearing hallmarks are particularly desirable and frequently demonstrate a finer degree of workmanship. Examples having copper covers and paktong bodies appear to be scarcer than those made entirely of the same metal.

Accumulations of dirt should be carefully removed with a mild soap, not detergent, and warm water applied with a soft brush; the parts then thoroughly dried. The original patina should be preserved at all costs. It is unnatural and undesirable to bring incense seals to a high polish, and buffing on a wheel is to be discouraged. Such a procedure generally produces an artificial brightness which cheapens the appearance of the object and destroys its history, the evidence of age and use.

It has been reported by collectors purchasing incense seals in the Far East during the past several decades that for the most part dealers sell them as obtained, often encrusted with a heavy patina. However, at the purchaser's request, they often are willing to have the corrosion removed and the seal buffed to brightness. Unfortunately, many examples purchased in the Far East in the past several decades have suffered from this process.

Chinese incense seals made of metal are cast or are soldered composites of several separate metal parts, cut or cast to shape and size. The soldering is a fine technique mastered by Chinese metal craftsmen to make the joints almost invisible. Repairs, if required, should be attempted only by technicians skilled in fine metalwork, perhaps a clockmaker or jeweler.

It is to be noted that with age, paktong becomes brittle and difficult to solder, making it almost impossible for any but a skilled craftsman to repair or replace missing parts. It is difficult to obtain paktong for shaping and replacing missing parts; in some instances nickel-silver has been used with moderate success.

Incense seals of paktong were made of several thicknesses of the metal, ranging from relatively thin sheets soldered together to others cast and quite thick. The latter generally demonstrate a finer quality of workmanship in the execution of the casting, engraving, and assembly, and consequently are more desirable. A characteristic often observed in the thicker paktong is a shiny appearance resembling gold or brass rather than silver, and may be mistaken for a light brass.

As well as can be determined, incense seals have not yet been "faked" or duplicated by modern makers. Nevertheless, the wise buyer must beware, for it is likely that as the number of collectors becomes greater than the supply, these esoteric devices may become lucrative subjects for reproduction.

JAPANESE *KŌBANDOKEI*

Relatively few examples of Japanese *kōbandokei* have survived, for two reasons. First, they were produced in limited numbers, and second, because as they were made of wood, they were more liable to damage and destruction. Surviving examples are often fragile, with parts charred, perhaps with other damage from misuse or unsatisfactory storage.

The potential purchaser should be aware that *jōkōban* are being made and sold in shops in Japan purveying accessories for Buddhist shrines. Examples of *jōkōban* that have been used for the purpose intended and that bear some age invariably show burnt scars, charred sections of the incense tray floor, and an overall evidence of handling and of the presence of powder, whether ash or incense, in all interstices.

Kōbandokei should be treated in the same manner as pieces of fine furniture. They require little if any polish, but should be protected from dampness, extreme changes of temperature, and dust.

A preservative used by many museums for preserving and polishing both metal and wooden objects is Renaissance Micro-Crystalline Wax/Cleaner/Polisher, marketed by Picreator Renaissance Products, 44 Park View Gardens, London NW4 2PN, England.

TYPES OF INCENSE SEALS

Shapes. *Geometrical:* square, rectangular, octagonal, hexagonal, sexagonal, round, cylindrical, oval, quarter-circular, double-square, double-circular.

Objects: bell, leaf, vase, fan, *ju-i* scepter, melon, lute, *ch'ing* (musical stone), eight-petaled lotus, spade money, cash (coin), gourd, Buddhist thunderbolt, peach, *shu* (pair of books).

Cover Designs. Seal-characters, landscapes, bats, double fish, phoenix, Kuan Yin, crane, trigrams, yin and yang, archaic poetic inscriptions, prunus blossoms, iris, grasses and other plants, bats, butterflies, swastika, etc.

Decoration. Landscapes, plants, meditating scholars, poetic inscriptions in archaic scripts, family scenes, leaf motifs, seal-characters.

Trail Patterns. Seal-characters representing happiness, double happiness, conjugal felicity, longevity, good luck, fecundity, prolonged years, etc.

Hallmarks. Identification of shops, makers and materials, descriptions of product, dates.

SALES DATA

Relatively few incense seals have been offered for sale at public auctions within the past several decades; generally they are sold in shops or in private sales. Following are typical entries from catalogues of auction sales held since the late 1960s. It should be noted that the catalogue information is not always accurate and that the information given here is not intended to be a comprehensive inventory of sales for this period.

Galerie Am Neumarkt, Zurich, Switzerland, March 4, 1969, Auktion IV.

Lot 163 FEUERUHR, Chinesische, 18 Jh. (Provinz FUKIENG). Boden signiert. Form eines Bootes mit Drachenkopf und -schwanz. Wandung durchloch zum Einstecken der Metallbolzen, die den Räucherstab tragen. Beim Verglimmen brennt der Faden mit den hängenden Kugeln durch, und das Geräusch des Herunterfallens auf den untenliegenden Teller zeigt das Ende der Gebetszeit an L. 21 cm. [Illustrated].

Ausserordentlich seltenes Objekt, vor allem in dieser nicht üblichen Kleinheit. Abgebildet und beschreiben in: MATHIEU PLANCHON "L'HORLOGE ET SON HISTOIRE." Ausgabe 1923, Fig. 132, Seite 226. Ebenfalls abgebildet im Katalog das Wuppertaler Museums: "UHREN IM WANDEL DER ZEIT."

[Estimate] Sw. Fr. 2600.

Galerie am Neumarkt, Zurich, March 6, 1970, Auktion XII.

Lot 140. CHINESISCHE FEUERUHR. 18 Jh. Form eines holzgeschnitzen Drachen, mit Goldlackmalerei (Fukieng-Lack). Diese Feueruhren dienten zur Zeitmessung der Dauer eines Gebetes, währenddem der über der eingelegten Zinnwanne liegende Räucherstab abbrannte. L 67 cm. [Illustrated]

Ähnliche Feueruhr, abgebildet und beschreiben in "Die Uhr, Lübke," Seite 100, Abb. 145. [Illustrated].

[Estimate] Sw. fr. 5000.

Sotheby Park Bernet, Los Angeles, California, May 7–8, 1973, Sale No. 76B.

Lot 134. JAPANESE KIRI WOOD INCENSE CLOCK *Kōdokei, 19th century*. The arched grid pattern above a square box containing fine ashes, raised on a stepped base with a square cabinet having two drawers containing the tools for arranging rows of incense to be burned. Height 17 inches. [Illustrated] Sold for $350.

[No estimate].

Auktionshaus Peter Ineichen, Zurich, Switzerland, October 26, 1973:

Lot 163. VERY RARE CHINESE FIRE-CLOCK, eighteenth Century. Boat-shaped with snail-shaped prow and stern. Extremely rich gold decorations "lacque-de-chine." Length 54.5 cm. [Illustrated].

[Estimate] Sw. fr. 3000.-/5000.-.

Auktionhaus Peter Ineichen, Zurich, June 5, 1973, Auktion IX. [In German and English]

Lot 175. FIRE-CLOCK, Province Fukien, China, approx. 1800. Carved wooden frame, boat-shaped, set on claw-supports, scale lacquer mounting. Fabulous animal with both head and tail gilt, the body to originally hold and support joss-sticks; in fixed intervals, these joss-sticks would burn a silk-thread with stringed balls, i.e., the falling balls would indicate the right time for the liturgy. L. 69.5 cm. [Illustrated].

[Estimate] Sw. fr. 2800.-/3800.-.

Auktionshaus Peter Ineichen, Zurich, October 25, 1976:

Lot No. 120. CHINESE FIRE-WATCH, c. 1760, shaped as a dragon, carved wood with lacquer decorations. Head, claws and tail gilt. The hollow body inside with tin-coating and 9 rings to originally keep the joss-sticks to burn, in fixed intervals, the silk thread stretched above and thus cause the falling of the little balls on the silk thread; the falling would indicate the beginning of a prayer or liturgy. Length 70 cm. [Illustrated].

[Estimate] Sw. fr. 1200.-/1600.-.

Sotheby's London, February 2, 1976:

Lot 128. A CHINESE FIRE CLOCK, the boat frame with wire supports for carry-ing the joss sticks, the terminals with carved dragon's head and tail, the lacquered body with gold scales, raised on dragon's feet, *length overall 625 mm. 19th Century*.

[Estimate] £200/300.

Sotheby's, London, February 18, 1982:

    Lot 54. A CHINESE PEWTER INCENSE CLOCK, comprising four sections, the square with compartment for spare incense, spoon and metal utensil, the second section comprising a tray for the ash bed, the third a brass perforated grid for impressing the seal, the lid with pierced brass cover to permit the smoke to escape, the various components engraved with undecipherable Chinese characters and foliage design, 82 mm. *square, 19th Century.*

[Estimate] £200–300.

    [Not sold]

    Lot 55. A CHINESE PEWTER INCENSE CLOCK, by *Yum Kuen Tai* signed on the underside of the base in Chinese *Hupeh Province, Yum Kuen Tai, Self Made,* comprising four sections, the square base with compartment for spare incense, with spoon, the second section comprising a tray for the ash bed, the third with perforated brass grid for impressing the seal, with presser, and with pierced cover to allow the smoke to escape, the corners with pierced decoration, 82 mm. *square, 19th Century.*

[Estimate] £150–250.

    [Sold for £308]

    Lot 56. A CHINESE INCENSE CLOCK IN THE FORM OF A RUJI [sic, *ju-i*] SCEPTRE comprising four sections, with metal trowel and utensil, the third section with perforated grid for the ash bed, the upper section pierced with stylized decoration, *length overall 260 mm. 19th Century.*

[Estimate] £250–350.

    [Sold for £440]

Robert C. Eldred Co., Inc., East Dennis, Massachusetts, August 23–26, 1982. "Oriental Art at Auction:"

    Item 1523. Rare copper and paktong incense clock in *ju-i* form, with archaic style calligraphy. Length 11½ in. Early 19th century.

[No estimate]

    Author's note: Made by Ting Yün, second half of the nineteenth century.

Nouveau Drouot – Salles Nos. 5 et 6, Paris, June 3 1983:

    Lot 262. Rare Incense clock in the form of a river steamer, japanned and gilt wood, late 19th century. Extremely rare. Most incense time-keepers of this kind are made in the form of dragons. The present example is one of a very small group made as paddle boats. Sticks of incense are laid lengthways in a pewter tray inside the body of the vessel. Across this a string or strings carrying small metal balls are stretched. As the incense stick burns down it will burn the string causing the balls to fall noisily into a plate set below the instrument. Long: 50 cm. [Illustrated].

[No estimate]

Christie, Manson & Woods, Ltd., London, March 7, 1985.

    Lot 127. A CHINESE PAKTONG FIRE CLOCK, in the form of a *ju-i* sceptre, the body of copper with a detachable lid pierced with an archaic Chinese inscription, the interior with a brass incense tray and perforated maze with corresponding damping cover, 19th Century. [Illustrated] 11½ in. [29 cm] long.

The shape of this particular form of fire clock derives from the sacred fungus, *ling-chih*, one of the Taoist emblems of longevity. It has been suggested that these clocks were employed in Taoist and Buddhist temples to measure intervals between the striking of the prayer bell.

[Estimate] £400–600.

Sold for £432.

Author's note: Illustrated. The template is broken and lacks part of the pattern.

Christie, Manson & Woods, Ltd., London, May 28 1986:

Lot 80. A CHINESE BRASS, COPPER AND PAKTONG FIRE CLOCK, in the form of a Ju' [sic, *ju-i*] sceptre, the body of copper with a detachable lid pierced with an archaic Chinese inscription, the interior with a brass incense tray and perforated maze with corresponding dampening cover. 19th Century. 11½ in, [29 cm] long. [Illustrated]. *Literature*

[Estimate] £300–500.

[Sold for £540]

Author's note: Illustrated in dismantled form depicting the five parts. The description and illustration of this example coincides with those of the incense seal sold in the sale of Christie, Manson & Woods sale of March 7, 1985, and may be the identical piece.

Antiquorum Auctioneers, Hong Kong, N. D. [1983?].
[Entries in French and English.]

Lot 80. CURIOUS CHINESE FIRE ALARM CLOCK, 19TH CENTURY.

Boat shaped, in sculptured wood decorated with scroll-pattern in lacquer. The prow represents a cock head and the stern his feather tail. The inside is delved, contains a metal recipient, graduated on the sides, and embrasures through which a silk thread was passed, with two metal balls at its ends.

Inside the metal recipient, one was used to put a mixture of powders, based on argil, Tibet-tree wood saw-dust to which was added some musk and gold dust. This mixture decayed slowly, without ever going out, showing by its position in the graduated recipient, the past time. By positioning the thread with the ball on a chosen hour, when the moment had arrived, the fire burned the thread, making the balls fall into a metal plate, placed underneath.

Length 570 mm. Width 60 mm. [Illustrated].

[Estimate] Hong Kong $3600–4500.
[Sw. fr. 1200.-/1500.-.]

Lot 81. CURIOUS CHINESE FIRE ALARM CLOCK, 19TH CENTURY.

Boat-shaped mounted on wheels, in sculptured wood decorated with lacquer showing family life, scenes, and scroll-pattern. The prow represents a dragon head, and the stern, his rolled tail. The inside is delved, contains a metal recipient graduated on the sides, and embrasures through which a silk thread was passed, with two metal balls at its ends. Inside the metal recipient, one used to put a mixture of powders, based on argil, Tibet tree wood saw-dust to which was added some musk and gold dust. The mixture decayed slowly, without ever going out, showing by its position in the graduated recipient the past time. By positioning the thread with the ball on a chosen hour, when the moment had arrived, the fire burned the thread, making the balls fall into a metal plate, placed underneath. [Illustrated].

Length: 480 mm. Width: 55 mm.

[Estimate]
Hong Kong $3600–4500.
Sw. frs. 1200.-/1500.-.

Christie's, London, July 14, 1986.

Lot 17. AN ORIENTAL BLACK LACQUER FIRE CLOCK in the elongated form of a bird with gilt decorations, the back hollowed for the fire tray, now lacking. 25½ in. [65 cm] long. [Illustrated].

[Estimate] £400–600.

Christie's, London, April 14, 1988.

Lot 129. A 19th century Chinese paktong and copper incense burner, the sides engraved with characters, with pierced cover and inner tray, the grill pierced with geometric motifs, inscribed with characters with scoop and cover, on ogee shaped feet – 3¼ in. [9.5 cm] wide.
Inventory No. 3149

[Estimate] £100–200.

*See Turner 2, pp. 154/155, no. 34. *The Time Museum*, vol. I, part 3, pp. 154–55. [Described on pp. 154–55 and illustrated in Figure 92. In the Time Museum catalogue, it is erroneously identified as a "Japanese Incense Timekeeper."]
[Sold for: £242 ($451.54)]

Lot 130. A fine 19th century lacquered wood Chinese incense timekeeper, in the form of a river paddle steamer, decorated in gilt with butterflies, moths, birds and flowers, the bow with winged female figurehead, the underside of the hull decorated with waves, the stern with raised deck enclosing a storage compartment, the main deck with funnel and mast enclosing a pewter inner tray with seven "u"-shaped cross wires to carry an incense stick, the underside with signature, the two main paddle wheels with additional bow and stern wheel supports – 8¼ in. [51.4 cm] long. Inventory No. 3389.

[Estimate] £500–800.

*A. J. Turner, *The Time Museum*, vol. I, part 3, *Time Measuring Instruments*, pp. 149–51, for other examples. [Illustrated in full color]
[Sold for: £4950 ($9,256.50)]

Appendix E

Select glossary

a wei	阿魏	Chan Hsi-yüan	詹希元
Ai Wu-lu	愛吾廬	chan t'an	旃檀
Amida	阿彌陀	chang	樟
Amida kōkaryō	阿彌陀公課料	Chang Chung-ting	張忠定
shizaichō	資財帳		
Amida Nyorai	阿彌陀如來	Chang Chün-heng	張鈞衡
an hsi hsiang	安息香		
An Te-hai	安德海	Chang Heng	張衡
Anpuku-ji	安福寺	Chang Hsüan	張萱
ansokukō	安息香	Chang Kuo-lao	張果老
aoshide	アオシデ	Chang Ti	章帝
Arita	有田	chang ts'ao	章草
ashiginu	絁	Chang Yüan-k'ai	張元愷
Ashikaga Yoshimasa	足利義政	Ch'ang-an	長安
		chao	朝
		Chao ch'eng ming hsin	肇城銘新
bakuryō	曝涼		
Birushina	毗盧遮那	Chao Fei-yen	趙飛燕
boku	木	ch'ao hsiao	炒硝
Bokumin kinkan	牧民金鑑	Chao Hsü	趙頊
bonji	梵字	Chao Mu-lan	趙木蘭
bōzu	坊主	Chao Yang	朝陽
bu	分	ch'en	辰（龍）
Bukkyō daijiten	佛教大辭典	Ch'en Ching	陳敬
bun	文	ch'en hsiang	沉香
bunkō	聞香	Ch'en Lang-wen	陳烺文
Bunzon	文存	chen nan	真南
butsudan	佛壇	ch'en su	沉速
		cheng	正
Ch'a Te-ch'ing	查德卿	Ch'eng Cheng-kung	成正恭
Ch'an	禪		

Cheng Ho	鄭和	ch'in	琴
Cheng Ku	鄭谷	chin	斤
ch'eng men lang	城門郎	chin kang ch'eng	金剛乘
Ch'eng Ti	成帝	chin kang chih	金剛智
Ch'eng-tu	成都	*Chin kang ching*	金剛經
Chen-yen	真言	chin shih	進士
chi	己	Ch'ing	清
Chi chiu chang	急就章	Ch'ing Huai yin tao	青淮陰道
chia	甲		
chia hsiang	甲香	*Ch'ing i lu*	清異錄
chia yin	甲寅	Ch'ing Li	慶曆
chiang chen hsiang	降真香	ch'ing lo	罄鑼
		ch'ing t'an	清談
chiang hsiang	降香	Chinju-fu	鎮守府
chiang huang	姜黃	chiu ts'eng Po-shan lu	九層博山爐
Ch'iao Chi	喬吉		
ch'ieh hu shih	挈壺氏	chōdō	朝堂
Chieh-yü	婕妤	Chōgen-ji	長源寺
ch'ien	錢	ch'ou	丑（牛）
chien	箋	Chou	周
chien	覵	chou	籀
Ch'ien Chai	潛齋	Chou Chia-chou	周嘉冑
Ch'ien Chai hsiang p'u shih i	潛齋香譜拾遺	Chou Ho	周賀
		Chou li	周禮
Chien Chen	鑑真	*Chou pei suan ching*	周髀算經
Ch'ien Han shu	前漢書	Ch'u	楚
chien hsiang	箋香	ch'u	初
chien yün hsiang	檢芸香	Chu En-hsi	朱恩錫
Ch'ien-fo-tung	千佛洞	*Ch'u hsüeh chi*	初學記
ch'ih	尺	ch'u keng	初更
chih chia hsiang	製甲香	Chu Shih	朱史
Chih Chih	至治	Chu Shu-chen	朱淑真
Chih Tun	支遁	Chu Yü-chin	朱域謹
Chih Yüan	至元	ch'ü	曲
Chihō bonrei roku	地方凡例錄	Ch'ü P'ing	屈平
		Ch'ü Yu	瞿祐
Chihō ochiboshū	地方落穗集	Ch'ü Yüan	屈原
Chi-lung	繼隆	chuan	篆
Ch'in	秦	chuan shu	篆書

chuan yen	篆言	Daizai-fu	大宰府天満宮
chüan	卷	Tenmangū	
Ch'üan Te-yü	權德輿	dattan no myōbō	達陀の妙法
Ch'üan Yüan	全元散曲	Deha	出羽
san ch'ü		Dengyō Daishi	傳教大師
Ch'üan-chou	泉州	dōdōji	堂童子
Chuang-tzu	莊子	Doryū Rankei	道隆蘭溪
chūdō no ichi	中燈之一	dōshi	童子
Ch'üeh shu	搉暑廨	dōtsukasa	堂司
hsieh		dōtsukasa	堂司宿所
Chūjō-hime	中將姫	shukusho	
Chung kuo jen	中國人名大辭		
ming ta tz'u	典	Edo	江戶
tien		ekishi	驛使
Ch'ung Ming	崇明	Enki	延喜
Chung wen ta	中文大辭典	*Enki onyōrō*	延喜陰陽寮式
tz'u tien		*shiki*	
Chung-chai chü	中齋居士	Enkyū	延久
shih		Eshin	惠心
Ch'ung-i	崇義		
Chung-kuo	中國	fa lu	發爐
Chung-li Ch'üan	鍾離權	fan hun hsiang	反魂香
Chung-shan ta	中山大辭典	Fan Shun-ch'en	范舜臣
tz'u tien i tzu	"一字"長	Fang Kan	方干
ch'ang pien	編	fang ku	仿古
Ch'un-k'un	春困	fen	分
Chū-sō	中倉	*Feng huang t'ai*	鳳凰臺上憶
chūya	中夜	*shang i ch'ui*	吹簫
Chūyūki	中右記	*hsiao*	
		Feng-shun	豐順
Dai Bukkyō jiten	大佛教辭典	fu lu	復爐
Dai Kanwa jiten	大漢和辭典	Fu Shih-chün	撫石軍
Daibutsu	大佛	Fudō	不動
Daigaku Zenji	大覺禪師	Fugen-in	普賢院
Daigo	醍醐	Fujiwara	藤原豐成
daijōshi	大導師	Toyonari	
daimyō	大名		
daishōya	大鐘屋	gaki	餓鬼
Daizai-fu	大宰府	ganju-dori	頷珠鳥

ganza	龕座	Hōjisan	法事讚
geisha	藝者	Hōjō Tokiyori	北條時賴
Genji	源氏物語	Hōju-in	寶壽院
monogatari		Hōki	寶龜
Genjikō	源氏香	Hoku-sō	北倉
Genjimon	源氏紋	*Ho-nan Ch'en*	河南陳氏香譜
Gen-Pei	原平	*shih hsiang*	
gonshosekai	權處世界	*p'u*	
Gotsuchimikado	後土御門	Ho-nan fu	河南府
goya	後夜	Hōnen Shōnin	法然上人
Guchū	具注	hōō-dori	鳳凰鳥
gyoku	玉	Horikawa	堀川
		Hōryō-ji	法隆寺
Hachiman	八幡	Hoshi Kisen	星喜仙
Hachiya-ryū	鉢屋流	Hōshō-ji	寶生寺
hai	亥（豬）	hōsō-ka	寶藏花
Hai-ch'ang	海昌	Hossōge	法相華
hainarashi	灰均	hsi	兮
Han	漢	*Hsi ch'ing ku*	西清古鑑
hana-kui-dori	花食鳥	*chien*	
Hanazono	花園	*Hsi ching tsa chi*	西京雜記
hango-kō	反魂香	Hsi Ning	熙寧
hangyoku	半玉	*Hsi yin hsüan*	惜陰軒叢書
Han-lin	翰林	*ts'ung shu*	
hanya	半夜	hsiang	香
Harima Nyūdō	播磨入道	*Hsiang ch'eng*	香乘
hasu	蓮	*Hsiang chien*	香筻
hasu	ハス	hsiang chuan	香篆
hei hsiang	黑香	hsiang chuan lu	香篆爐
Heian	平安	hsiang fu tzu	香附子
Heijō	平城	*Hsiang kuo*	香國
hi	火	hsiang lu	香爐
hideboku	火出木	*Hsiang lu*	香錄
Higashiyama	東山	*Hsiang lu t'u p'u*	香爐圖譜
himoto	火下	*Hsiang lu t'u*	香爐圖式
Hiraoka	枚岡	*shih*	
hitori	火取	hsiang p'an	香盤
hitsuji	未（羊）	hsiang pang	香棒
Hizen	肥前	*Hsiang p'u*	香譜

Hsiang pu	香部
Hsiang pu hsüan chü	香部選句
Hsiang shih	香史
hsiang yin	香印
hsiang yin lu	香印爐
Hsiang yin t'u k'ao	香印圖考
hsiao	曉
hsiao chuan	小篆
Hsiao hsüeh kan chu	小學紺珠
hsiao ling	小令
Hsiao Shan	筱山
Hsiao Tsung	孝宗
Hsieh Chin	謝晉
Hsin hsiu pen ts'ao	新修本草
hsin i hua	辛夷花
Hsin ts'uan hsiang p'u	新纂香譜
Hsing-chiang	星江
hsiu	宿
Hsiung P'eng-lai	熊朋來
hsü	戌 (狗)
Hsü Ch'i	徐琪
Hsü Hun	許渾
Hsü Lang-hsüan	許朗軒
Hsü Lü-t'ung	徐呂同
Hsüan	宣
Hsüan Te	宣德
Hsüan Tsang	玄奘
Hsüan Tsung	玄宗
Hsüan-ch'eng	宣城
Hsüan-chou	宣州
Hsüeh Chi-hsüan	薛季宣
Huai ch'an mo hsiang	淮産末香
Huakuninshu	百人一首

huang shu hsiang	黃熟香
huang t'an	黃檀
Huang Ti	黃帝
huang t'ung	黃銅
Huang-fu Hung-tse	皇甫洪澤
Hu-chou	湖州
hui hsiang	茴香
hui k'ao	彙考
Hui Tsung	徽宗
Hui-chou	惠州
hun	昏
Hung Ch'u	洪芻
Hung lou meng	紅樓夢
huo hsiang	藿香
Huo Jung	霍融
hwa-sung	火繩
hyaku-tsū-kiri-kami	百通切紙
hyang-jeon	香篆
i	亥 (豬)
i	鳰
I Ching	義淨
I ching	易經
i chu hsiang ti shih hou	一炷香的時候
I Tsung	懿宗
I yao	疑耀
Ichinomiya	一の宮
ihai	位牌
Ihara Saikaku	井原西鶴
I-hsing	宜興
Ike no bunka	池の文化
Imari	伊萬里
Inagaki	稲垣
inji	院士
Inoue	井上
inu	戌 (犬)

inushide	イヌシデ	jūrokinichi no ne	十六日の夜半
Ise-jingū	伊勢神宮	no toki	の時
Ishikawa	石川	jūyokka tora no	十四日寅の時
Ittoku	一德	toki	
ittokuka	一德火		
Izumoi	出雲井	Kaei	嘉永
Izumo-jinja	出雲神社	k'ai shu	楷書
Izumo-taisha	出雲大社	Kaisandō	開山堂
		kakae	抱え
		kakemono	挂物
jen	壬	Kakiemon	柿右衛門
jen	荏	kakubugyō	加供奉行
jih	日	Kamakura	鎌倉
jih ch'u	日出	kamidana	神棚
jih ju	日入	Kamiguchi	上口和時計保
jikikōki	食香鬼	Wadokei	存協會
jikikoku	時期刻	Hozon	
jikō no annai	時香の案内	Kyōkai	
jikō no hōkoku	時香の報告	Kaminochō	上町
jikō shoya made	時香初夜まで	kan	貫
nan sun nan	何寸何分	Kan Po	甘伯
bu		Kan Sung	甘松
jikōban	時香盤	Kanan	河南
jikoku fuda	時刻札	K'ang hsi	康熙
jimae	自前	Kanku	慼玖
jinchō	晨朝	Kannon	觀音
Jitchū	實忠	Kanzeon	觀世音大菩薩
jitchū-kō	十炷香	Daibosatsu	
Jizō	地藏	Kanzeon-ji	觀世音寺
jō	丈	*K'ao kung chi*	考工記
Jōgan	真觀	*k'ao kung tien*	考工典
Jōgon	淨嚴	Kao Lien	高濂
jōkōban	常香盤	kao pen	藁本
jōkōro	常香爐	karahana	唐花
Jōraku-ji	常樂寺	karahitsu	唐櫃
jōtōmyō	常燈明	Kasuka	日下
Jōun	承雲	Kawachi	河内
ju hsiang	乳香	*Kawachi-shi*	河内志
ju-i	如意	Kegon-shū	華嚴宗
Jun-chou	潤州	Keichō	慶長

Keiun	慶雲	kōgō	香盒
ken	間	Koguryŏ	高麗
ken	拳	kōin	香印
Kenchō-ji Rinzai-shū	建長寺臨濟宗	kōinban	香印盤
		kōinoukeban	香印御受盤
keng	庚	kōinza	香印座
keng hsiang	更香	*Koji ruien*	古事類苑
keta	氣多	kōkoku	香刻
keyaki	欅	koku	刻
Kichijō-ji	吉祥寺	komori	木守
Kichisaburo	吉三郎	kōmoto	香元
kikikō	聞香	Kōmyō	光明
Kimimaro	公麿	Konchi-ji	金地寺
Kimitaka Sangi	公貴參議	Kongōbu-ji	金剛峰寺
kin	金	Kongō-in	金剛院
Kinmei	欽明	Kōnin	光仁
kiri	桐	kōnoji	香野字
kirikuji kōrō	紇哩字香爐	kōro	香爐
kiroku	記錄	kōro makura	香爐枕
kitazashū no ichi	北座眾之一	*Kōshiki reigikai*	公式禮義解
kitazashū no ni	北座眾之二	kōsun	香寸
k'o	刻	Kōtoku	孝德
kō	香	*Kōtoku-ki*	孝德紀
Ko chih ching yüan	格致鏡原	kōuke	香受
		Kōya	高野
k'o chu	刻炷	kōya	後夜
kō o kiku	香を聞く	Kōyasan	高野山
kō o motte jōshiki to nasu	香をもって常式となす	ku wen	古文
		kua tieh mien mien	瓜瓞綿綿
kōan	公案	kuan huo	爟火
kōawase	香合せ	kuan kuei hsiang	官桂香
kōban	香盤	Kuan Yin	觀音
kōbandokei	香盤時計	Kuang Hsü	光緒
Kōbō Daishi	弘法大師	*Kuang ya*	廣雅
kōdō	香道	Kuang-te	廣德
kōe	香會	Kuan-hsiu	貫休
kōfuda	香札	kuei	癸
kōkago	香籠	kuei hsiang	桂香
kōgata	香型	kuei lou	晷漏

kuei piao	晷表	Liao	遼
kuei-ch'ou	癸丑	Li-chih	立之
Kujikkajō	九拾ヶ條	Lin	璘
Kūkai	空海	ling chih	靈芝
Kumikaō	組花押	ling mao hsiang	靈貓香
kumikō	組香	Ling-ling hsiang	零陵香
Kunashiri	國後	*Liu ching t'u*	六經圖
Kuo	過	Liu Jui-fen	劉瑞芬
Kuo Shou-ching	郭守敬	Liu Yü-hsi	劉禹錫
Kuo tzu chien	國子監	lo le hsiang	羅勒香
Ku-shan	鼓山	*Lou k'o ching*	漏刻經
kushi	駈士	*Lou k'o fa*	漏刻法
kusuri-dama	藥玉	*Lou shui chuan*	漏水轉渾天
kyaku	客	*hun t'ien i*	儀制
kyara	伽羅	*chih*	
Kyong Bok	京福	Lu	魯
Kyōto	京都	lu chu	爐主
Kyūshū	九州	lu hsia	爐下
		lu kan shih	爐甘石
lan hui	蘭蕙	Lu Kuei-meng	陸龜蒙
Lang-shan	狼山	*lu pu tsa lu*	爐部雜錄
Lan-pin	蘭賓	Lu Wa	魯瓦
Lao-tzu	老子	lung hsien	龍涎香
li	隸	hsiang	
li chih k'o	荔枝殼	lung nao hsiang	龍腦香
Li Ch'ing-chao	李清照	*Lung-shu p'u-sa*	龍樹菩
Li Chui	李錐	*ho hsiang*	薩和香方
Li Ho	李賀	*fang*	
Li Hou-chu	李後主		
Li Hsüeh-yü	李學裕	ma t'i hsiang	馬蹄香
li kuei i hsia lou	立晷儀下漏刻	Maeda	前田加賀經
k'o		Kagatsune	
Li Lan	李蘭	Mai Yü Wan	賣魚灣
Li Po	李白	makkō	抹香
Li Shun-feng	李淳風	Manaban	真南蠻
Li Ssu	李斯	manji	卍
Lin Te	麟德	mannentake	萬年茸
Li T'ieh-kuai	李鐵枴	man-su-hyang	萬歲香
Li Ts'ui	李璀	mao	卯（兔）
liang	兩	Mao Chin	毛晉

mao hsiang	茅香	nakama	中間
Mao Pang-han	毛邦翰	Namegawa Naga	並河永
matsu no jin	松のじん	nan mu	楠木
mei	妹	*Nan shih*	南史
Mei Yao-ch'en	梅堯臣	Nanbokuchō	南北朝
Mei-ch'i	梅溪	Nan-sō	南倉
Meiji	明治	Nan-t'ung	南通
Meng Chüeh An	夢覺庵妙高香	nanzashū no ichi	南座衆之一
miao kao	方	nanzashū no ni	南座衆之二
hsiang fang		ne	子（鼠）
Meng Liang lu	夢梁錄	nenbutsu	念佛
Meng Yüan-lao	孟元老	nerikō	練香
mi	巳	nichibotsu	日沒
Miao Wan	沙灣	nichimotsu	日沒
mien	綿	Nigatsudō	二月堂
mien mien kua	綿綿瓜瓞	*Nihon shoki*	日本書紀
tieh		*Nihongi*	日本紀
mikado	御門	Ninna-ji	仁和寺
Mikumari-jinja	水分神社	Ninsei	仁濟
Ming	明	Nintoku	仁德
Ming shih	明史	*Nintoku-ki*	仁德紀
Mi-tsung	密宗	nioi-bukuro	匂い袋
mo li hua hsiang	茉莉化香	nishiki	錦
Momoyama	桃山	nitchū	日中
Monmu	文武	Niu-ch'ang	牛廠
mu	墓	*Nōminshi goi*	農民史語彙
mu hsiang	木香	*Nung shu*	農書
mu ku	木鼓		
Mu-chou	睦州	Ō nyu	遠敷
Murashiki	紫式部	Ōchō	應長
Shikibu		Oda Nobunage	織田信長
Muromachi	室町	Oda Tokunō	織田得能
museirō	無聲漏	Ōgimachi	正親町
Mutō	無頭	ōi	大炊
Myōden	妙典	Oie-ryū	御家流
Myōon-ji	妙音寺	oiran	花魁
Myōshin-ji	妙心寺	Ōjin	應神
		Ōjin-ki	應神紀
Nabeshima	鍋島	Ōjukkō	黃熱香
naijin	內陣	ok-dung-an	玉燈盞

okiya	置屋	po hsiang	柏香
Ōkuni-jinja	大國神社	po k'o chuan hsiang	百刻篆香
Ōkunitama-jinja	大國魂神社		
omizutori	お水取り	po k'o hsiang yin	百刻香印
omizutori no gyōji	お水取りの行事	po k'o yin hsiang	百刻印香
Omuro	御室		
Ono Takeo	小野武夫	Po-shan	博山
osae	押	Po-shan hsiang lu	博山香爐
Oshichi	お七		
oshi-dori	鴛鴦	Po-shan lu	博山爐
Ou Lao	歐老	pu cheng shih	布政使
Ou-yang Hsiu	歐陽修	pu shih kuan	補事官
		Pu-k'ung chin-kang	不空金剛
pa chi hsiang	八吉象	*P'u-sa man*	菩薩蠻
pa hsien	八仙		
pa kua	八卦	Rakoku	羅國
pa pao	八寶	ranjatai	蘭奢待
Paekche	百濟	Reiun-ji	靈雲寺
pai ch'ien	白鉛	rengyōshū	練行衆
pai chih	白芷	reshi	靈芝
pai la	白鑞	rin	厘
p'an	盤	Rōben	良辨
Pan Chieh-yü	班婕妤	ro-jeon	爐殿
P'an Feng-t'ai	潘逢泰	rōkoku	漏刻, 漏剋
P'an Ku	盤古	roku	六
pang hsiang	棒香	rokuji	六時
pei chung hsiang lu	被中香爐	rokuji no raisan	六時の禮贊
P'ei wen yün fu	佩文韻府	rokuji raisan	六時禮贊
p'en hsiao	盆硝	Roku-jizō	六地藏
pi	璧	Ryūmei shō	龍鳴抄
pi hsieh	辟邪		
P'i Jih-hsiu	皮日休	Saichō	最澄
pi kuan	壁觀	*Saiikki*	西域記
p'i shuang	砒霜	Saikōin-ji	西光院寺
p'ien li	駢儷	Saimei	濟明
P'ing lan jen	憑欄人	Saiun	最雲
Po Chü-i	白居易	Sakai	酒井
po ho hsiang	百合香	samisen	三味線

san ch'ü	散曲		*Shih chi*	史記
san lai	三賴		shih chien ti tsu chi	時間的足跡
Sandai jitsuroku	三代實錄		*Shih ching*	詩經
Sanrō shukusho	參籠宿所		shih erh chih	十二支
Sanzenin-ji	三千院寺		Shih Huang	始皇
saru	申（猿）		Shih Kang	石港
Sasaki Dōyo	佐佐木道譽		*Shih lin kuang chi*	事林廣記
Satomi-ryū	里見流			
Sayama	狹山		*Shih t'ou chi*	石頭記
Sei Shōnagon	清少納言		*Shih yüan ts'ung shu*	適園叢書
Seiryō	清涼			
Seiwa	清和		Shikimi	樒
sen	錢		Shimizu Kannondō	清水觀音堂
Senju Kannon Bosatsu	千手觀音菩薩			
			Shimonoseki	下關
senkō	線香		shinchō	晨朝
senkōdokei	線香時計		Shingon	真言
Shaka	釋迦		shinkoku	辰刻
shaku	尺		Shino Sōshin	志野宗信
Shan chü hsin hua	山居新話		Shino-ryū	志野流
			Shinpen Kamakura shi	新編鎌倉志
shan li	山梨			
Shan Sheng	善勝			
shang	上		*Shinsen Bukkyō daijiten*	新撰佛教大事典
Shang	商			
Shang Tao	商道		Shinshū	真宗
Shan-wu-wei	善無畏		Shintō	神道
Shan-yang	山陽		Shirakawa	白河
Shao-nan	紹南		shishi	獅子
shen	申（猴）		Shizuoka	靜岡
Shen Li	沈立		Shō Kannon	小觀音
shen pien	神變		Shōgaku-ji	常樂寺
Shen Tsung	神宗		shogoya no toki	初後夜の時
Shen Yu-jen	沈有壬		shokō	小綱
sheng hsiang	生香		Shōmu	聖武
Sheng shih	笙氏		shosekai	處世界
shide	シデ		shosekai	小世界
shidebō	しでぼう		Shōsō-in	正倉院
shideboku	しで木		*Shōsō-in*	正倉院御物棚
shih	時			

gobutsubyō	別目錄	sugi	杉
betsumoku-		Sui	隋
roku		Suiko	推古
Shōsō-in	正倉院御物圖	Sūjin	崇神
gyobutsu	錄	Sūjin-ki	崇神紀
zuroku		sun	寸
Shōsō-in	正倉院寶物	Sung	宋
hōbutsu		Sung Chih-wen	宋之問
Shōtoku Taishi	聖德太子	Sung Ching	宋景
shou	壽	Su-Sung t'ai tao	蘇松臺道
Shou hu t'ang	受祜堂	Suzuka	鈴鹿
shou lu	手爐		
shoya	初夜	T'ien Li	天曆
shoya made nan	初夜まで何寸	ta chiao	大教
sun nan bu	何分	ta chuan	大篆
shoya no gyōhō	初夜の行法	ta huang	大黃
shu	書	Ta jih ching	大日經
Shu	蜀	Ta Ming	大明
shuang hsi	囍	Ta T'ang liu	大唐六典
shūka	集荷	tien	
Shumidan	須彌壇	Ta Yu Tung	大有洞
Shunie	修二會	Ta-fo-ssu	大佛寺
shushi	咒師	T'ai chi	太極
shusshi no annai	出仕の案內	T'ai p'ing yü	太平御覽
Silla	新羅	lan	
Soga no Iname	蘇我稻目	tai tan	待旦
Sono-ryū	曾野流	Taika	大化
soroban	算盤	Taima-dera	當麻寺
Soshi ryaku	祖師略	taimatsu shiki	松明式
Sōsō	雙倉	T'ai-po	太白
ssu	巳（蛇）	Taishō	大正
su ho hsiang	蘇合香	Tai-yü	黛玉
su hsiang	速香	Takamatsu	高松
Su Kuo	蘇過	Takauji	高氏
Su Shih	蘇軾	Takigawa Seijirō	瀧川政次郎
Su Sung	蘇頌	Takuan	澤庵
Su Tsung	肅宗	tameshi	試し
Su Tung-p'o	蘇東坡	t'an hsiang	檀香
Suenaga Tadao	末永雅雄	tan p'i	丹皮
Sugawara	菅原道真	Tan-chou	儋州
Michizane		T'ang	唐

Tan-yang	丹陽	Ting Wu	丁午
tao	道	Ting Yen	丁晏
T'ao Ku	陶穀	Ting Yüeh-hu	丁月湖
tao t'ai	道臺	Ting Yün	丁云
Tao te ching	道德經	Ting-chou	定州
tarumizu	垂水	Ting-chou kung	定州公庫印香
Tatebe-ryū	建部流	k'u yin	
tateki-age	叩き上げ	hsiang	
tatsu	辰（龍）	ting-tzu chih	丁子之香
Te Tsung	德宗	hsiang	
Tenchi	天智	to	鐸
Tendai	天台	Tōdai-ji	東大寺
teng hua	燈花	Tōjō-ji no	東城寺の藥師
Tenpaizan	天拜山	Yakushi	
Tenpyō	天平	tōka	唐花
Tenryaku	天曆	toki	時
Ti Jen-chieh	狄仁傑	toki no fuda	時の簡
Ti kung an	狄公案	toki no kui	時の杭
t'ieh mien ma ya	鐵面馬牙香	Tōkō-ji	東光寺
hsiang		tokonoma	床の間
T'ien hsiang	天香傳	Tokugawa	德川家康
chuan		Ieyasu	
T'ien kung fu	天公賦	tora	寅（虎）
T'ien kung k'ai	天工開物	tori	酉（雞）
wu		t'ou hu	投壺
T'ien wen chih	天文志	t'ou shih	鍮石
ting	鼎	*Tōyō monyō shi*	東洋文樣史
ting hsiang	丁香	Toyotomi	豐臣秀吉
Ting Huan	丁緩	Hideyoshi	
Ting Jih-ch'ang	丁日昌	Toyoura	豐浦
Ting Li-ch'eng	丁立誠	Ts'ai Wei-hsing	蔡蓮幸
Ting Li-chung	丁立中	tsao	棗
Ting Pao-chen	丁寶楨	Ts'ao Hsüeh-	曹雪芹
ting p'i	丁皮	ch'in	
Ting Ping	丁丙	Ts'ao Jui	曹叡
Ting Shen	丁申	Ts'ao Mo	曹沫
Ting Shou-	丁壽昌	*Ts'ao mu tien*	草木典
ch'ang		Ts'ao P'ei	曹丕
Ting Shou-ch'i	丁壽麒	ts'e ching jih	測景日晷
Ting Wei	丁謂	kuei	

Tso Ch'iu-ming	左丘明	Tz'u En	慈恩
Tsou Hsiang- t'an	鄒象潭	Tzu-chou	梓州
		Tz'u-li	慈利
tsu	度		
tsubonebugyō	局奉行	u	卯（兔）
Tsuchiura-shi Chūo	土浦市中央	uchū sahan	右重左半
		Ueda	上田
tsuketake	付立	Ujiamakō	宇治甘香
ts'un	寸	uma	午（馬）
Tsun sheng pa *chien*	遵生八牋	*Umezo no nikki*	梅園日記
		ungen	繧繝
Tsung cheng chai	宗正齋	urushi-kinpaku- eban	漆金箔繪盤
Tsung cheng chai tsao	宗正齋造	ushi	丑（牛）
		uta-awase	歌合せ
Ts'ung shu chi *ch'eng*	叢書集成		
		wajō	和上
tsung tzu	粽子	Wakasa	若狹
Tsutsui Eishun	筒井英俊	Wakasa-i	若狹井
Tu Fu	杜甫	wan	卍, 萬
tu huo	獨活	Wang An-shih	王安石
T'u Lung	屠隆	Wang Ch'ang- ling	王昌齡
Tu Mu	杜牧		
Tu Shen-yen	杜審言	Wang Chen	王禎
T'u shu chi *ch'eng*	圖書集成	Wang Chien	王建
		Wang Shu-wen	王叔文
Tu Tao-hsiu	杜道修	Wang Ting- hsiang	王定祥
Tu Tao-i	杜道義		
t'u ts'ao hsiang	土草香	Wang Wei	王維
Tu Tzu-sheng	杜子盛	Wang Ying-lin	王應麟
Tuan Ch'eng- shih	段成式	wei	未（羊）
		Wei Tan	韋誕
t'ung hu	銅壺	Wen Tsung	文宗
Tung t'ien *ch'ing lu*	洞天清錄	Wen-chou	溫州
		Whang Ryong	黃龍
Tung-ching *meng hua lu*	東京夢華錄	wo ch'ien	倭鉛
		Wu	武
Tun-huang	敦煌	wu	午（馬）
tzu	子（鼠）	wu	戊
tzu	字	wu-cheng	午正

Wu Cheng-chung	吳正仲
wu keng yin k'o	五更印刻
Wu Seng-jui	吳僧瑞
Wu Shan	五山
Wu Tsung	武宗
Wu Tzu-mu	吳自牧
wu yeh chuan hsiang	五夜篆香
wu yeh hsiang k'o	五夜香刻
wu yüeh chieh	五月節
Wu-lin	武林
Wu-chou	婺州
wu-tzu	戊子
yama	山
Yanagibashi	柳橋
yang	陽
Yang Chia	楊甲
Yang Hsiung	揚雄
Yang-yün shan-chuang ch'üan chi	養雲山莊全集
Yao	堯
yashiki	邸
Yeh shih hsiang p'u	葉氏香譜
Yeh Tao-lu	葉道陸
Yeh T'ing-kuei	葉廷珪
Yen Ch'ih-yüeh	顏持約
Yen Ch'ih-yüeh hsiang shih	顏持約香史
yen chuan	煙篆
yen hsiao	焰硝
yen nien	研碾
yen nien	延年
Yen Su	燕肅
Yen Yu	延祐
yen yün kung	煙雲供養

yang	
Yi Sampei	李參平
yin	陰
yin	印
yin	寅（虎）
yin hsiang	印香
Yin hsiang fang	印香方
Yin hsiang kung Fo fang	印香供佛方
Yin hsiang t'u p'u	印香圖譜
Yin Hsien	尹戚
yin yang	陰陽
Ying-chou	潁州
Ying-cheng	郢正
yōji no annai	用事の案内
yōkyoku	謠曲
Yonekawa-ryū	米川流
Yōrō	養老
Yōrō no gokurei	養老の獄令
Yōrō no kanshirei	養老の關市令
Yōrō no kōshikirei	養老の公式禮
Yōrō no kyūerei	養老の宮衛令
Yōrō no shokusei-ritsu	養老の職制律
Yoshida Hanbei	吉田半兵衛
yu	酉（雞）
Yü Wan	魚灣
Yü-li	寓澧
Yüan	元
Yüan Chen	元稹
Yüan Hu	鴛湖
Yüan P'u	袁浦
yüan shen	元參
Yüan Ti	元帝
Yü-chang	豫章
Yüeh-heng	嶽恆

Yu-hsiung	有熊	*Zenrin zōkisen*	禪林象器箋
yün hsiang	芸香	*Zōho den'en*	增補田園類說
Yung	永	*ruisetsu*	
		Zokki	續紀
Zen	禪	zukō	塗香
zenjō	禪定		

Appendix F

Romanization conversion table

WADE-GILES TO PINYIN

Wade-Giles	Pinyin	Wade-Giles	Pinyin	Wade-Giles	Pinyin	Wade-Giles	Pinyin
a	a	ch'ih	chi	chün	jun	hsing	xing
ai	ai	chin	jin	ch'ün	qun	hsiu	xiu
an	an	ch'in	qin	chung	zhong	hsiung	xiong
ang	ang	ching	jing	ch'ung	chong	hsü	xu
ao	ao	ch'ing	qing			hsüan	xuan
		chiu	jiu	en	en	hsüeh	xue
cha	zha	ch'iu	qiu	erh	er	hsün	xun
ch'a	cha	chiung	jiong			hu	hu
chai	zhai	ch'iung	qiong	fa	fa	hua	hua
ch'ai	chai	cho	zhuo	fan	fan	huai	huai
chan	zhan	ch'o	chuo	fang	fang	huan	huan
ch'an	chan	chou	zhou	fei	fei	huang	huang
chang	zhang	ch'ou	chou	fen	fen	hui	hui
ch'ang	chang	chou	zhou	feng	feng	hun	hun
chao	zhao	ch'ou	chou	fo	fo	hung	hong
ch'ao	chao	chu	zhu	fou	fou	huo	huo
che	zhe	ch'u	chu	fu	fu		
ch'e	che	chü	ju			i	yi
chen	zhen	ch'ü	qu	ha	ha		
ch'en	chen	chua	zhua	hai	hai	jan	ran
cheng	zheng	ch'ua	chua	han	han	jang	rang
ch'eng	cheng	chuai	zhuai	hang	hang	jao	rao
chi	ji	ch'uai	chuai	hao	hao	je	re
ch'i	qi	chuan	zhuan	hei	hei	jen	ren
chia	jia	ch'uan	chuan	hen	hen	jeng	reng
ch'ia	qia	chüan	juan	heng	heng	jih	ri
chiang	jiang	ch'üan	quan	ho	he	jo	ruo
ch'iang	qiang	chuang	zhuang	hou	hou	jou	rou
chiao	jiao	ch'uang	chuang	hsi	xi	ju	ru
ch'iao	qiao	chüeh	jue	hsia	xia	juan	ruan
chieh	jie	ch'üeh	que	hsiang	xiang	jui	rui
ch'ieh	qie	chui	zhui	hsiao	xiao	jun	run
chien	jian	ch'ui	chui	hsieh	xie	jung	rong
ch'ien	qian	chun	zhun	hsien	xian		
chih	zhi	ch'un	chun	hsin	xin	ka	ga

Wade-Giles	Pinyin	Wade-Giles	Pinyin	Wade-Giles	Pinyin	Wade-Giles	Pinyin
k'a	ka	liang	liang	nien	nie	pu	bu
kai	gai	liao	liao	nien	nian	p'u	pu
k'ai	kai	lieh	lie	nin	nin	sa	sa
kan	gan	lien	lian	ning	ning	sai	sai
k'an	kan	lin	lin	niu	niu	san	san
kang	gang	ling	ling	no	nuo	sang	sang
k'ang	kang	liu	liu	nou	nou	sao	sao
kao	gao	lo	luo	nu	nu	se	se
k'ao	kao	lou	lou	nü	nü	sen	sen
kei	gei	lu	lu	nuan	nuan	seng	seng
k'ei	kei	lü	lü	nueh	nüe	sha	sha
ken	gen	luan	luan	nung	nong	shai	shai
k'en	ken	lüan	lüan			shan	shan
keng	geng	lüeh	lüe	o	e	shang	shang
k'eng	keng	lun	lun	ou	ou	shao	shao
ko	ge	lung	long			she	she
k'o	ke			pa	ba	shen	shen
kou	gou	ma	ma	p'a	pa	sheng	sheng
k'ou	kou	mai	mai	pai	bai	shih	shi
ku	gu	man	man	p'ai	pai	shou	shou
k'u	ku	mang	mang	pan	ban	shu	shu
kua	gua	mao	mao	p'an	pan	shua	shua
k'ua	kua	mei	mei	pang	bang	shuai	shuai
kuai	guai	men	men	p'ang	pang	shuan	shuan
k'uai	kuai	meng	meng	pao	bao	shuang	shuang
kuan	guan	mi	mi	p'ao	pao	shui	shui
k'uan	kuan	miao	miao	pei	bei	shun	shun
kuang	guang	mieh	mie	p'ei	pei	shuo	shuo
k'uang	kuang	mien	mian	pen	ben	so	suo
kuei	gui	min	min	p'en	pen	sou	sou
k'uei	kui	ming	ming	peng	beng	ssu	si
kun	gun	miu	miu	p'eng	peng	su	su
k'un	kun	mo	mo	pi	bi	suan	suan
kung	gong	mou	mou	p'i	pi	sui	sui
k'ung	kong	mu	mu	piao	biao	sun	sun
kuo	guo			p'iao	piao	sung	song
k'uo	kuo	na	na	pieh	bie		
		nai	nai	p'ieh	pie	ta	da
la	la	nan	nan	pien	bian	t'a	ta
lai	lai	nang	nang	p'ien	pian	tai	dai
lan	lan	nao	nao	pin	bin	t'ai	tai
lang	lang	nei	nei	p'in	pin	tan	dan
lao	lao	nen	nen	ping	bing	t'an	tan
le	le	neng	neng	p'ing	ping	tang	dang
lei	lei	ni	ni	po	bo	t'ang	tang
leng	leng	niang	niang	p'o	po	tao	dao
li	li	niao	niao	pou	bou	t'ao	tao
lia	lia	nieh	nie	p'ou	pou		

Wade-Giles	Pinyin	Wade-Giles	Pinyin	Wade-Giles	Pinyin	Wade-Giles	Pinyin
te	de	ts'ai	cai	ts'ui	cui	wei	wei
t'e	te	tsan	zan	tsun	zun	wen	wen
teng	deng	ts'an	can	ts'un	cun	weng	weng
t'eng	teng	tsang	zang	tsung	zong	wo	wo
ti	di	ts'ang	cang	ts'ung	cong	wu	wu
t'i	ti	tsao	zao	tu	du		
tiao	diao	ts'ao	cao	t'u	tu	ya	ya
t'iao	tiao	tse	ze	tuan	duan	yai	yai
tieh	die	ts'e	ce	t'uan	tuan	yang	yang
t'ieh	tie	tsei	zei	tui	dui	yao	yao
tien	dian	tsen	zen	t'ui	tui	yeh	ye
t'ien	tian	ts'en	cen	tun	dun	yen	yan
ting	ding	tseng	zeng	t'un	tun	yin	yin
t'ing	ting	ts'eng	ceng	tung	dong	ying	ying
tiu	diu	tso	zuo	t'ung	tong	yu	you
to	duo	ts'o	cuo	tzu	zi	yü	yu
t'o	tuo	tsou	zou	tz'u	ci	yüan	yuan
tou	dou	ts'ou	cou			yüeh	yue
t'ou	tou	tsu	zu	wa	wa	yün	yun
tsa	za	tsuan	zuan	wai	wai	yung	yong
ts'a	ca	ts'uan	cuan	wan	wan		
tsai	zai	tsui	zui	wang	wang		

PINYIN TO WADE-GILES

Pinyin	Wade-Giles	Pinyin	Wade-Giles	Pinyin	Wade-Giles	Pinyin	Wade-Giles
a	a	bo	po	chi	ch'ih	da	ta
ai	ai	bou	pou	chong	ch'ung	dai	tai
an	an	bu	pu	chou	ch'ou	dan	tan
ang	ang			chu	ch'u	dang	tang
ao	ao	ca	ts'a	chua	ch'ua	dao	tao
		cai	ts'ai	chuai	ch'uai	de	te
ba	pa	can	ts'an	chuan	ch'uan	deng	teng
bai	pai	cang	ts'ang	chuang	ch'uang	di	ti
ban	pan	cao	ts'ao	chui	ch'ui	dian	tien
bang	pang	ce	t'se	chun	ch'un	diao	tiao
bao	pao	cen	ts'en	chuo	ch'o	die	tieh
bei	pei	ceng	ts'eng	ci	tz'u	ding	ting
ben	pen	cha	ch'a	cong	ts'ung	diu	tiu
beng	peng	chai	ch'ai	cou	ts 'ou	dong	tung
bi	pi	chan	ch'an	cu	ts'u	dou	tou
bian	pien	chang	ch'ang	cuan	ts'uan	du	tu
biao	piao	chao	ch'ao	cui	ts'ui	duan	tuan
bie	pieh	che	ch'e	cun	ts'un	dui	tui
bin	pin	chen	ch'en	cuo	ts 'o	dun	tun
bing	ping	cheng	ch'eng			duo	to

Pinyin	Wade-Giles	Pinyin	Wade-Giles	Pinyin	Wade-Giles	Pinyin	Wade-Giles
e	o	huan	huan	leng	leng	nian	nien
en	en	huang	huang	li	li	niang	niang
er	erh	hui	hui	lia	lia	niao	niao
		hun	hun	lian	lien	nie	nieh
fa	fa	huo	huo	liang	liang	nin	nin
fan	fan			liao	liao	ning	ning
fang	fang	ji	chi	lie	lieh	niu	niu
fei	fei	jia	chia	lin	lin	nong	nung
fen	fen	jian	chien	ling	ling	nou	nou
feng	feng	jiang	chiang	liu	liu	nu	nu
fo	fo	jiao	chiao	long	lung	nü	nü
fou	fou	jie	chieh	lou	lou	nuan	nuan
fu	fu	jin	chin	lu	lu	nüe	nüeh
		jing	ching	lü	lü	nuo	no
ga	ka	jiong	chiung	luan	luan		
gai	kai	jiu	chiu	lüan	lüan	ou	ou
gan	kan	ju	chü	lüe	lüeh		
gang	kang	juan	chüan	lun	lun	pa	p'a
gao	kao	jue	chüeh	luo	lo	pai	p'ai
ge	ko	jun	chün			pan	p'an
gei	kei			ma	ma	pang	p'ang
gen	ken	ka	k'a	mai	mai	pao	p'ao
geng	keng	kai	k'ai	man	man	pei	p'ei
gong	kung	kan	k'an	mang	mang	pen	p'en
gou	kou	kang	k'ang	mao	mao	peng	p'eng
gu	ku	kao	k'ao	mei	mei	pi	p'i
gua	kua	ke	k'o	men	men	pian	p'ien
guai	kuai	kei	k'ei	meng	meng	piao	p'iao
guan	kuan	ken	k'en	mi	mi	pie	p'ieh
guang	kuang	keng	k'eng	mian	mien	pin	p'in
gui	kuei	kong	k'ung	miao	miao	ping	p'ing
gun	kun	kou	k'ou	mie	mieh	po	p'o
guo	kuo	ku	k'u	min	min	pou	p'ou
		kua	k'ua	ming	ming	pu	p'u
ha	ha	kuai	k'uai	miu	miu		
hai	hai	kuan	k'uan	mo	mo	qi	ch'i
han	han	kuang	k'uang	mou	mou	qia	ch'ia
hang	hang	kui	k'uei	mu	mu	qian	ch'ien
hao	hao	kun	k' un			qiang	ch'iang
he	ho	kuo	k'uo	na	na	qiao	ch'iao
hei	hei			nai	nai	qie	ch'ieh
hen	hen	la	la	nan	nan	qin	ch'in
heng	heng	lai	lai	nang	nang	qing	ch'ing
hong	hung	lan	lan	nao	nao	qiong	ch'iung
hou	hou	lang	lang	nei	nei	qiu	ch'iu
hu	hu	lao	lao	nen	nen	qu	ch'ü
hua	hua	le	le	neng	neng	quan	ch'üan
huai	huai	lei	lei	ni	ni	que	ch'üeh

Pinyin	Wade-Giles	Pinyin	Wade-Giles	Pinyin	Wade-Giles	Pinyin	Wade-Giles
qun	ch'ün	sui	sui	ya	ya	zuan	tsuan
		sun	sun	yai	yai	zui	tsui
ran	jan	suo	so	yan	yen	zun	tsun
rang	jang			yang	yang	zuo	tso
rao	jao	ta	t'a	yao	yao		
re	je	tai	t'ai	ye	yeh		
ren	jen	tan	t'an	yi	i		
reng	jeng	tang	t'ang	yin	yin		
ri	jih	tao	t'ao	ying	ying		
rong	jung	te	t'e	yong	yung		
rou	jou	teng	t'eng	you	yu		
ru	ju	ti	t'i	yu	yü		
ruan	juan	tian	t'ien	yuan	yüan		
rui	jui	tiao	t'iao	yue	yüeh		
run	jun	tie	t'ieh	yun	yün		
ruo	jo	ting	t'ing				
		tong	t'ung	za	tsa		
sa	sa	tou	t'ou	zai	tsai		
sai	sai	tu	t'u	zan	tsan		
san	san	tuan	t'uan	zang	tsang		
sang	sang	tui	t'ui	zao	tsao		
sao	sao	tun	t'un	ze	tse		
se	se	tuo	t'o	zei	tsei		
sen	sen			zen	tsen		
seng	seng	wa	wa	zeng	tseng		
sha	sha	wai	wai	zha	cha		
shai	shai	wan	wan	zhai	chai		
shan	shan	wang	wang	zhan	chan		
shang	shang	wei	wei	zhang	chang		
shao	shao	wen	wen	zhao	chao		
she	she	weng	weng	zhe	che		
shen	shen	wo	wo	zhen	chen		
sheng	sheng	wu	wu	zheng	cheng		
shi	shih			zhi	chih		
shou	shou	xi	hsi	zhong	chung		
shu	shu	xia	hsia	zhou	chou		
shua	shua	xian	hsien	zhu	chu		
shuai	shuai	xiang	hsiang	zhua	chua		
shuan	shuan	xiao	hsiao	zhuai	chuai		
shuang	shuang	xie	hsieh	zhuan	chuan		
shui	shui	xin	hsin	zhuang	chuang		
shun	shun	xing	hsing	zhui	chui		
shuo	shuo	xiong	hsiung	zhun	chun		
si	ssu	xiu	hsiu	zhuo	cho		
song	sung	xu	hsü	zi	tzu		
sou	sou	xuan	hsüan	zong	tsung		
su	su	xue	hsüeh	zou	tsou		
suan	suan	xun	hsün	zu	tsu		

Bibliography

Manuscripts and published works in East Asian languages

Aiga, Tetsuo. *Tōdaiji omizutori*. Tokyo: Shogaku kan. 1985.

Asahina, Teiichi. "Jikōban ni tsuite" [On the Time Measuring Incense Board]. *Yamato Bunka Kenkyū* [Research on the Culture of Yamato]. Vol. 2, no. 3, June 29, 1955. Pp. 19–34.

Tokei. Tokyo: Kokuritsu hakubutsukan. 1970

Buddhist Canon. See *Takakusa, J.*

Bukkyō daijiten [Buddhist Dictionary]. 1932.

Chang Chün-heng. *Shih-yüan ts'ung shu*. 1914.

Chang Hsüan, *Iyao* (Ts'ung shu chi ch'eng edn).

Chang T'ing-yü *et al. Ming shih*.

Chang Yü-shu, ed. *P'ei wen yün fu* [Encyclopedia of Phrases and Allusions Arranged According to Rhyme]. 1711.

Ch'en Ching, compiler, c. 1322. Hsin tsuan hsiang p'u.

Ch'en Meng-lei and Chiang T'ing-hsi, *et al.*, compilers. *Ku chin t'u shu chi ch'eng*. 1725. *Shanghai T'u-shu chi-ch'eng chü*. 100 vols, plus index. 1884, 1934. (See Chiang T'ing-hsi.)

Ch'en I and Ch'en Hao, c. 1168 *Hsin tsuan hsiang p'u* (Newly Compiled Book of Incense). Also known as *Ho-nan Ch'en shih hsiang p'u* [also *Ho-nan Ch'en shih i shu*] (Handbook on Incense by the Ch'en Family in Honan).

Ch'en Ching. *Hsin tsuan hsiang p'u* [Newly Compiled Handbook of Incense] c. 1322. Also known as *Ho-nan Ch'en shih hsiang p'u* [Handbook on Incense by the Ch'en Family in Honan]. Included in Chang Chün-heng, *shih-yüan ts'ung-shu*,

as well as in the *Hsiang ch'ien chou* [Comprehensive Account of Incense] by Chou Chia-chou (fl. 1580–1650). N.D. [Early fourteenth century].

Chiang, T'ing-hsi. *Ch'in ting ku chin t'u shu chi ch'eng*. 800 vols. in 102 cases. 1933. (See Ch'en Meng-lei, *Ku chin*.)

Chiou Li. [Record of the Rites of (the) Chou (Dynasty)]. Translated by E. Biot, *Le 'Tcheou-Li' ou 'Rites Jes Tseou.'* 3 vols. Paris: Imprimerie Nationale, 1851.

Chou Chia-chou. *Hsiang ch'eng* [Comprehensive Account of the Incense]. c. 1618–1641.

Chou-pei suan ching [Arithmetical Classics of the Gnomon and the Circular Paths of Heaven). Sixth century B.C.–ante A.D. 80.]

Chu Ch'i-feng. *Tz'u t'ung*. 1204–1206.

Chu Hsi. *Ho-nan Ch'eng shih i shu* [Remaining Records of Discourses of the Ch'eng Brothers of Honan]. 1168.

Ch'üan T'ang shih. Peking: Chung-hua shu-ch. 1960.

Damrong Rajanubhab, Prince. *Nithan Borannakhadi* [Stories of Ancient Times]. Khlang Witthaya. 1950.

Domon, Ken. *Domon Ken "Todaiji"*. Tokyo: Heibon-sha. 1973

Hashimoto, Monpei. *Nihon nojikōku-seido* [Measurement of Time in Japan]. Hanawa-shobo. Revised and enlarged. N.D. [Twentieth century].

Hirano, Mitsu. *Mji Tokyo Tokeit Ki*. N.D. *Tokei no romansu*. Nagoya. 1957.

Hirashi, Yoshio. "Karakuri Giemon no tokei – Nagasaki tokei." *Tsubomi*. No. 3. 1976.

Hirayama, Kiyotsugu. *Rikihō oyobi jihō* [The Calendar and Time Measurement in Japan]. Tokyo. N.D.

Hisanori, Wada. *On Yeh Thing-Kuei, Hsiang Lu* [Catalogue of Incense]. N.D.

Ho-nan Ch'en shih hsiang p'u [Handbook of Incense by the Ch'en Family in Honan]. Early fourteenth century.

Hsiang pu hui k'ao. In *Ch'in ting t'u shu chi ch'eng*.

Hsin hsiu pen ts'ao. c. 956.

Hsiung P'eng-lai. *Hsiang p'u*. 1322.

Hsüeh Chi-Hsüan. *Hsiao hsüeh kan chu*. 1270.

Hung Ch'u. *Hsiang p'u* [Catalogue of Aromatics]. Late eleventh or early twelfth century.

Kentaro, Yamada. *Tōzai kōyaku shi* [History of Aromatics East and West]. Tokyo. 1957.

Kōryō no michi. 1977.

Kita Seiro. *Bai en nikki* [Diary of Bai-en Kita Seiro]. *Nihon zuihitsu-taisei* [Compilation of Japanese Essays]. 3rd term, vol. 12. Yoshikawa-kobun-kan. N.D. [Twentieth century].

Kunitomo, Hideo. *Tokei no hanashi* [Discourses on Timepieces]. Tokyo: Seibundo shinkosha. 1948.

Kyōdōhakubutsukan, Ota-ku. *Toki wo shiru: Koyomi to wadokei* [Japanese Calendars and Clocks]. Oota-ward Folk Museum. 1986.

Li Ho. *Li Ch'ang-chi ko shih*. 1760.

Liu Hsien-chou. "Chung-kuo tsai chi shih ch'i fang mien ti fa ming" [Chinese Inventions in Horological Engineering]. *Ch'ing hua ta hsüeh chi hsieh kung ch'eng hsüeh ao*. New series, vol. 4, no. 1. 1956.

Liu Jui-fen. *Yang-yün-shan-chuang ch'uan chi*. 1893.

Lu Erh-K'uei *et al.*, compilers. *Tz'u yuan* [Encyclopedia]. Shanghai: Com. Press. 1915.

Mao Pang-han. (fl. 1170). *Shih lin mang chi* [Guide Through the Forest of Affairs]. Encyclopedia.

Mitsuo, Hirano. *Tokei no romansu* [The Romance of the Timepiece]. Privately printed. 1957.

Meiji Tōkyō tokeitō [Clock Towers of the Meiji Era in Tokyo]. Tokyo. 1958.

Morohashi, Tetsuji, ed. *Dai Kanwa jiten* [The Great Chinese-Japanese Dictionary]. 13 vols. Taipei. Reprint c. 1970.

Nakajima, Koji. "[Incense Clock]." *Omi Bunko*. No. 237 Go. 1983.

Nam Pyŏng-ch'ŏl. *Uigi chipsŏl*. 1860.

Nishimura, Hyobu. *Karakusa: Nihon no monyō*. Kyoto: Shuppansha. 1974.

Ogawa, Tomozo. *Teikoku Kassoku* [Practical Method for Measurement of the Fixed Time by the European Horloge]. Kobe. 1838.

Onodera Koji. "[Incense Clock]." *Kurashi no Tetyoo*. Summer issue, 1970.

Po Shou-i. "Sung shih I-szu-lan chiao-t'u-ti hsiang-liao mao-i." *Yü kung*. Vol. 7, no. 4, April 1937. Pp. 47–77.

Shen Li-chih [Li-chih]. *Hsiang p'u* [Aromatic Catalogue]. Ante 1073. See Ch'eng I and Ch'eng Ho, *Hsin-tsuan hsiang-p'u*.

Shōsōin gyobutsu zuroku. Tokyo. 1928.

Shōsōin hōbutsu. Tokyo. 1960.

Ssu k'u ch'üan shu tsung mu. 1782.

Ssu-ma Ch'ien. *Shih Chi* [Historical Records].

Sui Shu-sen, ed. *Ch'üan Yüan Yän san-ch'ü* [Complete Yüan-Dynasty *san-ch'ü*]. Reprint Taipei: Chung-hua shu-chü, 1969.

Takabayashi, Hyoe. *Tokei no hanashi* [Story of the Timepiece]. Tokyo: Privately printed. 1924.

Tokei hattatsushi [The Development of Time Measurement]. Tokyo: Toyo Shuppansha. 1927.

Takakusa, S. and Watanabe, Ko, eds. *The Tripitaka in Chinese (Taishō Shinshū Daizōkyō)*. Tokyo: Society for the Publication of the Taishō Edition of the Tripitaka. 1928.

Takakusa, J. and Ono, G. *The Tripitaka in Chinese. Picture Section*. Tokyo: Society for the Publication of the Taishō Edition of the Tripitaka. 1933.

Takigawa, Seijirō. "Kawachi no minka shiyō no jikōban" [The Jikōban Used in Households in Kawachi]. *Yamato Bunka*

Kenkyū. Vol. 7, no. 4, April 1962. Pp. 8–10.

Ting Yün [Ting Yüeh-hu]. *Hsiang lu t'u p'u* [Perhaps also known as *Yin hsiang t'u p'u*]. 2 vols. Shanghai [?]: Privately printed by Yüan Wu-lu. N.D. [1881–1882].

Toktaga and Ouyang Hsüan. *Sung Shih*. K'ai ming edition. 1345.

Tso Kuei. *Pai ch'uan hsüeh hai* [Hundred Rivers Oceans of Learning]. Shanghai. 1921.

Tuan Ch'eng-shih. "Tseng chu shang jen lien chü." *T'ang shih po ming chia ch'uan chi*. Ante 863.

T'u Lung. *Hsiang chien*. N.D. [Sixteenth century].

Watanabe, Soshu. *Tōyō monyō shi*. N.D.

[Twentieth century].

Yabuuchi, Kiyoshi. "Chūgoku no tokei" [Timekeeping Instruments in Ancient China]. *Kagaku-shi kenkyū* [Japanese Journal of the History of Science]. Vol. 19, July 1951. Pp. 19–25.

"Kōban" [Incense Trays Used As Timepieces]. *Hiroaka*. No. 11, 1962. Pp. 8–10.

Yamaguchi, Ryūji. *Nihon no tokei: Tokugawa Jidai no Wadokei no Kenkyū* [The Clocks of Japan. A Study on the antique Japanese clocks of the Tokugawa Shogunate]. Tokyo: Nihon Hyoron-sha Publishing Co., Ltd. 1942–1950.

Yamamoto, Reiko. *Hana no kaori to onna no kurashi*. 1987. Pp. 51–63.

Manuscripts and published works in Western languages

An Introduction to Thailand. Bangkok: Local Affairs Press. 1966.

"Analysis of Tutenag." *London Journal of Arts and Sciences*. Vol. 5, 1823. Pp. 48–49.

Armstrong, Robert Cornell. *Buddhism and Buddhists in Japan*. New York: The Macmillan Company. 1927.

Atchley, E. G., Cuthbert F. *A History of the Use of Incense in Divine Worship*. London: Longmans, Green and Company. 1909.

Ayres, Lew. *Altars of the East*. Garden City, NY: Doubleday and Company, Inc. 1956.

Baedeker's Japan. Engelwood Cliffs, NJ: Prentice Hall, Inc. 1987.

Bailey, L. H. *The Standard Cyclopedia of Horticulture*. New York: The Macmillan Company. 1947.

Ball, James Dyer. *Things Chinese*. Hong Kong: Kelly & Walsh. 1903.

Ball, Katherine M. *Decorative Motives of Oriental Art*. London: John Lane, The Bodley Head, Ltd., and New York: Dodd, Mead and Company. 1927.

Bayer, Theophilus Sigefried Regiomontani. *De Horis Sinicis et Cyclo Horario Commentationes Accedit Eiusdem Auctoris Parergon Sinicum de*

Calendariis Sinicis ubi Etiam Quaedam in Doctrina Temiorum Sinica Emendantur. Petrograd: Typis Academiae Scientiarum. 1735.

Beardsley, Richard K., John W. Hall and Robert E. Ward. *Village Japan*. University of Chicago Press. 1959.

Bedini, Silvio A. *The Scent of Time: A Study of the Use of Fire and Incense for Time Measurement in Oriental Countries*. Transactions of the American Philosophical Society, New Series, vol. 53, part 5, August 1963.

"Holy Smoke: The Oriental Fire Clocks." *The New Scientist*. Vol. 21, no. 380, February 27, 1964. Pp. 537–39.

"Oriental Concepts in the Measure of Time. The Role of the Mechanical Clock in Japan and China." *The Study of Time II*. Edited by J. T. Fraser and N. Lawrence. New York: Springer-Verlag, 1975. Pp. 451–84.

"Identification of Incense Seals." *Arts of Asia*. Vol. 18, no. 3, May–June 1988. P. 10.

Bhattacharyya, Benoytosh. *An Introduction to Buddhist Esotericism*. Oxford: Humphrey Milford, Oxford University Press. 1932.

Sadhanamala. Baroda: Oriental Institute. 1968.

Bhirasri, Silpa. *Thai Wood Carvings*. Bangkok: The Fine Arts Department. 1963.

Biot, E. '*Tcheou-Li ou* 'Rites des Tstou.' Paris: Timprimerit Nationale, 1851.

Bonnin, Alfred. *Tutenag and Paktong*. Oxford University Press. 1924.

Bramsen, William. "Japanese Chronology and Calendars." *Transactions of the Asiatic Society of Japan*. Vol. 37 Supplement, 1910. Pp. 1–128.

Brinkley, Captain Frank, ed. *Japan Described and Illustrated by the Japanese*. Boston: J. B. Millet Company. 1901.

Japan, Its History, Arts and Literature. 12 vols. London: T. C. and E. C. Jack. 1903–1904.

Broderick, James, S. J. *Saint Francis Xavier 1506–1552*. London: Burns and Oates. 1958.

Burkhardt, V. R. *Chinese Creeds and Customs*. 3 vols. Hong Kong: The South China Morning Post. 1955.

Burnouf, Eugene. *Le Lotus de le Bonne Loi*. 2 vols. New edition. Paris: Librairie Orientale et Americaine. 1925.

Bushell, Stephen W. *Chinese Art*. 2 vols. London: Wyman & Sons, Ltd. for H. M. Stationery Office. 1906.

Cao Xueqin (Ts'ao Hsüeh-ch'in). *The Story of the Stone*. 5 vols. Volume 1. *The Golden Days*. Translated by David Hawkes. London: Penguin Books. 1973.

Casal, U. A. "Incense." *Transactions of the Asiatic Society of Japan*. 3rd series, vol. 3, 1954. Pp. 46–73.

Chamberlain, Basil Hall. *Things Japanese*. London: J. Murray. 1902.

Chamberlain, Basil Hall and W. B. Mason. *A Handbook For Travellers In Japan Including The Whole Empire From Yezo to Formosa*. 7th edition. London: John Murray. 1903.

Chambers, Sir William. *Traité des édifices, meubles et habits de la Chine*. Paris: Chez le Sieur, Le Rouge. 1757.

Chapuis, Alfred, Loup, G., and De Sassure, L. *Relations de l'horlogerie Suisse avec la Chine*. *La Montre Chinois*. Neuchatel: Attinger Frères. N.D. [1919].

Charannes, E., "Le cycle turc des douze animaux," *Tóung Pao*, vol. 7, 1906.

Chiang, Monlin. *Tides From the West*. New Haven: Yale University Press. 1947.

Chiang, Yee. *Chinese Calligraphy: An Introduction to Its Aesthetic and Technique*. 2nd edition. London: Methuen & Co., Ltd. 1954.

The Chinese Repository, Vol. 7, March 1839. Article 4.

Chou Yi-liang. "Tantrism in China." *Harvard Journal of Asiatic Studies*. Vol. 8, 1945. Pp. 241–332.

Churchill, Awnsham. *A Collection of Voyages and Travels, Some Now First Printed from Original Manuscripts, Others Now First Published in English ... 6 vols*. 3rd edition. London: For H. Lintot, etc. 1744–1746.

Clement, E. W. and Bramsen, William. *Transactions of the Asiatic Society of Japan*. Vol. 37. Supplement. 1910.

Cobbold, George A. *Religion in Japan: Shintoism-Buddhism-Christianity*. London: Society for Promoting Christian Knowledge. 1905.

Collis, Maurice. *The Land of the Great Image. Being Experiences of Friar Manrique in Arakan*. New York: Alfred A. Knopf. 1958.

Combridge, John. "Chinese Sexagenary Calendar Cycles." *Antiquarian Horology*. Vol. 5, no. 4, September 1966. P. 134.

"Hour Systems in China and Japan." *Bulletin of the National Association of Watch and Clock Collectors*. Vol. 18, no. 4, August 1976. Pp. 236–38.

Couling, Samuel. *Encyclopedia Sinica*. Shanghai: Kelly & Walsh. 1917.

Cox, Warren E. *Pottery and Porcelain*. 2 vols. New York: Crown Publishers. 1944.

Cronin, Vincent. *The Wise Man From the West*. London: Rupert Hart-Davies. 1955.

Crump, James I. *Songs From Xanadu. Studies in Mongol-Dynasty Song-Poetry*

Crump James I. (*contd.*)

[*San-ch'ü*]. Michigan Monographs in Chinese Studies No. 47. Ann Arbor: Center for Chinese Studies, University of Michigan. 1983.

Curzon, George, Marquess of Kedleston. *Leaves From A Viceroy's Notebook.* London: Macmillan and Co. 1927.

Daito Shuppansha, *Japanese-English Buddhist Dictionary*, Tokyo. Compiled 1979.

Dastur, J. F. *Useful Plants of India and Pakistan.* Bombay: D. B. Taraporevala Sons & Co., Ltd. 1969.

Dechelette, J. "Le swastika ou croix gammé et les symboles derivés de la roüe." *Manuel d'archéologie préhistorique.* Vol. 2. 1983. Pp. 454–464.

D'Elia, Pasquale M., S. J., ed. *Fonte Ricciane. Documenti originali concernenti Mattao Ricci e la storia delle prime relazioni tra l'Europa e la Cina (1579–1615).* 3 vols. Rome: Libreria dello Stato. 1942–1949. Volume I. 1942.

De Lubac, Henri, S. J. *Aspects du Bouddhisme. Amida.* Paris: Editions du Seuil. 1954.

Dennys, Nicholas Belfield. *The Folklore of China and its affinity with that of the Aryan and Semitic Races.* London: Trubner & Co. 1876.

De Poncin, Gontran. *From a Chinese City.* Garden City, NY: Doubleday and Company, Inc. 1957.

De Visser, M. W. *Ancient Buddhism in Japan. Sutras and Ceremonies in Use in the Seventh and Eighth Centuries A. D. and Their History in Later Times.* 2 vols. Leiden: E. J. Brill. 1935.

Diels, H. *Antike Technik.* Leipzig/Berlin: Teubner. 1914.

Doré, Henri. *Recherches sur les superstitions en Chine.* 18 vols. Shanghai: Imprime de la Mission Catholique a l'Orphelinat de T'ou-sé-wé. 1911–1938.

Douglas, Robert K. *China.* Boston: D. Lothrop and Company. 1885.

Society In China. London: A. D. Innes & Co. 1894.

Dubs, Homer H. "Han 'Hill Censers.'" *Studia Serica Bernard Karlgren*

Dedicata. Copenhagen: Ejnar Munksgaard. 1959. Pp. 259–64.

Du Halde, Jean Baptiste. *The General History of China, Chinese-Tartary, &c.* London: J. Watts. 1736.

Dye, Daniel Sheets. *Chinese Lattice Designs.* New York: Dover Publications. 1974.

Earle, Alice Morse. *Sun-Dials and Roses of Yesterday.* New York: The Macmillan Company. 1902.

Eberhard, Wolfram. *Lexicon Chinesischer Symbole.* Koln: Eugen Diederichs Verlag. 1983.

Eitel, Ernest J. *Hand-Book of Chinese Buddhism Being A Sanskrit-Chinese Dictionary With Vocabularies of Buddhist Terms . . .* Hong Kong: Lane, Crawford & Co. 1888.

Erskine, William Hugh. *Japanese Festivals and Calendar Lore.* Tokyo: Kyo Bun Kwan. 1933.

Everett, Thomas H. *The New York Botanical Garden Illustrated Encyclopedia of Horticulture.* 10 vols. New York: Garland Publishing, Inc. Vol. 10. 1980.

Eyries, Jean Baptiste Benoit. *Voyage à Peking* [*Voyage de Golownin, officier de la marine Russe, 1824*]. Paris: Dondey-Dupre. 1827.

Favier, Alphonse. *Peking: Histoire et description.* Lille: Desclée, De Brouwer et Cie. 1900.

Ferrand, Gabriel, trans. *Voyage de Sulayman suivi de remarques par Abu Zayd Hasan.* Paris. 1922.

Forestry of Japan. Ministry of Agriculture and Forestry, Forestry Agency. 1935.

Fraissinet, Edouard. *Le Japon contemporain.* Paris: Librairie Hachette. 1857.

Freed, Stanley A. and Freed, Ruth S. "Origin of the Swastika." *Natural History.* Vol. 89, no. 1, January 1980. Pp. 68–74.

Frodsham, J. D., trans. *Goddesses, Ghosts, and Demons. The Collected Poems of Li He (790–816).* London: Anvil Press Poetry, Ltd. 1983.

Fu, James S. "The Mirror and the Incense in the Tale of Genji and the Dream of the Red Chamber." *Tamkang Review* (Tamsui, Taipei). Vol. 10. 1983. Pp. 199–209.

Fu Shen C. Y. T., Marilyn W. Fu, Mary G. Neill and Mary Jane Clark. *Traces of the Brush. Studies In Chinese Calligraphy*. New Haven: Yale University Art Gallery. 1977.

Fujii, Keiichi and Perkins, P. D., trans. *Gonin Onna (Five Women) by Ihara Saikaku*. N.D.

Fyfe, Andrew, M. D. "Analysis of Tutenag, or the White Copper of China." *Edinburgh Philosophical Journal*. Vol. 7, 1822. Pp. 69–71. See also *London Journal of Arts and Sciences*. [Letter to the editor concerning Fyfe] Vol. 5, 1823. Pp. 18–19.

Gallagher, Louis J., S. J. *China In the Sixteenth Century. The Journals of Matteo Ricci 1583–1610*. New York: Random House. 1953.

Gardner, C. T. "On the Chinese Race: their Language, Government, Social Institution, and Religion." *The Journal of the Ethnological Society of London*. Vol. 2, 1870. Pp. 26–27.

Garrido-Atienza, Miguel. *Los Alquezares de Santafe*. Granada: F. Reyes. 1893.

Gaubil, A. *Traité de l'astronomie Chinoise*. Paris: Rollin. 1732.

Gibbs, Sharon. *Greek and Roman Sundials*. New Haven: Yale University Press. 1976.

Giles, Herbert A. *A Chinese Biographical Dictionary*. London/Shanghai: Bernard Quaritch. 1898.

Glick, Thomas F. "Medieval Irrigation Clocks." *Technology and Culture*. Vol. 10, no. 3, July 1969. Pp. 424–28.
Irrigation and Society in Medieval Valencia. Cambridge: Harvard University Press. 1970.
Gobernaçion. Valencia, Archivo del Reino de Valencia. 2377, 46th band of 1485.

Goblet-Alviella, Albert-Joseph. *Migration des symboles*. Paris: E, Leroux. 1891.

Golovnin, Captain [Vasilif Michael]. *Memoirs of A Captivity in Japan During the Years 1811, 1812, and 1813*. 3 vols. London: For H. Colburn & Co. 1824.

Greg, Robert Philips. "On the Meaning and Origin of the Fylfot and Swastika." *Archeologia*. Vol. 48, part 2, 1885. Pp. 293–326.

Griffis, William Eliot. 1902. *Corea: the Hermit Nation*. New York: C. Scribner & Sons.

Guitton, Robert. 1958. *Quand sonne l'heure*. Brive: de Michel de Montaigne.

Hamel, Hendrik. "An Account of the Shipwreck of a Dutch Vessel on the Coast of the Isle of Quelpaert, Together With a Description of the Kingdom of Korea." *Transactions of the Royal Asiatic Society of Great Britain and Ireland*. Vol. 9, 1918. Pp. 99–128.

Harada, Yoshito. "The Interchange of Eastern and Western Culture as Evidenced in the Shō-sō-in Treasures." *Memoirs of the Research Department of the Tōyō Bunko*. Vol. 2, 1939. Pp. 55–78.

Harvey, Edwin D. *The Mind of China*. New Haven: Yale University Press, 1933.

Hearn, Lafcadio. *In Ghostly Japan*. Boston: Little, Brown & Co. 1890.
Glimpses of Unfamiliar Japan. 2 vols. Boston: Houghton Mifflin & Co. 1894. Reprint, New York: Ames Press, Inc. 1969.
Japan, An Attempt At Interpretation. New York: Macmillan Company. 1904.
A Japanese Miscellany. Boston: Little, Brown & Co. 1908.
Out of the East. Boston: Houghton Mifflin Company. 1914.
Exotics and Retrospectives. London: Kegan Paul, Trench, Trubner. 1918.

Hedin, Sven. *Jehol, City of Emperors*. London: Kegan Paul, Trench & Trubner, Ltd. 1932. Reprint, New York: E. P. Dutton & Company, Inc. 1933.

Hilton-Simpson, M. W. "Further Notes on Time-Measurement for Irrigation in the Aures." *Geographical Journal*. Vol. 62, 1924. P. 430.
"The Influence of Its Geography on the People of the Aures Massif, Algeria." *Geographical Journal*. Vol. 59, 1922. Pp. 27–29.

Hirth, Frederick and Rockhill, W. W., trans. *Chau Ju-kua: His Work on the Chinese and Arab Trade in the Twelfth and Thirteenth Centuries, entitled Chu-fan-chi*. St. Petersburg: Imperial Academy of

Sciences. 1911. Reprint, New York: Paragon Book Reprint Corp. 1966.

History of Japanese Science and Technology. Tokyo. 1962.

Hoffmann, J. J. *A Japanese Grammar.* Leiden: A. W. Synthoff. 1868.

Holland, Clive. *Old and New Japan.* London: J. M. Dent & Co. 1907.

Hommel, Rudolph P. *China At Work.* New York: John Day & Co. 1937.

Hora, Bayard, ed. *The Oxford Encyclopedia of Trees of the World.* Oxford University Press. 1981.

Hough, Walter. "Timekeeping By Light and Fire." *The American Anthropologist.* Vol. 6, April 1893. Pp. 207–8.

Huc, M. [Évariste Régis]. *A Journey Through the Chinese Empire.* 2 vols. New York: Harper and Brothers. 1859.

Hummel, Arthur W., ed. *Eminent Chinese of the Ch'ing Period (1644–1912).* 2 vols. Washington, DC: Government Printing Office. 1944.

Ihara, Saikaku. *Five Women Who Loved Love.* Translated by Theodore de Bary. Rutland, VT and Tokyo: Charles E. Tuttle Company. 1956.

Imazeki, Rokuya. "Mannentake, A Happy Fungi." *Natural Science and Museum (Shizen Kagakuto Hakubutsu-kan).* Vol. 6, no. 1, January 1935. Pp. 11–15.

"Studies on Ganoderma of Nippon." *Bulletin of the Tokyo Science Museum.* No. 1, March 1939. Pp. 29–51.

Inouye, Jukichi. *Sketches of Tokyo Life.* Yokohama: Torandio. N.D.

Introduction to Thailand. N.D.

Ishida, Mosaku and Gunichi Wada. *The Shōsōin. An Eighth Century Treasure-House.* Tokyo/Osaka/Moji: The Mainichi Newspapers. 1954.

Iwaki, Masao. "A Report of Experiments on Ancient Fire-Making Techniques." *Japanese Studies in the History of Science.* Vol. 16, 1977. Pp. 91–93.

Iwao, Hino. "Ganoderma lucidum." *Botany and Zoology.* Vol. 5, no. 5, May 1937. Pp. 1–8.

Japan The Official Guide. Tokyo: Japan Travel Bureau. 1955.

Jeon, Sang-woon. *Science and Technology in*

Korea. Traditional Instruments and Techniques. Cambridge, MA: The MIT Press. 1974.

Jisaburo Ohwi. *Flora of Japan.* Edited by Frederick G. Meyer and Egbert H. Walker. Washington, DC: Smithsonian Institution. 1984.

John, W. D. and Katherine Coombes. *Paktong. The Non-Tarnishable Chinese "Silver" Alloy Used for "Adam" Firegrates and Georgian Candlesticks.* Newport, England: The Ceramic Book Company. 1970.

Jünger, Ernst. *Das Sanduhrbuch.* Frankfurt-am-Main: Vittorio Klostermann. 1957.

Karlgren, B. "The Book of Documents (*Shu Ching.*)" *Bulletin of the Museum of Far Eastern Antiquities* (Stockholm). Vol. 22, no. 1. 1950.

Kates, George N. *Chinese Household Furniture.* New York: Harper & Brothers. 1948.

Kim Yong-Woon. "Origins of Time-keeping Mechanisms: Similarities in China and Korea." *Korea Journal,* Vol. 15, no. 7, August 1975. Pp. 4–10.

Knox, George William. *Japanese Life in Town and Country.* New York: G. P. Putnam & Sons. 1905.

Kwo Da-Wei. *Chinese Brushwork: Its History, Aesthetics, and Techniques.* Montclair, NJ: Allanheld & Schram. 1981.

Kwock, H. and McHugh, Vincent, trans. *Old Friends From Far Away.* Berkeley, CA: North Point Press. N.D. [Twentieth century].

Laufer, Berthold. *Chinese Pottery of the Han Dynasty.* Rutland, VT: Charles E. Tuttle Company. 1962.

Leathart, Scott. *Trees of the World.* New York: A & W Publishers, Inc. 1977.

Legge, J. *The Texts of Confucianism, translated: Part II, the I Ching.* Oxford University Press. 1899.

Lewis, Norman. *A Dragon Apparent in Indo-China.* London: Cape. 1951.

J. D. Frodsham. Arthur Waley. London: George Allen and Unwin, Ltd. 1950.

Liebert, Gosta. *Iconographic Dictionary of the Indian Religions. Hinduism-*

Buddhism-Jainism. Leiden: E. J. Brill. 1976.

Lin, Shuen-fu. *The Transformation of the Chinese Lyrical Tradition. Chiang K'uei and Southern Sung Tz'u Poetry.* Princeton University Press. 1978.

Ling, Trevor. *A Dictionary of Buddhism. Indian and South-East Asian.* Calcutta and New Delhi: K. P., Bagchi & Company. 1981.

Linschoten, Jan Huyghen Van. *John Huighen Van Linschoten: His Discourses of Voyages Into Ye Easte and West Indies.* 4 vols. London: I. Wolfe. 1598.

Liu Hsien-chou, "On the Chinese Invention of Time-keeping Apparatus." *Actes du VIIIe Congres International d'Histoire des Sciences.* Vinci/Paris: Hermann & Cie. 1956. Pp. 329–49.

Loewenstein, Prince John. *Swastika and Yin-Yang.* China Society Occasional Papers. New Series, no. 1. London: China Society. 1942.

Lübke, Anton. "Ein Reformer der chinesischen Astronomie und Zeitmessung II." *Die Uhr.* Nr. 24, 1966. Pp. 13–20.

Magalhaens, P. Gabriel de, S. J. *Nouvelle relation de la Chine.* Paris: C. Barbin. 1668.

Magowan, Dr. D. J. "On Chinese Horology, With Suggestions of the Form of Clocks Adapted for the Chinese Market." Part 4, Communications. *Report of the Commissioner of Patents for the Year 1851. Senate Document 118.* Washington, DC: U.S. Printing Office. 1852. Pp. 335–342.

"Modes of keeping time known among the Chinese." *The Chinese Repository* (Canton). Vol. 20, July 1951. Pp. 426–32.

Mahabharata. The Song Celestial. London: Kegan Paul, Trench, Trubner. 1890.

[Manrique, Sebastião]. *Travels of Fray Sebastião Manrique 1629–1644.* Translated by C. Eckford Luard and H. Hosten. N.D.

Maspero, H. "Communautés et moines bouddhistes chinois au IIe et IIIe

siècles." *Bulletin de l'École Française d'Extréme-Orient.* Tome X, 1910. Pp. 222–32.

"L'astronomie chinoise avant les Han." *T'oung Pao.* Vol. 26, 1929. Pp. 267–356.

"Les instruments astronomiques des Chinois au temps des Han." *Mélanges Chinois et bouddhiques* (Brussels). 1939. Pp. 183–370.

Mathews, Robert Henry. *Dictionary of Chinese Dates.* N.D.

Matsuoka, Asako. *Sacred Treasures of Nara in "Shōsō-in" & "Kasuga Shrine."* Tokyo: The Hokuseido Press. 1935.

Mayers, William Frederick. *The Chinese Reader's Manual.* Shanghai: Presbyterian Mission Press. 1924.

Mikami, Tsugio. *The Art of Japanese Ceramics.* New York and Tokyo: Weatherhill/Heibonsha. 1972.

Mody, N. H. N. *A Collection of Japanese Clocks.* London: Kegan Paul, Trench, Trubner & Co., Ltd. 1932, Reprint, Rutland, VT: Charles E. Tuttle Company, Inc. 1967.

Moncrieff, C. C. Scott. *Irrigation in Southern Europe.* London. 1868.

Monier-Williams, Sir Monier. *Brahmanism and Hinduism; or, Religious Thought and Life in India, As Based on the Veda and Other Sacred Books of the Hindus.* London: John Murray. 1891.

Morgan, Harry T. *Chinese Symbols and Superstitions.* South Pasadena, CA: P. D. and Ione Perkins. 1942.

Murasaki, Lady Shikibu. *The Tale of Genji.* New York: Random House, Modern Library edition. 1960.

Nakamura, Julia. "Burning Tradition." *Asian-American Magazine.* January-February 1989. Pp. 70–72.

Navarrete, Martin Fernandez, O. P. *Histoire general des Voyages.* "Description de la Chine." Livre II. Paris. 1748.

An Account of the Empire of China, Historical, Political, Moral and Religious . . . Translated by Abbé Prevost. In *Churchill's Collection of Voyages and Travels.* London: For Henry Lintot and John Osborn. N.D.

Needham, Joseph. "The Wilkins Lecture. The Missing Link in Horological History: a Chinese Contribution." *Proceedings of the Royal Society*. A, vol. 250, 1959. Pp. 147–79.

"Time and Eastern Man. The Henry Myers Lecture 1964." Occasional Paper No. 21. *Royal Anthropological Institute of Great Britain & Ireland*. 1965.

"Astronomy in Ancient and Medieval China." *Philosophical Transactions of the Royal Society of London*. Vol. 276, 1974. Pp. 67–82.

Needham, Joseph *et al*. *Science and Civilisation in China*. 7 vols. Cambridge University Press. Vol. 3, 1959; vol. 4, Part II, 1965, Part III, 1971; vol. 5 Part II, 1974; vol. 5 Part III 1976.

Needham, Joseph, Wang Ling and Derek J. Price. *Heavenly Clockwork. The Great Astronomical Clocks of Medieval China*. 1959. Cambridge University Press. Reprint. 1986.

Needham, Joseph, Lu Gwei-Djen, John H. Combridge and John S. Major. *The Hall of Heavenly Records. Korean Astronomical Instruments 1380–1780*. Cambridge University Press. 1986.

Newark Museum. *Catalogue of the Tibetan Collection*. Part 2. Newark: Newark Museum. 1950.

Nieuhoff, John. *An Embassy From the East India Company of the United Provinces to the Grand Tartar Cham, Emperor of China, Delivered By Their Excellencies Peter de Goyer & Jacob de Keyser*. London: By the author, i.e., Ogilby. 1673.

Noguchi, Yone. *Emperor Shōmu and the Shōsōin*. 2 vols. Ginza, Tokyo: Kyo Bun Kwan. 1941.

Northeastern Asiatic Art. Toledo, OH: The Toledo Museum of Art. 1942.

"Notices in Natural History." *The Chinese Repository*. Vol. 7, p. 1839. P. 598.

Nouvelle Biographie générale depuis les temps plus reculés jusqu'a nos jours. 45 vols. Paris: Firmin Didot, Fils et Cie. 1863.

[Odoricus of Pordenane]. *Travels of Friar Odoricus Into China and the East*. London: Macmillan Company. 1928.

Ogawa Tomozo. *Teikoku Kassoku* [Practical Method of the Fixed Time by the European Horologe]. 1838.

Ohwi, Jisaburo. *Flora of Japan*. Edited by Frederick G. Meyer and Egbert H. Walker. Washington, DC: Smithsonian Institution Press. 1965.

Okakura, Yoshisaburo. *The Life and Thought of Japan*. London: J. M. Dent. 1913.

150 Chinese Poems from the Great Dynasties. San Francisco: North Point Press. 1980.

Owen, Stephen. *The Poetry of the Early T'ang*. New Haven/London: Yale University Press. 1977.

The Great Age of Chinese Poetry. The High T'ang. New Haven/London: Yale University Press. 1981.

The Oxford English Dictionary. Oxford University Press. 1986.

Piggot, Joan R. "Hierarchy and Economics in Early Medieval Todaiji." *Court and Bakufu in Japan. Essays in Kamakura History*. New Haven/London: Yale University Press. 1982. Pp. 45–77.

Planchon, Matthieu. "L'Heure en Chine par le soleil l'eau et le feu." *La Nature*. 23 année, Tome II, 1895 deuxieme semestre. Pp. 247–50.

L'Horloge: son histoire rétrospective, pittoresque et artistique. Paris: Henri Laurens. 1898.

[Polo, Marco]. *The Travels of Marco Polo*. London: J. M. Dent. 1925.

Poncins, Gontran de. *From A Chinese City*. Garden City, NY: Doubleday and Company. 1957.

Pound, Ezra. *The Classic Anthology Defined by Confucius*. Cambridge: Harvard University Press. 1954.

Prevost, Abbé, trans. *Churchill's Collection of Voyages and Travels*. London. N.D.

Priestley, J. B. *Man and Time*. London: Aldus Books, Limited. 1964.

Pugsley, Edwin. "Letter to Brooks Palmer." *Bulletin of the National Association of Watch and Clock Collectors*. Vol. 7, no. 7, February 1953. Pp. 321–24.

Ramayana: The Egyptian Priest's Story of Creation and the Spirit World. Everett Washington. 1926.

Read, Bernard. "Chinese Medicinal Plants." *Natural History Bulletin*. Peiping: French Bookstore. 1936.

Reischauer, Edwin O. *Ennin's Travels in T'ang China*. New York: Ronald Press. 1955.

Robertson, J. Drummond. *The Evolution of Clockwork With a Special Section on The Clocks of Japan*. London: Cassell & Company, Ltd. 1931.

Rockhill, William Woodville. "Laws and Customs of Korea." *The American Anthropologist*. Vol. 4, 1891. P. 183.

The Land of the Lamas. Notes on a Journey Through China, Mongolia and Tibet. New York: The Century Co. 1891.

Diary of a Journey Through Mongolia and Tibet in 1891 and 1892. Washington, DC: Smithsonian Institution. 1894.

The Journey of William of Rubruik To the Eastern Parts of the World, As Narrated By Himself. London: For the Hakluyt Society. 1900.

Rodrigues Tçuzu, João. *Historia de Igreja do Japão*. Edited and annotated by João de Amaral Abranches Pinto, in *Colecção Noticias de Macau* No. 14. Macao: Noticias de Macau. 1956.

Roerich, Georg N. *Trails To Inmost Asia. Five Years of Exploration with the Roerich Central Asian Expedition*. New Haven: Yale University Press. 1931.

Roggendorf, Joseph. "In memoriam, Robert Hans Van Gulik." *Monumenta Nipponica*. Vol. 23, nos. 1–2, 1967. Pp. i–vii.

Rufus, W. Carl. "Astronomy in Korea." *Transactions of the Royal Asiatic Society, Korea Branch*. Vol. 26, 1936. Pp. 12–14.

Rufus, W. Carl and Won-Chul Lee. "Marking Time in Korea." *Popular Astronomy*. Vol. 44, 1936. Pp. 252–57.

"Så visade rökelseuret dygnets löpp i 1500-talets Kina . . ." *Ur Optik* (Stockholm). March 1983. P. 13.

Sachse, Julius F. "Horologium Achaz (Christophorus Schissler, Artifex)." *Proceedings of the American Philosophical Society*. Vol. 34, no. 147, 1895. Pp. 21–30.

Sarton, George. *Introduction to the History of Science*. 2 vols. Huntington, NY: Robert Krieger. 1975.

Satō, Masahiko. *Kyoto Ceramics*. Translated and adapted by Anne Ono Towle and Usher P. Coolidge. New York and Tokyo: Weatherhill/Shibundo. 1973.

Satow, Sir Ernest. *A Diplomat in Japan: An Inner History From Personal Experiences of the Overthrow of the Shōgun Rule*. London: Seeley, Service & Co., Ltd. 1921.

Sauvaget, J. "*Akhbar al-Sin wa'l-hind.*" *Relation de la Chine et de l'Inde, redigée en 851. Texte établi, traduit et commenté*. Paris: Belles Lettres. 1948.

Schafer, Edward. "Rosewood, Dragon's Blood, and Lac." *Journal of the American Oriental Society*. Vol. 77, no. 2, April–June 1957. Pp. 129–36.

The Golden Peaches of Samarkand. A Study of T'ang Exotics. Berkeley, CA: University of California Press. 1963.

The Shore of Pearls. Berkeley, CA: University of California Press. 1970.

Schipper, K. M. "Taoism, the Liturgical Tradition." First International Conference on Taoist Studies. Bellagio, Italy. 1968.

Schliemann, Henry. *Ilios: The City and Country of the Trojans*. New York: Harper and Brothers. 1881.

Schurhammer, Georged. *Epistolae S. Francisci Xaverii*. N.D. [Twentieth century].

Scott, A. C. *The Flower and the Willow World. The Story of the Geisha*. New York: The Orion Press. 1960.

Scott-Moncrieff, C. C. *Irrigation in Southern Europe*. London. 1860.

Seely, F. A. "The Development of Time-Keeping In Greece and Rome." *The American Anthropologist*. Vol. 1, January 1888. P. 49 (Comment by Otis Tufton Mason).

Sei Shōnagon. *Pillow Book*. Translated by Arthur Waley. New York: The Grove Press. 1960.

Semedo, Alvarez, S. J. *Imperio de la China*. Madrid. 1642.

The History of the Great and Renowned Monarchy of China. London: E. Tyler. 1655.

Sei Shōnagon. *Pillow Book.* Translated by Arthur Waley. New York: The Grove Press. 1960.

[Shōsōin Office]. *Treasures of the Shōsōin.* Tokyo: Asahi Shimbun Publishing Co. 1965.

Siebold, P. F. von. *Nippon; Archiv zur Beschreibung von Japan,* 5 vols. N.D. [nineteenth century].

Soothill, William E. and Hodus, L. *A Dictionary of Chinese Buddhist Terms, with Sanskrit and English Equivalents and a Sanskrit-Pali Index.* London: Kegan Paul. 1937.

Steele, John. *The I-Li; or, Book of Etiquette and Ceremonial.* London: Probsthain & Co. 1917.

Stevens, John. *Sacred Calligraphy of the East.* Boulder, CO and London: Shambhala Publications, Inc. 1981.

"People's Buddhism and Japan's Calligraphy Renaissance." *Japan Society Newsletter.* Vol. 31, no. 10, May 1984. Pp. 2–4.

Stimpson, George W. *Nuggets of Knowledge.* New York: George Sully and Company. 1928.

Stoddard, John L. *Stoddard's Lectures.* 3 vols. Chicago: G. L. Shuman & Co. 1925.

Street, A. and Alexander, W. *Metals in the Service of Man.* Harmondsworth, England: Penguin Books. 1960.

[Sulaiman]. *Voyage du marchand arabe Sulaiman en Inde et en Chine rédigé en 851 suivi de remarques par Abu Zayd Hasan (vers 916).* French translation by Gabriel Ferrand. *Bois de Mille A. Karpelés.* Les classiques de Orient, no. 7. Paris: Bossard. 1922. Longue critique par P. Pelliot. *T'oung Pao.* t. 21, 1922. Pp. 399–413. And in *Isis.* Vol. 6, no. 17, part 2, 1924. Pp. 146.

Suzuki, Seitaro. "The Fires and the Weather." *Journal of the Department of Agriculture.* Kyushu Imperial University. Vol. 2, no. 1, March 20, 1928.

Takeo, Tamura. "Incense Clock of Ibaraki Prefecture." *Folklore of Ibaraki.* No. 6, 1967. Pp. 56–58.

"The Jōkōban of the Southern Region of Ibaraki Prefecture." *Koten Tokei Tsūshin.* Vol. 108, January 1988. Pp. 4–7.

"Jōkōban Made of Metal (Bronze)." *Koten Tokei Tsūshin.* Vol. 108, July 1988. Pp. 1–4.

Tardy. "Origine de la mesure du temps. La Chine." *La France horlogére.* New series, no. 145, November 1957. Pp. 125–30.

Taylor, Bayard. *Japan In Our Day.* New York: Charles Scribner's Sons. 1893.

Thomas, Edward. "The Indian Swastika and Its Western Counterparts." *Numismatic Chronicle.* Vol. 20, 1880. Pp. 18–48.

Thunberg, Charles Pierre. *Voyage de C. P. Thunberg au Japon, Per le Cap de Bonne-Esperance, les Iles de la Sonde, &c.* Paris: Chez Benoit Dandre, Garnery & Obre. Vol. 3. 1796.

Travels in Japan, and Other Countries. Philadelphia: J. J. Cruikshank. 1801.

Thurston, Edgar. *Ethnographic Notes in Southern India.* Madras: Superintendent, Government Press. 1906.

Tiffany, Osmond, Jr. *The Canton Chinese, or the American's Sojourn in the Celestial Empire.* Boston and Cambridge: James Munroe and Company. 1849.

Treasures of the Shōsōin. Tokyo: Asahi Shimbun Publ. Co. 1965.

Trigault, P. Nicolas, S. J. *De Christiana Expeditione Apud Sinas.* Vienna/Augsburg: Apud Christoph Mangium. 1615.

Turner, Anthony J. *The Time Museum.* Vol. I. *Time Measuring Instruments.* Part 3. "Water-clocks, Sand-glasses, Fire-clocks." Rockford, IL: The Time Museum. 1984.

Van Beek, Gus W. "Frankincense and Myrrh." *The Biblical Archeologist Reader.* Vol. 2. Garden City, NY: Doubleday and Company, Inc. 1964. Pp. 99–126.

Van Gulik, Robert H. *Siddham. An Essay on the History of Sanskrit Studies in China and Japan.* 2 vols. Delhi: Jayyed Press. 1954.

Judge Dee At Work. New York: Charles Scribner's Sons. 1967. "Five Auspicious Clouds."

The Haunted Monastery and The Chinese Maze Murders. New York: Dover Publications, Inc. 1978. "The Chinese Maze Murders," pp. 110–321.

"Remarks On My Judge Dee Novels." (Typewritten manuscript, 7 pp.). Mugar Memorial Library, Boston University. 1966.

[Vitruvius]. *Les Livres d'Architecture de Vitruve corrigez et traduite nouvellement en François, avec des notes et de figures.* Edited by Claude Perrault. Paris. 1684.

Waddell, L. Austine. *Lhasa and Its Mysteries*. London: J. Murray. 1905.

Waley, Arthur, ed. *The Book of Songs*. London: Allen & Unwin. 1954.

The Poetry and Career of Li Po 701–762 A.D. London: George Allen and Unwin, Ltd. 1950.

Trans. *The Tale of Genji. A Novel in Six Parts . . . by Lady Murasaki*. New York: The Modern Library. 1960.

Waseda, Eisaku. "Incense Ceremonies." *The Tourist*. July 1919. Pp. 14–23.

Wasson, R. G. *Soma, The Divine Mushroom of Immortality*. New York: Harcourt, Brace & World. 1968.

Watson, Burton, trans. and ed. *The Columbia Book of Chinese Poetry From Early Times to the Thirteenth Century*. New York: Columbia University Press. 1984.

Watson, Richard, Bishop of Llandaff. *Chemical Essays*. Cambridge: Printed for T. & S. Merrill. 1781–1789.

White, W. C. and P. M. Millman. "An Ancient Chinese Sundial." *The Journal of the Royal Astronomical Society of Canada*. Vol. 32, no. 9, November 1938. Pp. 417–30.

Whitlock, Herbert P. "Salutations and Inscriptions in Jade." *Natural History*. February 1942. Pp. 108–14.

Willetts, William. *Chinese Art*. 2 vols. London: Penguin Books. 1958.

Williams, C. A. S. *Outlines of Chinese Symbolism*. Peiping: Customs College Press. 1931.

Encyclopedia of Chinese Symbolism and Art Motives. New York: The Julian Press. 1960.

Williams, S. Wells. *The Middle Kingdom. A Survey of the Geography, Government, Literature, Social Life, Arts and History of the Chinese Empire and Its Inhabitants*. 2 vols. New York: Charles Scribner's Sons. 1904.

Wilson, Ernest Henry. *A Naturalist In the Far East*. London: Methuen & Co. 1913.

Wilson, Thomas. "The Swastika." *Annual Report of the Board of Regents of the Smithsonian Institution . . . for the Year Ending June 30, 1894. Report of the U.S. National Museum*. Washington: Government Printing Office. 1896.

Wins, Alphonse. *L'Horloge à travers les âges*. Paris/Mons: Librairie Edouard Champion. 1924.

Wu Chi-yu. "Le séjour de Kouan-hieou au Houa chan et le titre du recueil des ses poèmes." *Mélanges publies par l'Institut des Hautes Etudes Chinoises*. Vol. 2, 1969. Pp. 158–78.

Yamaguchi Ryūji. 1949. "Japanese Clocks and Their History." *The American Horologist and Jeweler*, April 1949. Pp. 71–86.

Timekeepers. Tokyo: Iwanami, Ltd. 1956.

Yetts, W. Perceval, trans. *The Cull Chinese Bronzes*. London: Courtauld Institute of Art. 1939.

Yule, Henry and Cordier, Henri. *The Book of Marco Polo, The Venetian, Concerning The Kingdoms and Marvels of the East*. 2 vols. St. Helier: Armorica Book Co., 1871. Reprint, Amsterdam: Philo Press. 1975.

Figure 1. Chinese equatorial sundial, Ming Dynasty, sixteenth or seventeenth century. Fire-gilt bronze, measuring 132 cm high, 76.5 cm wide and dial 63.5 cm in diameter. Dial is read on both sides, upper surface during summer months and lower surface in winter. It is identical in overall design to a sundial made of marble which is in the Hall of Supreme Harmony in Beijing.

Figure 2. Marble pedestal sundial on the terrace of Kuo tzu chien (Imperial University), Peking.

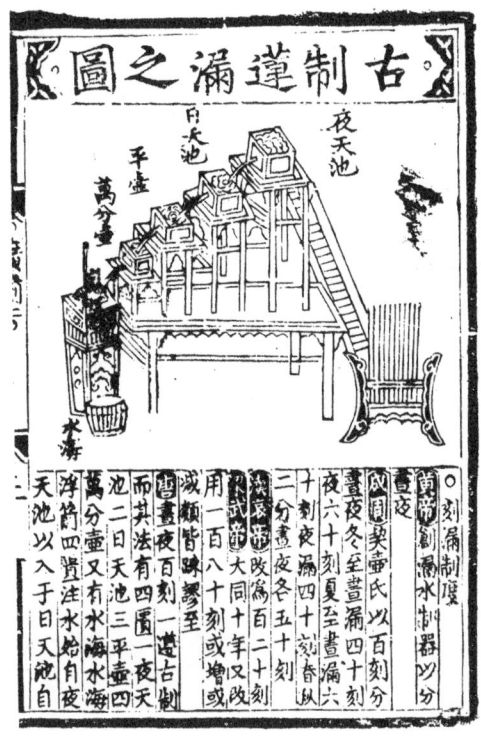

Figure 3. One of the oldest known representations of Chinese clepsydrae. Polyvascular inflow clepsydra from the *Shih lin kuang chi* encyclopedia.

Figure 4. Community clepsydra at Canton, erected in 1316.

Figure 5. *Po-shan hsiang lu*. Late Eastern Chou or early Western Han, fourth to third century B.C. Bronze inlaid with gold and silver with turquoise and carnelian insets throughout, 17.93 cm high, 10 cm wide.

Figure 6. Timetelling with incense sticks inserted in a bed of wood ash in a *ting*.

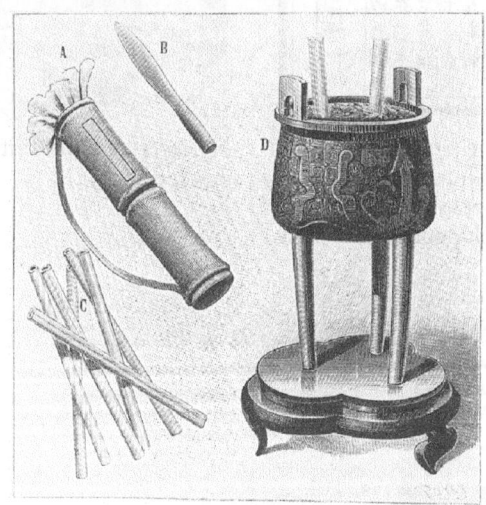

Figure 7. A timetelling device utilizing incense sticks inserted in a *ting*.

Figure 8. Knotted match-cord used for timetelling.

Figure 9. Interior of Cho Lon's Fukien temple in French Indochina in 1952 with a worshipper lighting a large incense coil suspended from the ceiling.

Figure 10. Incense coil equipped with alarm bell suspended on special stand with bronze basin with alarm bell attached.

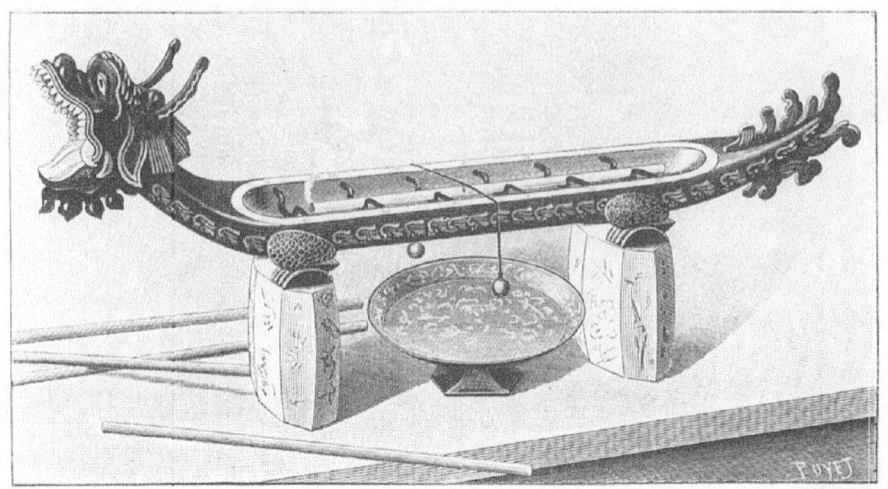

Figure 11. Dragon boat alarm shown with its pedestals and basin, drawn from an example noted as being in the Louvre.

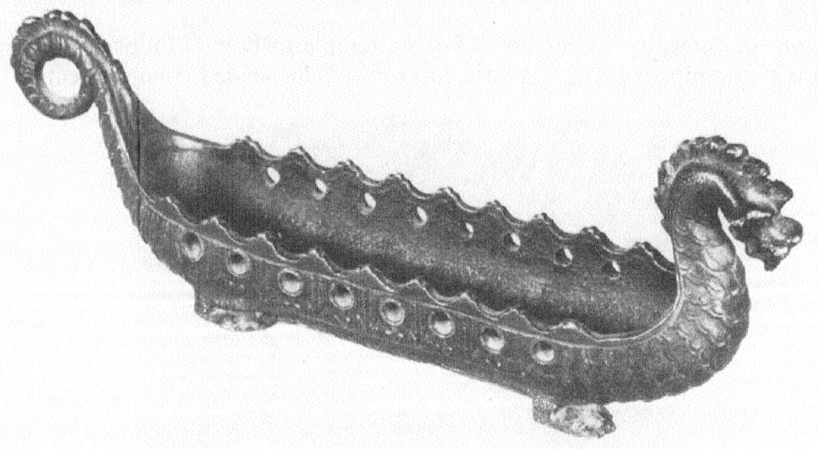

Figure 12. Cast bronze dragon boat alarm, with unusual features.

Figure 13. Drawing of the Dragon Boat
Festival in southern China.

Figure 14. Dragon boat alarm with large figurehead featuring a dragon with
heavy mane resembling a lion.

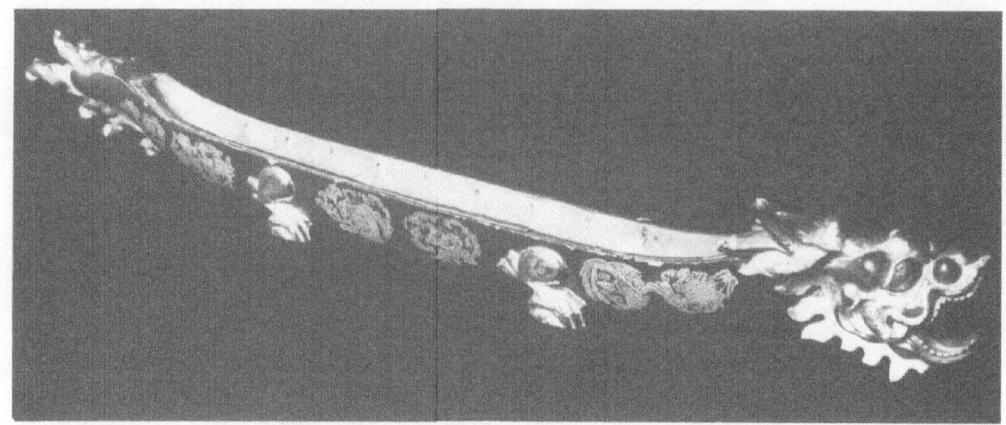

Figure 15. Dragon boat alarm with elongated body, carved and gilded dragon head, tail "feathers," and feet, body finished in black lacquer with six decorative designs on each side in red and gilt, pewter liner, 70.8 cm long, 4.6 cm to 9.5 cm high, and 5.4 cm wide.

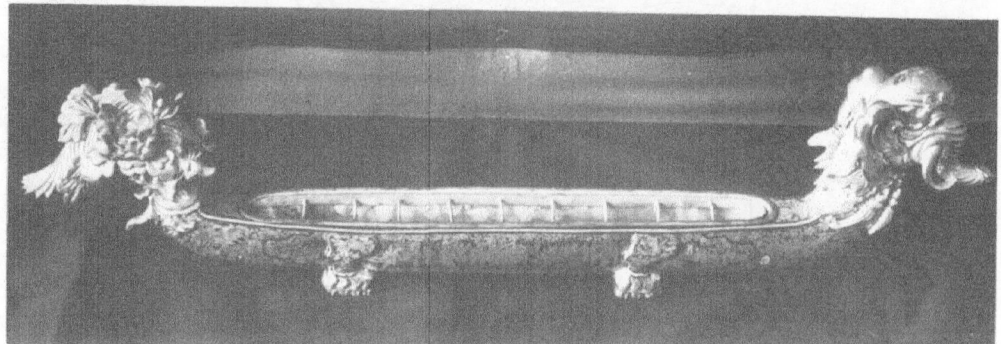

Figure 16. Dragon boat alarm owned by the eighteenth-century kings of Spain, one of two examples preserved in the royal palace at Aranjuez, where the sovereigns prepared themselves for going to sea.

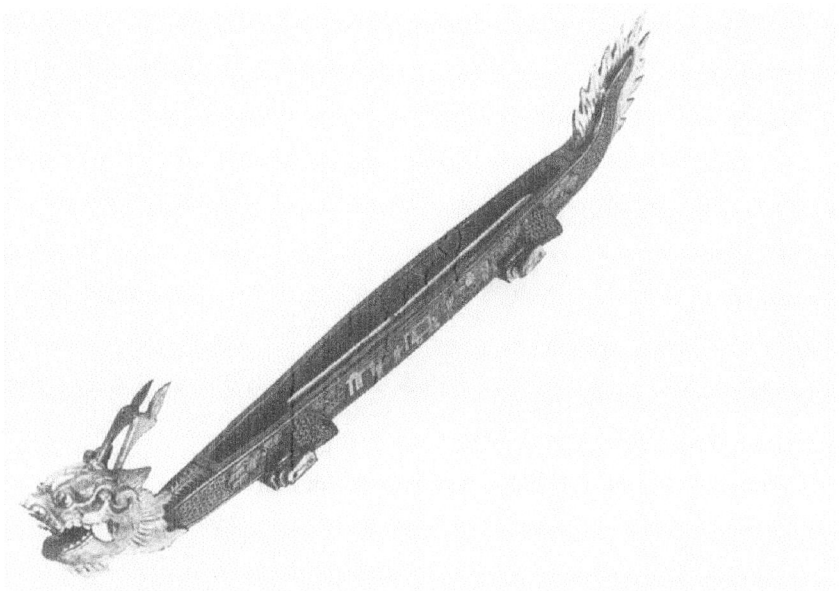

Figure 17a. Dragon boat alarm. Eighteenth or early nineteenth century. Wood, lacquered in black and gold, with paintings in red and gold. Pewter liner with 9 cross-wires. 67.3 cm overall.

Figure 17b. Dragon boat alarm. Eighteenth or early nineteenth century. Wood, lacquered in black and gold, with paintings in red and gold. Semi-seated figures fore and aft. Pewter liner with 9 cross-wires. 67.31 cm overall.

Figure 18. Dragon boat alarm featuring a box in its mid section and with wheels hidden in the feet. Black lacquer with designs in gilt.

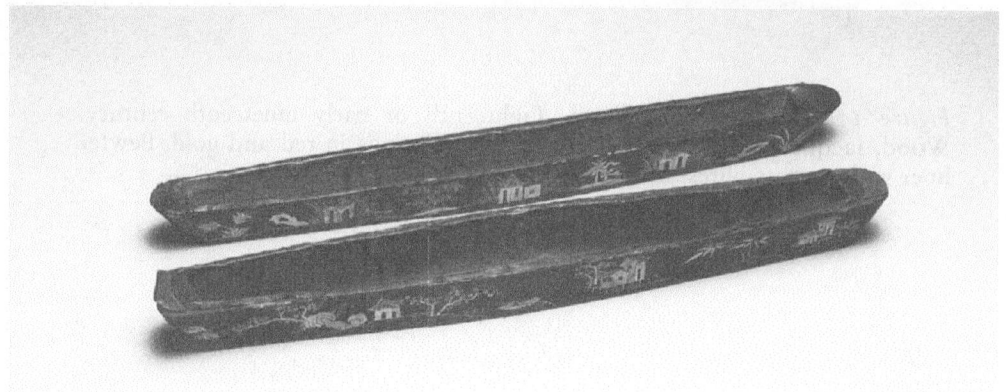

Figure 19. "Boat" alarms, without dragon figurehead and tail. Probably eighteenth century.

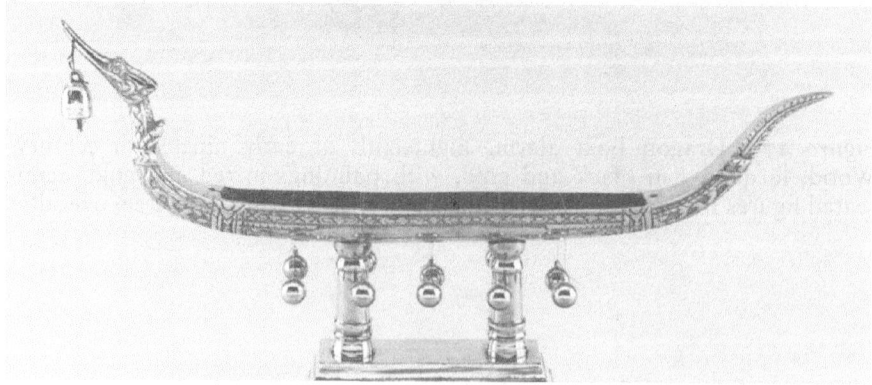

Figure 20. Dragon boat alarm, believed to be from Thailand, probably nineteenth century. Represents the royal barge of the king of Thailand. The bird figurehead is a "sacred swan" or *hansa*, a symbol of power.

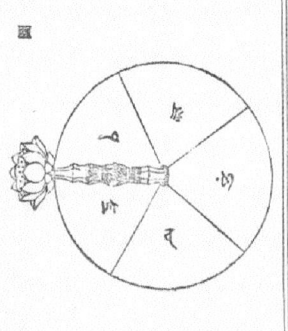

Figure 21. Tantric scripture in the Buddhist Canon on "The [Incense] Seal of Avalokitésvara Bodhisattva."

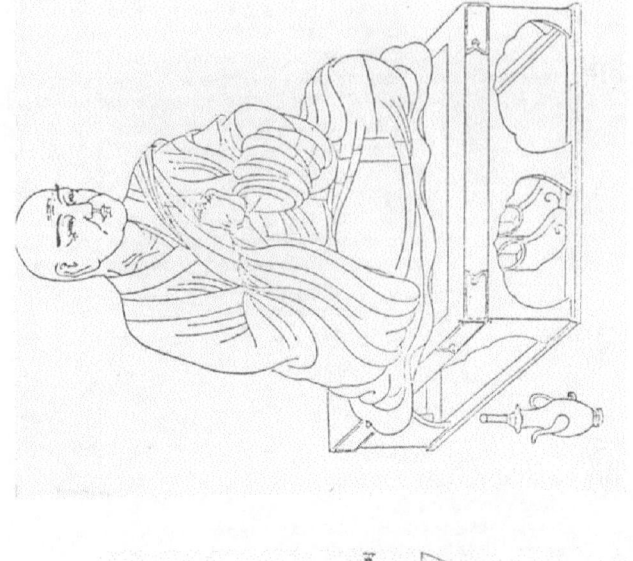

Figure 23. Portrait of the Buddhist monk Amoghavajra (A.D. 705–774).

Figure 22. Portrait of the Tantric master Subhakarasimha (A.D. 637–735) holding an incense burner.

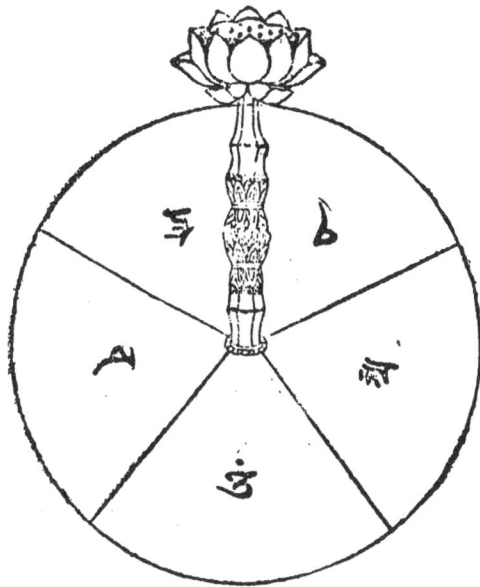

Figure 26. Enlarged detail of cover of "The [Incense] Seal of Avalokitésvara Bodhisattva" featuring the five Siddham seed-syllables of the *mantra* and the finial in the form of a stemmed eight-petaled lotus.

Figure 27. Enlarged detail of the incense trail of "The [Incense] Seal of Avalokitésvara Bodhisattva" forming the Siddham character *hríḥ*.

Figure 28. "Siddham" incense seal, cast brass with copper edging, three tiers, 15.5 cm in diameter and 20.3 cm in height to the base of the finial, which is lacking. The present finial is a later addition and incorrect.

Figure 29. The "Siddham" incense seal dismantled to display its parts. The cover features the Siddham seal-characters of the mantra *oṃ*, *va*, *jra*, *dha*, and *rma*. The original finial, in the form of a stemmed eight-petaled lotus, is lacking. Note that template forms the incense trail in the seal character *hríḥ*.

Figure 30. "The [Incense] Seal of Avalokitésvara Bodhisattva," or *kirikuji kōro*.

Figure 31. Utensils for forming the incense trail of "The [Incense] Seal of Avalokitésvara Bodhisattva." Made of heavy bronze in two parts, measuring 29.5 cm in diameter.

Figure 32. Underside of tamper revealing projections for forming the incense trail.

Figure 33. Template of "The [Incense] Seal of Avalokitésvara,"
with channels forming the Siddham character *hríḥ* through which
the powdered incense is sifted to form the incense trail. Underside
of the template showing incense trail pattern in reverse.

Figure 34. The burning incense pattern of "The [Incense] Seal of Avalokités-vara," known in Japan as a *kirikuji kōro*, in the Hōju-in Temple of the Shingon sect in Totsuka-ku, Yokohama. 30.6 cm in diameter.

Figure 35. *Hsiang p'an*, bronze, late Chou or Han Dynasty.

Figure 36. *Kōinza* in its place on the *ganza* seen from above.

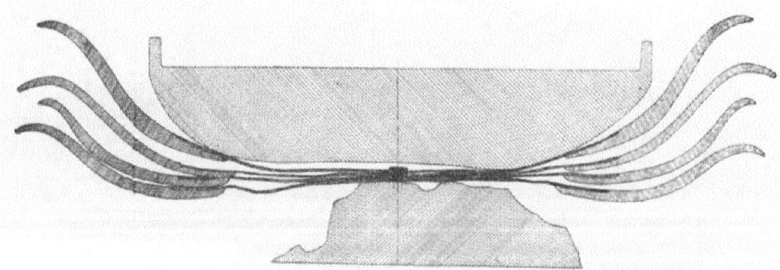

Figure 37. Schematic drawing of the arrangement of the *kōinza* in place on the *ganza*.

Figure 38. The *Kōinza* in place on the *ganza* in the Imperial Treasury.

Figure 39. Detail of painted plate of the
ganza depicting the phoenix.

Figure 40. The *shitsuban* viewed from above.

Figure 41. "Everlasting spring and long life incense seal."

Figure 42. Three incense seals – "Life as long as cotton strands" incense seal, and *fu* and *shou* incense seals.

Figure 43. "The Hundred Graduations Incense Seal."

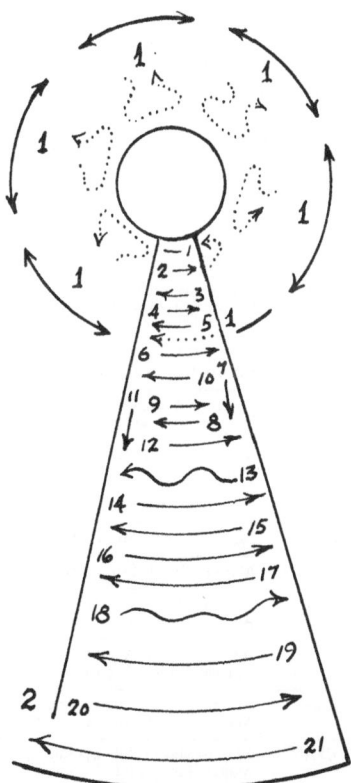

Figure 44. Overlay on "The Hundred Graduations Incense Seal.'

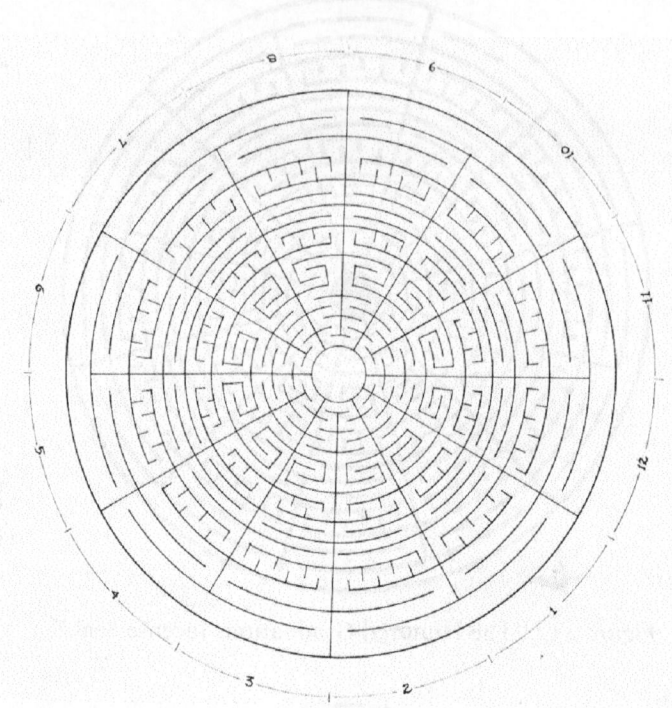

Figure 45. Diagram of "The Hundred Graduations Incense Seal" with explanatory notations added.

Figure 46. Table of Unified Data for "The Five Watch Aromatic Notch Timekeeper."

1. Seal	2. Number of notches (k'o)	3. Diameter	4. Length	5. Dates covered, according to the text (incorrect dates italicized)	6. Dates covered, according to emendations made in the text
First Seal	60	3.3″	2′7.5″	Dec 2–Jan 9	Dec 3–Jan 9
Seal A	59 58	3.2″	2′7″	Jan 10–Jan 23, Nov 21–Dec 3 Jan 24–Feb 1, Nov 11–Nov 20	Jan 10–Jan 23, Nov 21–Dec 2 Jan 24–Feb 1, Nov 11–Nov 20
Seal B	57 56	3.2″	2′6″	Feb 2–Feb 9, *Jan 31–Feb 8* Feb 10–*Feb 17, Oct 19*–Nov 2	Feb 2–Feb 9, Nov 3–Nov 10 Feb 10–Feb 16, Oct 27–Nov 2
Seal C	55 54	3.2″	2′5″	*Feb 16–Feb 22, Oct 21–Oct 26* Feb 23–*Mar 1, Oct 14*–Oct 20	Feb 17–Feb 22, Oct 21–Oct 26 Feb 23–Feb 28, Oct 14–Oct 20
Seal D	53 52	3″	2′4″	Mar 1–Mar 5, Oct 7–Oct 13 Mar 6–Mar 11, Oct 1–Oct 6	Mar 1–Mar 5, Oct 7–Oct 13 Mar 6–Mar 11, Oct 1–Oct 6
Seal E	51	2.9″	2′3″	Mar 12–Mar 17, Sep 26–*Oct 1*	Mar 12–Mar 17, Sep 26–Sep 30
Middle Seal	50	2.8″	2′2.5″	Mar 18–Mar 23, Sep 21–Sep 25	Mar 18–Mar 23, Sep 21–Sep 25
Seal F	49	2.8″	2′2″	Mar 24–Mar 28, Sep 15–Sep 20	Mar 24–Mar 28, Sep 15–Sep 20
Seal G	48 47	2.7″	2′1.5″	Mar 29–*Apr 1, Sep 9*–Sep 14 Apr 4–Apr 11, Sep 3–Sep 8	Mar 29–Apr 3, Sep 9–Sep 14 Apr 4–Apr 11, Sep 3–Sep 8
Seal H	46 45	2.6″	2′ .5″	Apr 12–Apr 17, Aug 27–Sep 2 Apr 18–Apr 23, Aug 19–Aug 26	Apr 12–Apr 17, Aug 27–Sep 2 Apr 18–Apr 23, Aug 19–Aug 26
Seal I	44 43	2.5″	1′9.5″	Apr 24–Apr 30, Aug 12–Aug 18 May 1–May 8, Aug 4–Aug 11	Apr 24–Apr 30, Aug 12–Aug 18 May 1–May 8, Aug 4–Aug 11
Seal J	42 41	2.4″	1′8.5″	May 9–May 18, Jul 25–Aug 3 *May 20*–Jun 1, Jul 11–Jul 24	May 9–May 18, Jul 25–Aug 3 May 19–Jun 1, Jul 11–Jul 24
Last seal	40	2.3″	1′7.5″	*Jun 3*–Jul 10	Jun 2–Jul 10

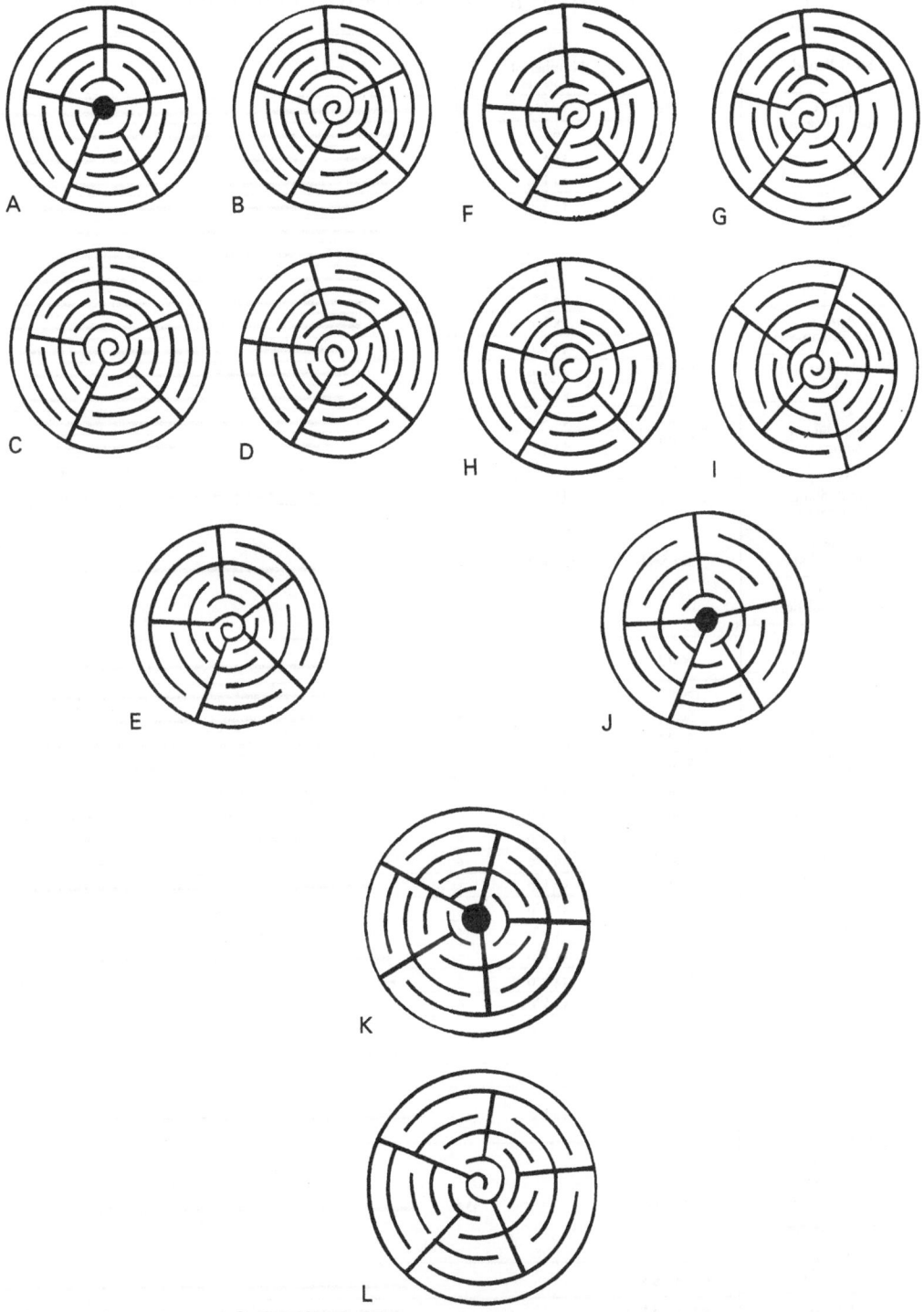

Figure 47. "The Five Watch Aromatic Notch Timekeeper." First seal – seals
A, B, C, D, E, F, G, H, I and last seal.

Figure 48. Chart for the "Five Night Seal-Character Aromatic Notch Timekeeper."

Solar Period	Seal	Number of Notches	Days used	Night Time
Feb. 5 (Beginning of Spring)	Seal B	57 k'o	8	
	Seal B	56 k'o	(7)	
Feb. 19 (Rain Waters)	Seal C	55 k'o	(6)	
	Seal C	54 k'o	(6)	
March 5 (Excited Insects)	Seal D	53 k'o	(5)	
	Seal D	52 k'o	6	
	Seal E	51 k'o	6	
March 20 (Vernal Equinox)	Middle Seal	50 k'o	6	
	Seal F	49 k'o	5	
April 5 (Clear Brightness)	Seal G	48 k'o	(6)	
	Seal G	47 k'o	(8)	
April 20 (Grain Rains)	Seal H	46 k'o	6	
	Seal H	45 k'o	6	
May 5 (Beginning of Summer)	Seal I	44 k'o	7	
	Seal I	43 k'o	8	
May 21 (Small filling [of Grain])	Seal J	42 k'o	10	
	Seal J	41 k'o	14	
June 6 (Grain in Ear) June 21 (Summer Solstice) July 7 (Slight Heat)	Last Seal	40 k'o	39	
July 7 (Slight Heat)	Seal J	41 k'o	14	
July 23 (Great Heat)	Seal J	42 k'o	10	
	Seal I	43 k'o	8	
Aug. 7 (Beginning of Autumn)	Seal I	44 k'o	7	
	Seal H	45 k'o	8	
Aug. 23 (Abiding Heat)	Seal H	46 k'o	7	
	Seal G	47 k'o	6	
Sept. 8 (White Dew)	Seal G	48 k'o	6	
	Seal F	49 k'o	6	
Sept. 23 (Autumnal Equinox)	Middle Seal	50 k'o	5	
	Seal E	51 k'o	(5)	
	Seal D	52 k'o	(6)	
Oct. 8 (Cold Dew)	Seal D	53 k'o	7	
	Seal C	54 k'o	7	
Oct. 23 (Frost Falls)	Seal C	55 k'o	6	
	Seal B	56 k'o	(7)	
Nov. 7 (Beginning of Winter)	Seal B	57 k'o	(8)	
	Seal A	58 k'o	10	
Nov. 22 (Slight Snow)	Seal A	59 k'o	(12)	
Dec. 7 (Heavy Snow) Dec. 21 (Winter Solstice) Jan. 6 (Slight Chill)	First Seal		(38)	
Jan. 6 (Slight Chill)	Seal A 59 k'o		13	
Jan. 21 Great Chill	Seal A 58 k'o		9	
	Seal B	57 k'o	8	

Figure 49. *Shou* (long life) seal, and *fu* (blessings) seal.

Figure 50. "Greatly Elaborated Incense Seal."

Figure 51. Square incense seal, bronze, two tiers, probably eighteenth century, with cover perforated with writings in archaic scripts and poetic inscriptions inlaid in silver on all four sides. 11.4 cm square and 8.5 cm high.

Figure 52. Square incense seal dismantled showing all parts. Six large characters on cover read "Great wealth and honor and a long life." The smaller characters on either side are "[Madein] the first month of Spring of the *chia tzu* Year [1744?]."

Figure 53. A paktong incense seal in its original gift box for presentation.

Figure 54. Lao Tzu, the founder of Taoism, on a journey shown bearing a *ju-i* scepter. Wood carving, probably early nineteenth century.

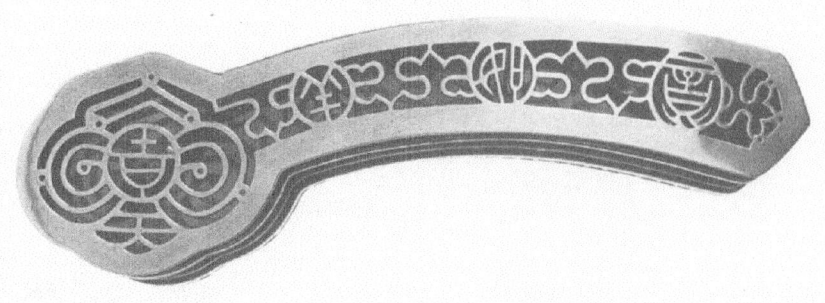

Figure 55. Incense seal in the form of a *ju-i* scepter, paktong, two tiers, separate copper gallery base, length 20.6 cm, width 5 cm, height 6.6 cm. The characters perforated on the cover are "good luck," "scepter," and "wishes to be fulfilled."

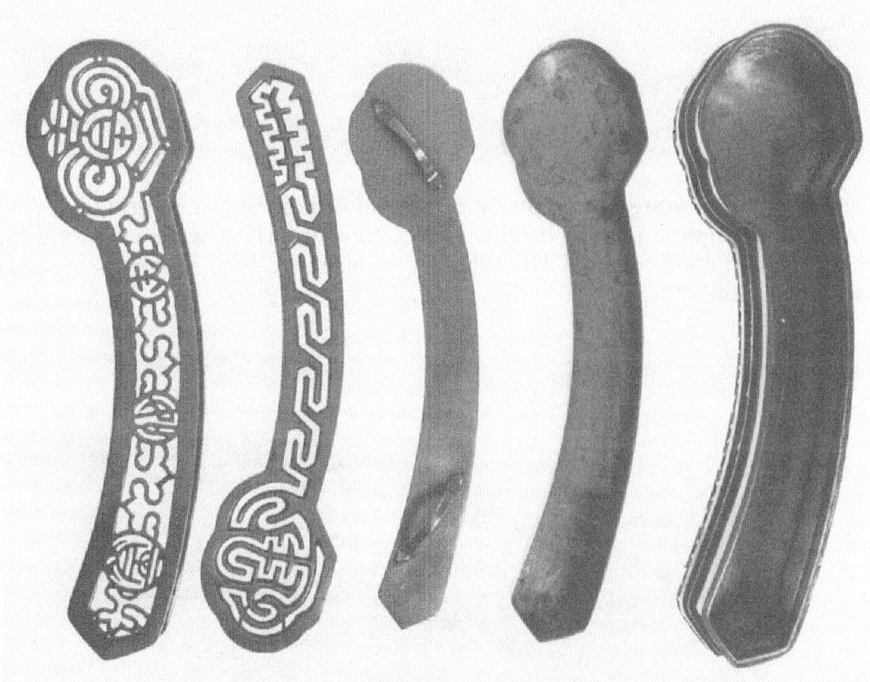

Figure 56. *Ju-i* scepter incense seal dismantled. Hallmark reads *Kawamoto* or "River route." The incense trail pattern also suggests the "*ju-i* scepter" form, with the seal-characters *yen nien* which may be interpreted to read "May there be long life." The *yen* may be symbolized by the long scroll instead of being spelled out.

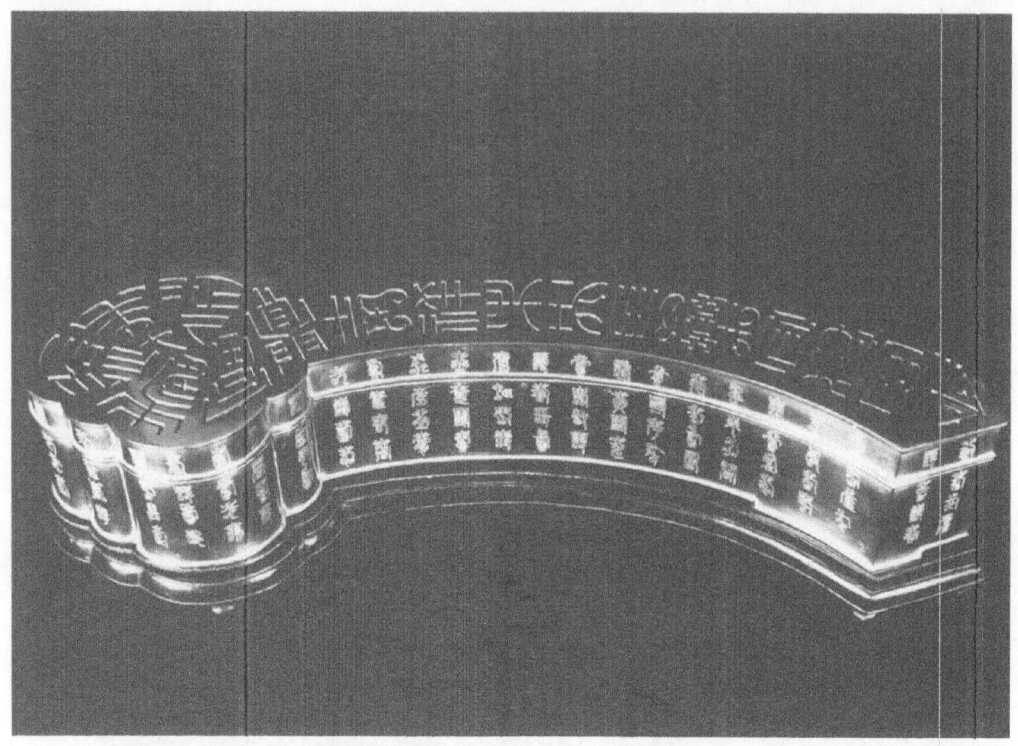

Figure 57. Elaborate *ju-i* scepter incense seal of paktong, measuring 53.3 cm in overall length, 14 cm in height. The cover is inscribed with a blessing for long life and good fortune. The sides are emblazoned with hundreds of script variants of the character *shou* (long life). Teakwood stand.

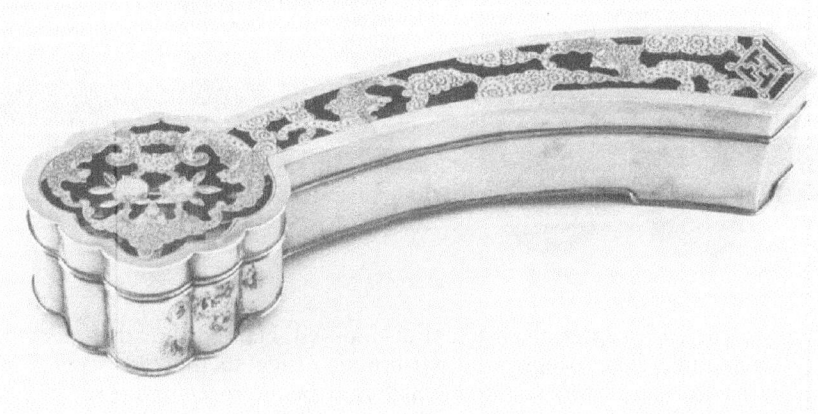

Figure 58. *Ju-i* scepter form, single tier with cover featuring sculptured designs, paktong with copper trim, 29.5 cm long, 7 cm wide and 5.4 cm high.

Figure 59. Paktong incense seal, four tiers, base of each tier decorated with perforated designs, 7.3 cm square and 9.2 cm high.

Figure 60. Incense seal with three tiers, heavy paktong, separate perforated rounded base, lattice patterned cover featuring seal characters for "wealth" and "longevity" in center, sides plain, 9.8 cm square, base 11.4 cm square, 10.8 cm high. Tamper and template only.

Figure 61. Incense seal with two tiers, thin paktong with copper beading on all edges, footed, with swastika design on feet, delicately made. Coffin-shaped cover perforated with "broken ice" lattice work, all four sides engraved – butterfly and grasses on overhang of cover, poetic inscriptions in archaic script on first tier, flowers, landscape with figures on bottom tier. No hallmark. Brass bases of tiers. 8.9 cm long, 6 cm wide, 8.6 cm high.

Figure 62. Urn shaped incense seal, footed, with projecting handles, copper beading and brass grid in cover, archaic characters around cover, pewter, 13.2 cm in diameter (bowl 9 cm diameter) and 11.6 cm high.

Figure 63. Rectangular incense seal, thick paktong, single tier, with separate base. Length 18.4 cm, width 10.2 cm, and height 11.4 cm. The cover is perforated with designs of prunus blossoms and "broken ice" and the "double happiness" character appears on either side with a stylized character for "riches" at the center.

Figure 64. Octagonal incense seal, paktong with copper inlay in cover and trim, 9 cm in diameter, 5.9 cm in height. The cover is decorated at its center with the circle of *yang* and *yin* surrounded by the eight trigrams derived from the "Book of Changes" (*I ching*).

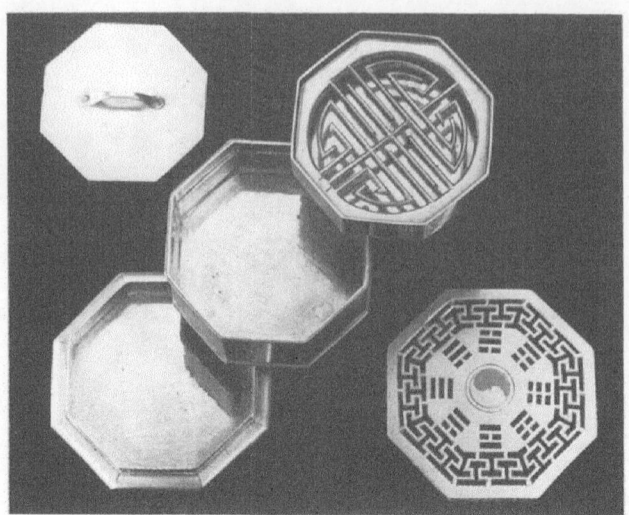

Figure 65. The component parts of the octagon shaped incense seal. The template is a stylized version of the character for "longevity."

Figure 66. Round incense seal, paktong with copper trim, 3 tiers, footed, 8.6 cm in diameter, 8.9 cm in height. Cover perforated with overall swastika pattern and central medallion featuring the seal-character for "wealth." Outer surfaces of middle tier inscribed with poetic phrases in archaic scripts and floral branches and a bird on lower tier. The maker's signature appears on the middle tier. No hallmark.

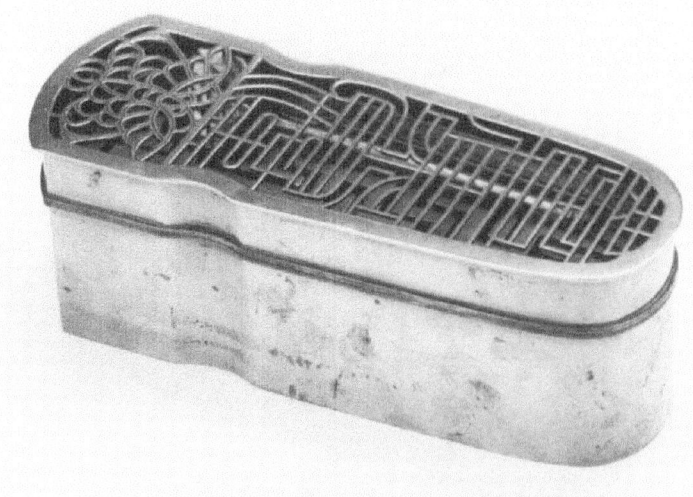

Figure 67. Incense seal in form of a lute (*ch'in*), paktong. The cover is perforated with the design of a butterfly and the characters for "worldly success and long life."

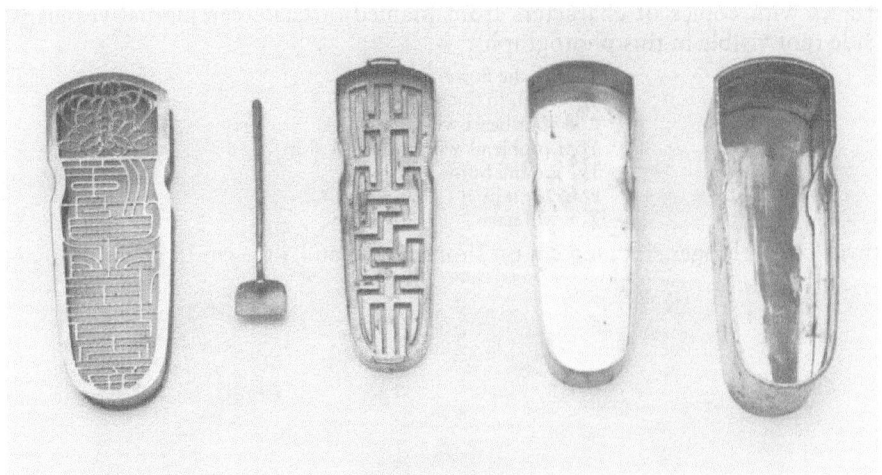

Figure 68. Lute incense seal dismantled. The template design is "intertwining hearts."

Figure 69. Incense seal in the form of a quarter circle. These were often produced in sets of four and assembled to form a complete circle. Heavy cast paktong, three tiers, footed, cover perforated with prunus blossom and "broken ice" pattern and egg-and-dart bordering. Copper lining and bases of all sections and copper template and tamper. Inscribed on all surfaces with copies of characters from (named) archaic ceremonial vessels. On the front side (not visible in this photograph):

> You see the flowers
> You listen to the bamboo
> And your heart will be at peace.
> Your problems will be cleared away.
> The ground burns
> Fragrant music
> You will have ...

Length 14 cm on longer side, 6.7 cm on shorter side, and 11.1 cm in height.

Figure 70. Quarter-circle incense seal dismantled.

Figure 71. Square incense seal, heavy paktong, three tiers (but lacking tier), footed, cover perforated with lattice pattern featuring a crane and floral motifs in a central medallion. Sides of upper tier inscribed with designs of archaic ceremonial vessels and seals, and the lower tier inscribed with wise and poetic inscriptions in *li* script. 7 cm square, 5.1 cm in height. No hallmark.

Figure 72. Incense seal in shape of a fan, thin paktong with copper trim, footed, single tier. Cover of copper with decorative bluing perforated with designs of flowers, melon, *ju-i* scepter, and engraved with two human figures. Template and bases appear to be of brass. Length 12.5 cm on longer side, 4.4 cm on shorter side, 4.6 cm in height. No hallmark.

Figure 73. Incense seal of pewter, three tiers and cover, measuring 7.6 cm square and 11.4 cm high. Copper beading on rims, brass cover grid featuring seal-character for "double happiness." All four sides decorated with engravings of landscapes, birds, flowers, and poetic inscriptions in different Chinese script forms.

Figure 74. Pewter seal dismantled. Template with *shou* (long life) seal-character.

Figure 75. Shown for comparative purposes, another square pewter incense seal, with three tiers and decorations similar to the foregoing. Measuring 7.6 cm square and 11.4 cm high.

Figure 76. Enlargement of hallmark on underside of square pewter incense seal.

Figure 77. Large pewter incense seal, three tiers and cover with overhang containing grid panel made of brass with filigree perforations and featuring the character for "double happiness" in the center. Footed, hallmark. 26.3 cm square, 29.2 cm high. Lowermost tier double the height of other tiers. Made in Swatow, seventeenth or eighteenth century.

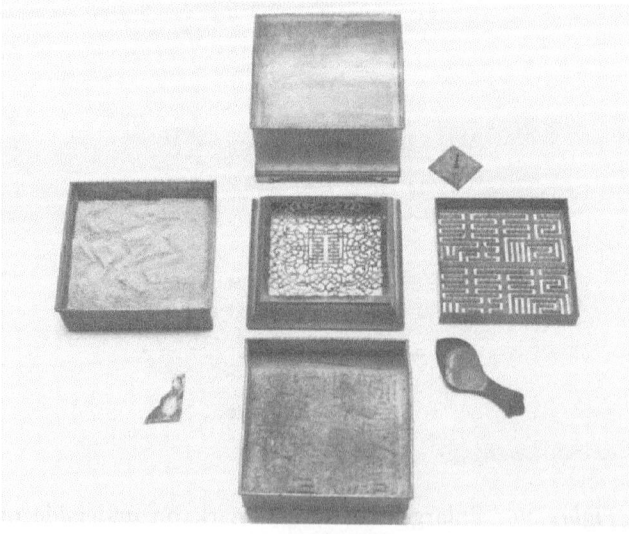

Figure 78. Large pewter incense seal dismantled to show parts and utensils – template, wooden tamper a later replacement, scoop shovel. Pewter template 23.2 cm square, slightly smaller in diameter than the incense tray. The design includes the characters for "blessings" and "long life" twice.

Figure 79. Hallmark of large pewter incense seal, reading "The shop is in Ch'ao Yang [the classical name of Swatow] Yen I Ho Old Shop." Within the double oval the four characters read "Real material superior pewter."

Figure 80. *Ju-i* scepter incense seal entirely of copper or bronze, length 20.6 cm, width 5.1 cm, height 6.7 cm. Identical to a design in Ting Yün's memorial volume, with hallmark. The cover design reads "Great worldly success and pursuit of a long life," and includes the "Moon Lake" signature.

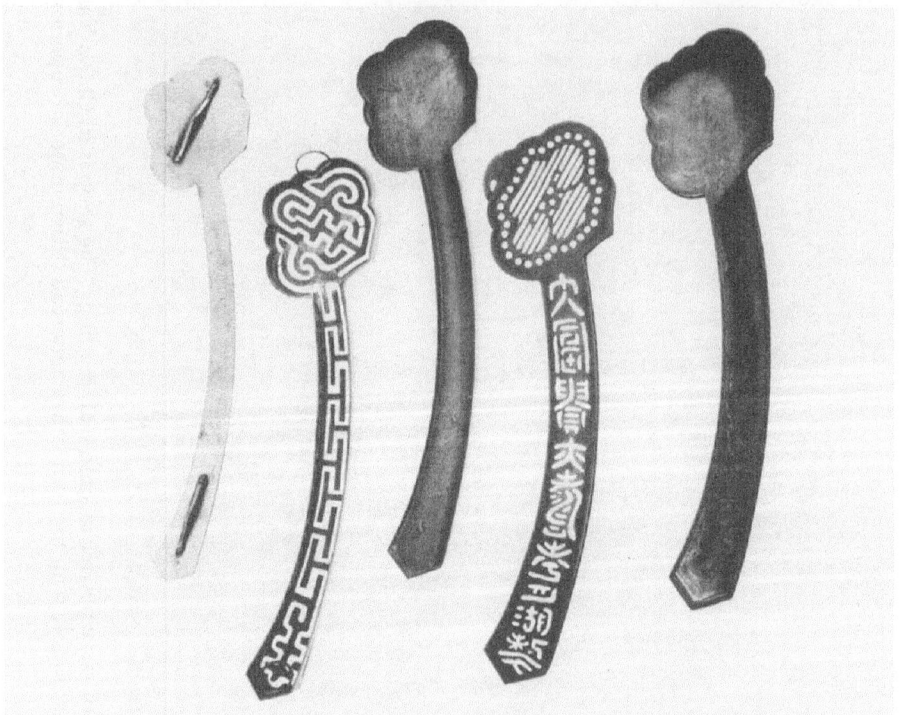

Figure 81. *Ju-i* scepter incense seal dismantled. The character *chi* meaning "auspicious" or "lucky" appears at either end of the template; the stem is merely a decorative pattern.

Figure 82. Hall-mark cast in underside of *ju-i* scepter, reading "Yüeh Hu (Moon Lake), in imitation of the ancients. Made in Shih Kang (Stone Harbor) by Li Hsüeh-yü."

Figure 83. The foregoing incense seal depicted in wood block prints from Ting Yün's memorial volume.

Figure 84. Incense seal in the form of a *ju-i* scepter. All parts made of paktong except for copper cover and copper beading on edges. The inscription on the cover reading counterclockwise around the end of the scepter then along the stem reads "Reading a passage from the 'Book of Changes' (*I ching*), playing a tune on the lute [*ch'in*], and sitting [meditating?] to purge the heart of envy and meanness. Made by Moon Lake." Overall length 29.2 cm, height 4.1 cm. Made by or for Ting Yün.

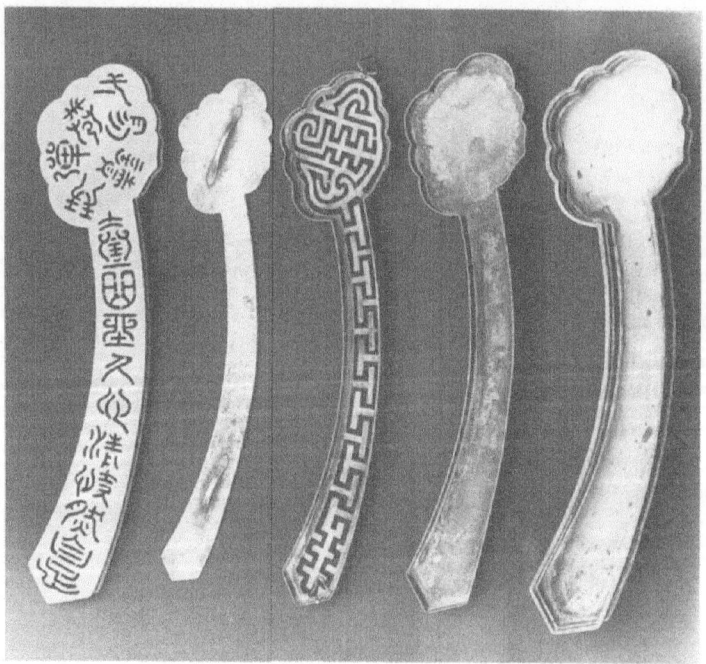

Figure 85. The foregoing *ju-i* scepter incense seal dismantled

Figure 86. Incense seal in the form of a *ju-i* scepter. All parts made of copper. Overall length 23.1 cm, width 5.7 cm, height 3.8cm. Made by or for Ting Yün.

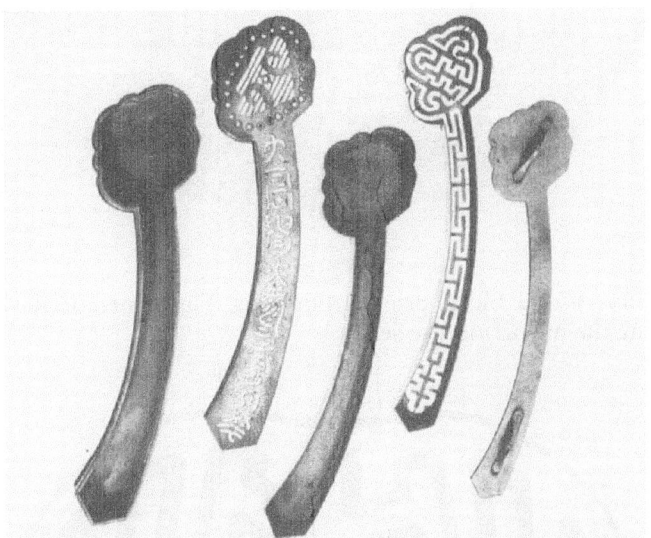

Figure 87. *Ju-i* scepter incense seal dismantled. The inscription of the cover reads. "Great honors and long life" and "Made by Moon Lake." The inscription formed by the template reads "Auspicous."

Figure 88. Wood block prints of the same *ju-i* scepter incense seal from Ting Yün's memorial volume.

Figure 89. Wood block prints from Ting Yün's memorial volume depicting the melon incense seal.

Figure 90. Wood block prints from Ting Yün's memorial volume illustrating the melon incense seal and its seal-character template.

Figure 91. Incense seal in the shape of a melon, one tier, of cast bronze or copper, template and bottoms of incense tray and storage compartment made of cast brass. May be eighteenth century. Length 10.2 cm, width 8.3 cm, height 5.1 cm. Unusual cover pierced with figures of butterflies and melons and the inscription *kua tieh mien mien*, which may be rendered "May your family line grow and prosper" or "The melon vine is long and spreading," reflecting the first line of Mao No. 237 [Waley 240] from *Shih ching* (The Book of Poetry).

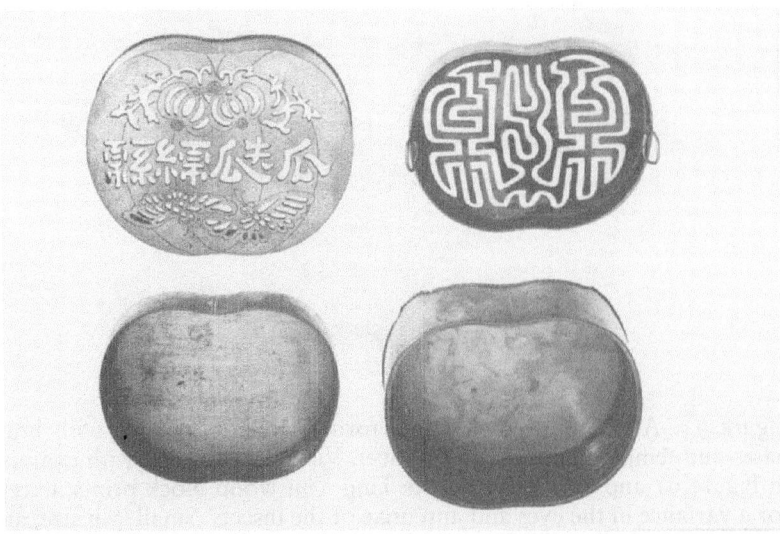

Figure 92. Melon incense seal dismantled. The template consists of the word *mien* meaning "to continue," or "spread," joined in Siamese-twin fashion.

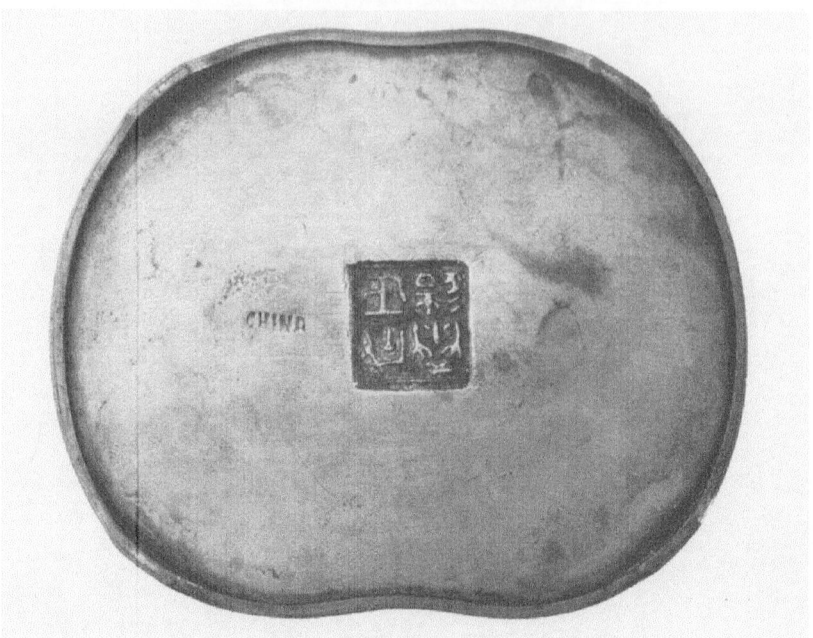

Figure 93. Hallmark on underside of melon incense seal.

Figure 94. Another incense seal in form of melon, copper with brass
bases and template, tamper and shovel. Virtually identical with example
in Figure 91 and as appears in the Ting Yün wood block prints, except
for a variance in the eyes and antennae of the insects. Smaller in size, and
variation of hallmark. Footed, length 8.4 cm, width 8.1 cm, height
4.4 cm.

Figure 95. Hallmark of second melon incense seal. It is the same name as that shown in Figure 91 except that the form of the seal is different. It reads "Ch'eng Cheng-kung."

Figure 96. A third incense seal in melon form, paktong with copper beading, footed. It is similar to the two preceding examples but with different designs. Length 9.8 cm, width 7.9 cm, height 4.1 cm.

Figure 97. Incense seal in form of a lute, single tier, copper, cover with archaic inscriptions, footed. 16.5 cm long, 5.4 cm wide, 4.4 cm high. No hallmark.

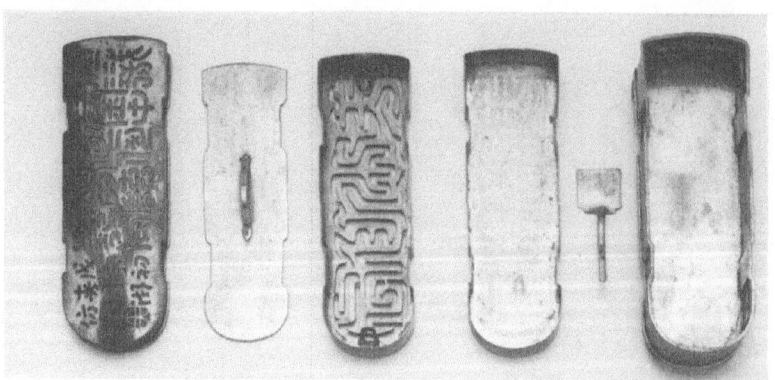

Figure 98. Lute seal dismantled.

Figure 99. Designs of lute incense seal from Ting Yün's memorial volume. The inscription is "But to know the lute (ch'in) is to know how to produce sound from its strings. Moon Lake – imitating the ancient. First month of spring 1878." The caption for the template design is "Untroubled mind at ease."

Figure 100. Incense seal in form of a leaf, single tier, paktong with copper trim on edges, footed, perforated inscription on cover, and "Second month of autumn, 1878, signed 'Moon Lake' and sealed 'Ting Yün.' Includes tamper, no shovel, 12.4 cm long, 6.7 cm wide, 4.4 cm high.

Figure 101. Wood block prints from Ting Yün's memorial volume, of the cover and template of leaf incense seal. The caption for the template design (not shown) is "Green sky."

Figure 102. Bell-shaped incense seal, paktong, two tiers, with copper beading and with gallery stand. Length 8.9 cm, width 7.3 cm, and height 6.3 cm.

Figure 103. Bell-shaped incense seal dismantled. The template design is the character for "happiness."

Figure 104. Wood block prints of bell incense seal from Ting Yün's memorial volume.

Figure 105. Incense seal in form of double *pi*, paktong with copper trim. Length 13.3 cm, width 9.8 cm, height 6.8 cm. Seal-characters perforated in the cover are stylized forms of the characters for "wealth and glory."

Figure 106. Wood block prints from Ting Yün's memorial volume illustrating the double *pi* incense seal and its template. Although the design of the template is the same, that of the cover in the block print is captioned "Linked pearls, joined *pi* disks." The seal-character formed by the template (not shown) is the word "heart" used twice and intertwined to indicate "intertwining hearts."

Figure 107. "Five Auspicious Clouds" incense seal, paktong, with copper beading, measuring 11.4 cm in diameter and 7 cm in height. Cover features prunus blossoms and "broken ice" design. Unsigned, but probably produced by or for Ting Yün and illustrated in his memorial volume.

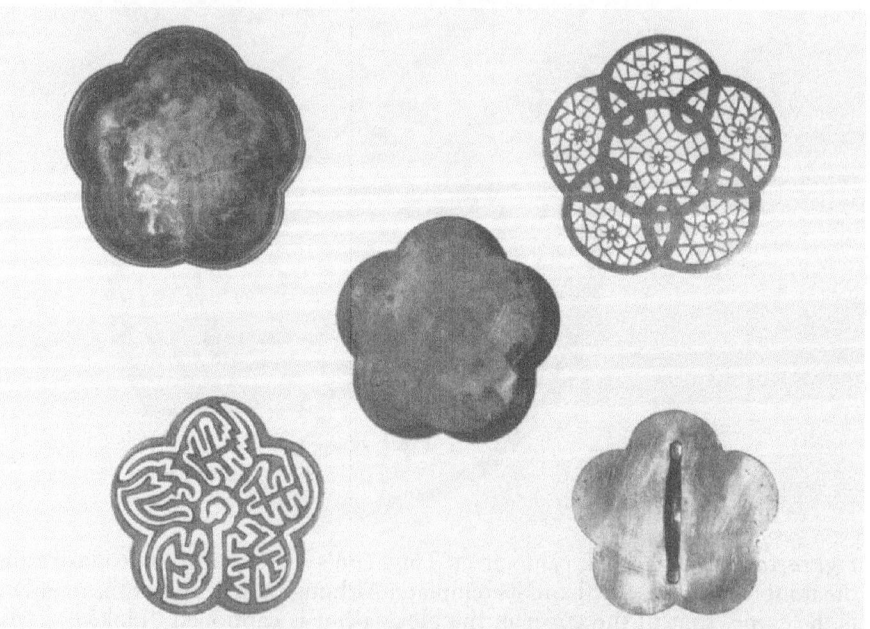

Figure 108. "Five Auspicious Clouds" incense seal dismantled. The template reads "How many lives [lifetimes] before I obtain my flowers" which appears to be a Zen Buddhist *kōan* or other philosophical question for a disciple. The words "my flowers" are replaced with a flower design at the center.

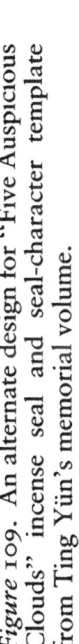

Figure 110. "Tower of Emptiness." Incense seal template design from Ting Yün's memorial volume. The words of the caption are the same as those formed by the seal-characters.

Figure 109. An alternate design for "Five Auspicious Clouds" incense seal and seal-character template from Ting Yün's memorial volume.

Figure 111. *Jōkōban* with swiveling incense tray upon a two-drawer cabinet made of cryptomeria or cypress with dominant grain. Trimmed in black lacquer. Drawers provide storage for tools and incense supply. Measures 19.7 cm square, 29.2 cm in overall height, and the upper section is 11.43 cm deep.

Figure 112. The *jōkōban* dismantled, with utensils, including tray for ash base, template, and leveling bar. The bottom of the incense box is heavily charred. Note the platform with pivot on which the incense tray swivels to form the incense trail of the swastika with double crampons.

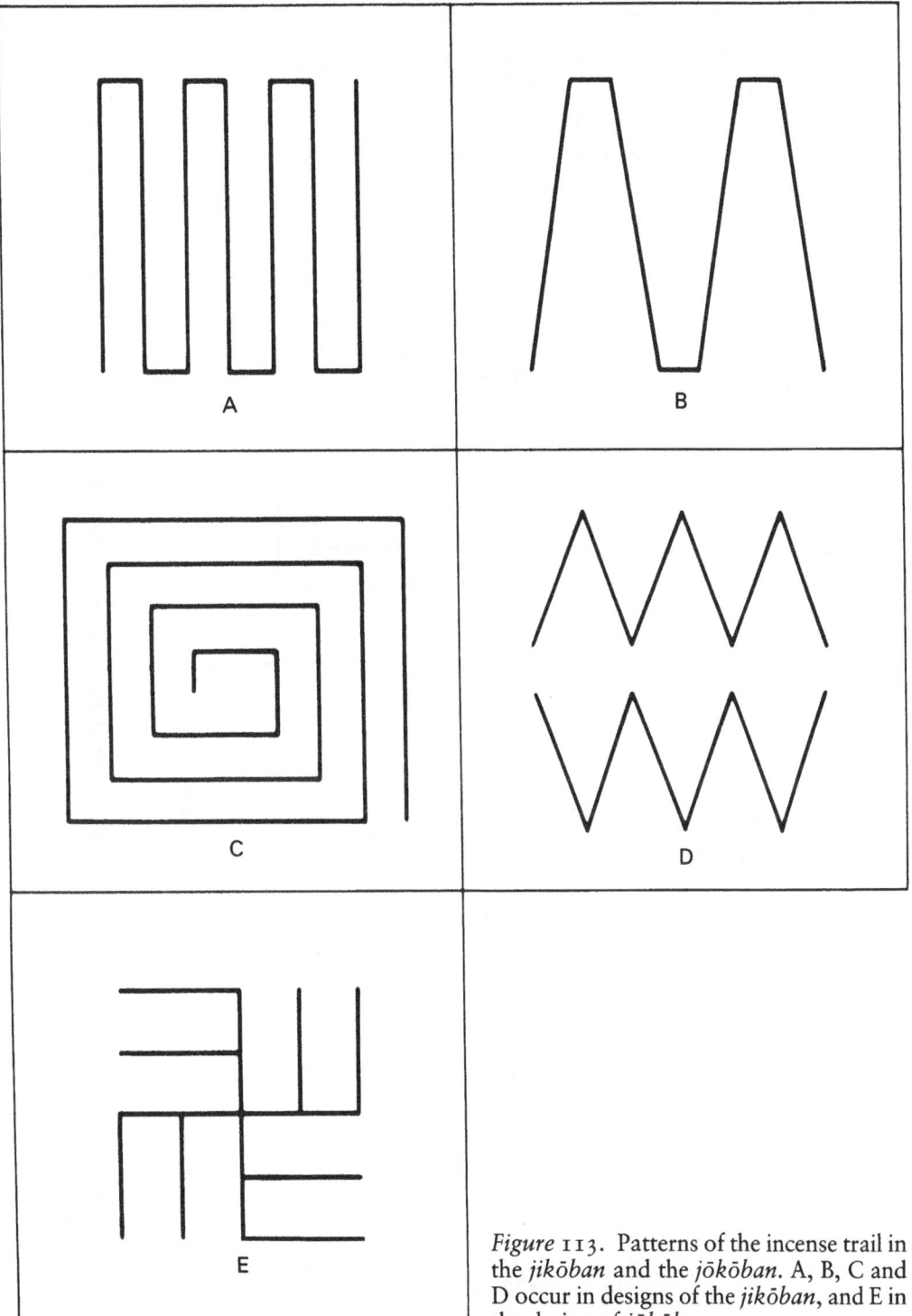

Figure 113. Patterns of the incense trail in the *jikōban* and the *jōkōban*. A, B, C and D occur in designs of the *jikōban*, and E in the design of *jōkōban*.

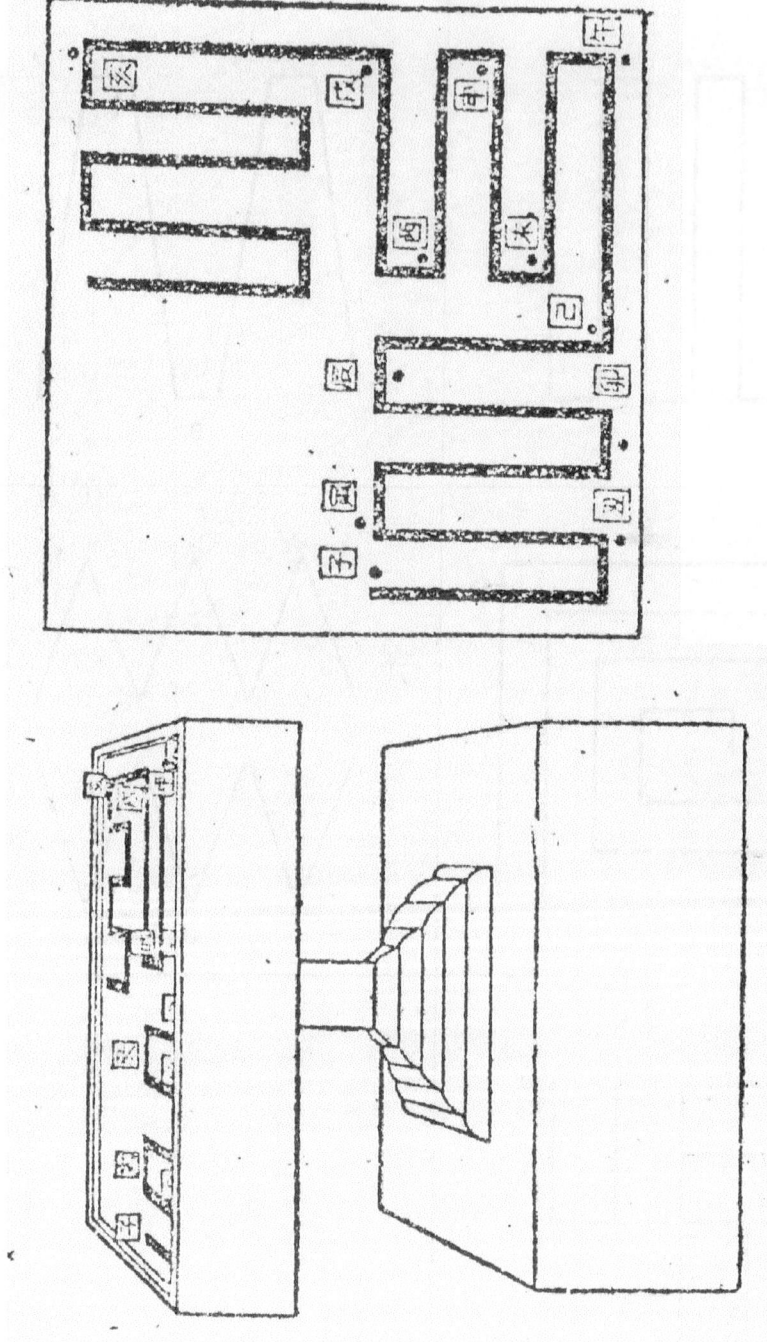

Figure 114. Sketch illustrating the manner in which the incense trail is formed in the *jōkōban* in the shape of the swastika.

Figure 115. *Jōkōban* of wood lacquered black, 33.3 cm square and 46 cm high. Each storage drawer inscribed with the personal name "Kangoro Isuka" and "Kaminochō Shinjuku," the name of a district in Tokyo.

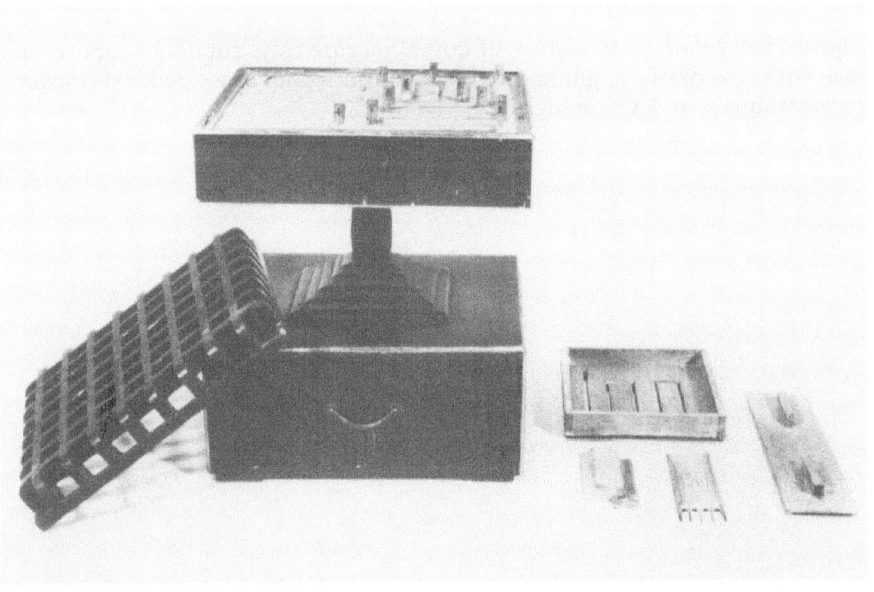

Figure 116. *Jokōban* shown dismantled with utensils.

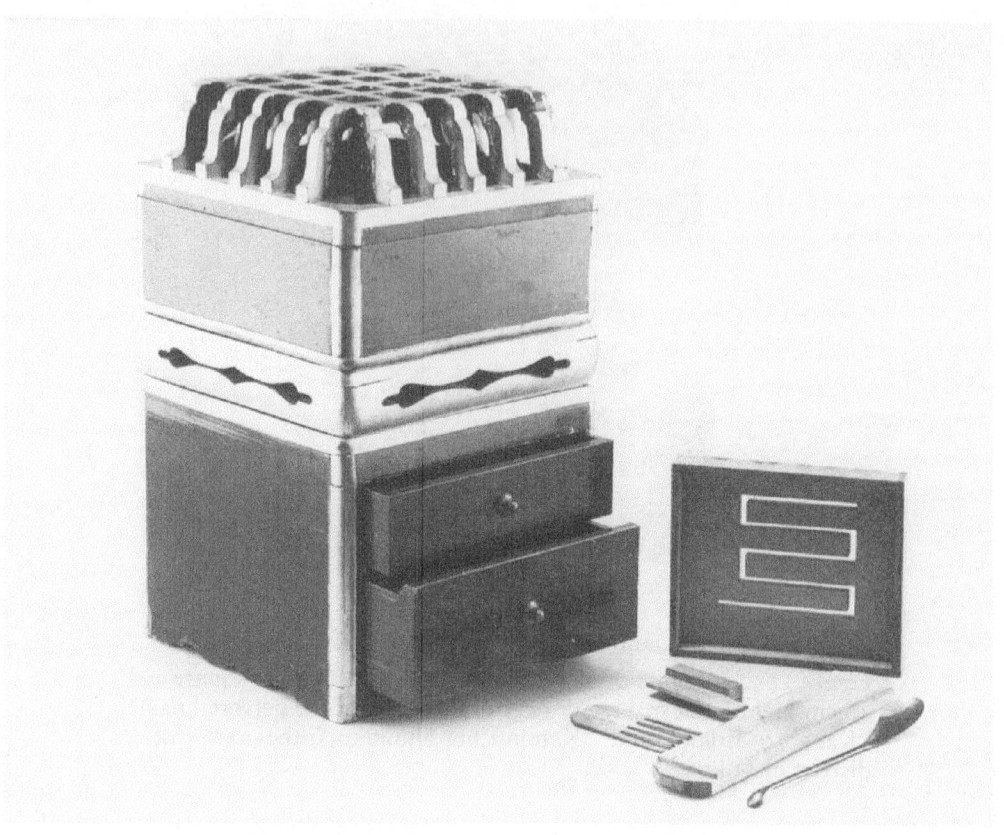

Figure 117. *Jōkōban* consisting of cover, incense tray, rotating support, and base with two drawers, gilt and red lacquer on wood, brass pulls on drawers, 19.7 cm square and 33 cm high.

Figure 118. Book illustration of a Buddhist novice preparing a *jōkōban* in the Kichijō-ji temple in Komagome. Seventeenth century, by Yoshida Hanbei. Wood cut.

Figure 119. *Jōkōban*, decorated with Buddhist symbols. Nara National Museum.

Figure 120. Large elaborate *jikōban* with pedestal, measuring 30.8 cm square, 30.5 cm high, with inside dimensions of incense tray 27.9 cm square and 6 cm deep. Lower section is assembled from more than 140 cubical units interlocked with plain lateral strips of graduated length, kept properly spaced by means of wooden pins inserted at intervals. The whole may be disassembled by removal of a wedge. Each side of the incense box consists of a decoratively carved panel, each featuring ocean waves, clouds, lotus and artemisia blossoms. The four smaller panels at the base of the pedestal feature waves and clouds. The bottom of the incense box is heavily charred. The incense trail is in the form of seven parallel lines connected at right angles at alternate ends. On a bed of compacted wood ash from 3.2 cm to 3.8 cm in depth, is laid an incense trail 0.5 cm in width, 0.8 cm in depth, and 172.7 cm in length.

Figure 121. The large *jikōban* with latticed cover removed exposing the template.

Figure 122. *Jikōban* of extremely elaborate form. Nara National Museum.

Figure 123. *Jikōban* of Arita porcelain. In the Sanzenin-ji temple at Ohara, Japan.

Figure 124. *Jikōban* in use during the water-drawing ceremony at the Nigatsudō at Nara.

Figure 125. Incense pattern tray of the Nigatsudō *jikōban* viewed from the upper side.

Figure 126. Enlargement of previous figure showing indicators for time intervals.

Figure 127. Underside of pattern grid tray of the Nigatsudō *jikōban*.

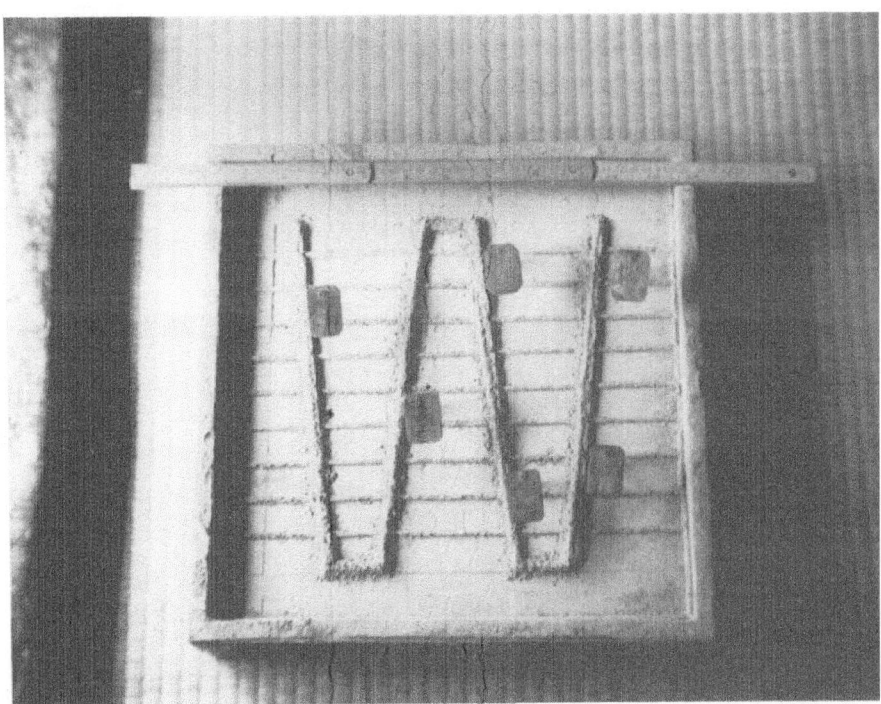

Figure 128. The Nigatsudō *jikōban* shown with the incense trail, and the character *kanji* used to mark the time intervals.

Figure 129. *Senkōdokei* or geisha house timekeeper. The device is made with nineteen openings, and two drawers provide space for new and used incense sticks.

Figure 130. View of *senkōdokei* from above.

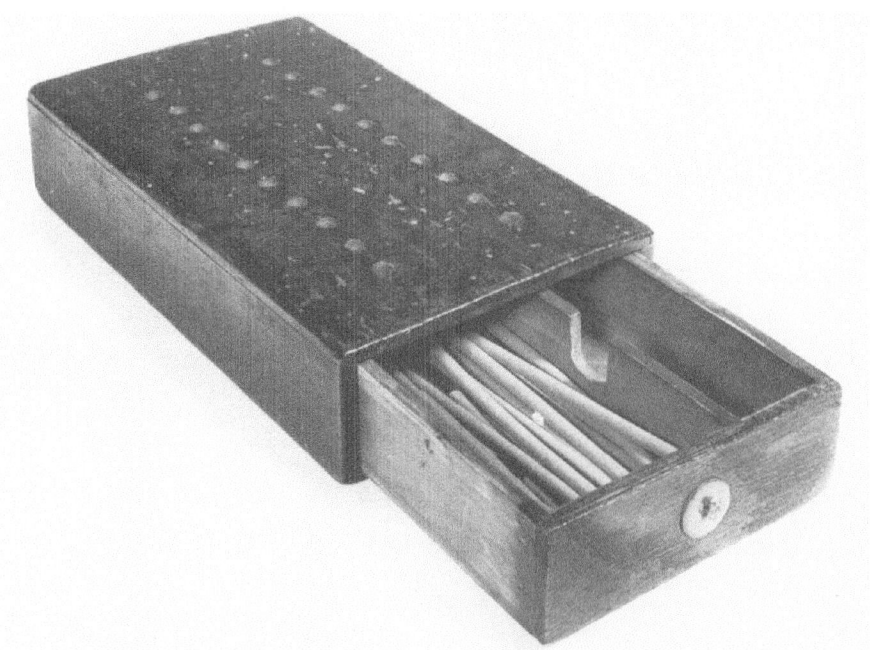

Figure 131. Another form of the geisha timekeeper, with compartmented drawer for storage of new and used incense sticks. Made of cryptomeria wood, 21.6 cm long, 12.1 cm wide, 5.1 cm high, with 20 openings.

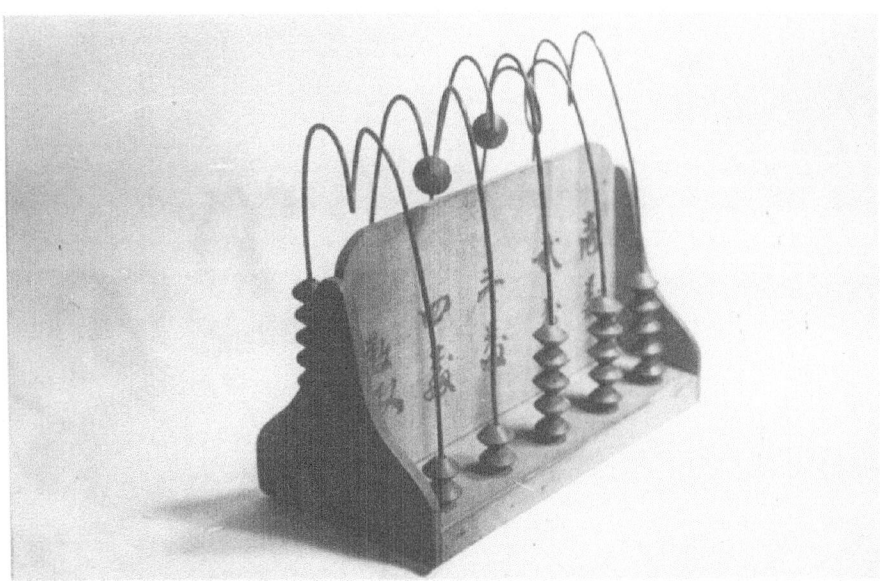

Figure 132. Geisha house *soroban* for calculating number of incense sticks consumed for the time of each *oiran*.

Index